# Springer Series in Statistics



**Springer**
*New York*
*Berlin*
*Heidelberg*
*Barcelona*
*Budapest*
*Hong Kong*
*London*
*Milan*
*Paris*
*Santa Clara*
*Singapore*
*Tokyo*

# Springer Series in Statistics

*Andersen/Borgan/Gill/Keiding:* Statistical Models Based on Counting Processes.
*Andrews/Herzberg:* Data: A Collection of Problems from Many Fields for the Student and Research Worker.
*Anscombe:* Computing in Statistical Science through APL.
*Berger:* Statistical Decision Theory and Bayesian Analysis, 2nd edition.
*Bolfarine/Zacks:* Prediction Theory for Finite Populations.
*Borg/Groenen:* Modern Multidimensional Scaling: Theory and Applications
*Brémaud:* Point Processes and Queues: Martingale Dynamics.
*Brockwell/Davis:* Time Series: Theory and Methods, 2nd edition.
*Daley/Vere-Jones:* An Introduction to the Theory of Point Processes.
*Dzhaparidze:* Parameter Estimation and Hypothesis Testing in Spectral Analysis of Stationary Time Series.
*Fahrmeir/Tutz:* Multivariate Statistical Modelling Based on Generalized Linear Models.
*Farrell:* Multivariate Calculation.
*Federer:* Statistical Design and Analysis for Intercropping Experiments.
*Fienberg/Hoaglin/Kruskal/Tanur(Eds.):* A Statistical Model: Frederick Mosteller's Contributions to Statistics, Science and Public Policy.
*Fisher/Sen:* The Collected Works of Wassily Hoeffding.
*Good:* Permutation Tests: A Practical Guide to Resampling Methods for Testing Hypotheses.
*Goodman/Kruskal:* Measures of Association for Cross Classifications.
*Gouriéroux:* ARCH Models and Financial Applications.
*Grandell:* Aspects of Risk Theory.
*Haberman:* Advanced Statistics, Volume I: Description of Populations.
*Hall:* The Bootstrap and Edgeworth Expansion.
*Härdle:* Smoothing Techniques: With Implementation in S.
*Hart:* Nonparametric Smoothing and Lack-of-Fit Tests.
*Hartigan:* Bayes Theory.
*Heyde:* Quasi-Likelihood And Its Application: A General Approach to Optimal Parameter Estimation.
*Heyer:* Theory of Statistical Experiments.
*Huet/Bouvier/Gruet/Jolivet:* Statistical Tools for Nonlinear Regression: A Practical Guide with S-PLUS Examples.
*Jolliffe:* Principal Component Analysis.
*Kolen/Brennan:* Test Equating: Methods and Practices.
*Kotz/Johnson (Eds.):* Breakthroughs in Statistics Volume I.
*Kotz/Johnson (Eds.):* Breakthroughs in Statistics Volume II.
*Kotz/Johnson (Eds.):* Breakthroughs in Statistics Volume III.
*Kres:* Statistical Tables for Multivariate Analysis.
*Küchler/Sørensen:* Exponential Families of Stochastic Processes.
*Le Cam:* Asymptotic Methods in Statistical Decision Theory.

*(continued after index)*

Jeffrey S. Simonoff

# Smoothing Methods in Statistics

With 117 Figures

Jeffrey S. Simonoff
Department of Statistics and Operations Research
Leonard N. Stern School of Business
New York University
44 West 4th Street
New York, NY 10012-1126 USA

Library of Congress Cataloging-in-Publication Data
Simonoff, Jeffrey S.
Smoothing methods in statistics / Jeffrey S. Simonoff.
p. cm. – (Springer series in statistics)
Includes bibliographical references and indexes.
ISBN 0-387-94716-7 (hard:alk. paper)
1. Smoothing (Statistics) I. Title. II. Series.
QA278.S526 1996
519.5′36–dc20 96-11742

Printed on acid-free paper.



Production managed by Hal Henglein; manufacturing supervised by Jeffrey Taub.
Camera-ready copy prepared from the author's TeX files.
Printed and bound by Edwards Brothers, Inc., Ann Arbor, MI.
Printed in the United States of America.

9 8 7 5 4 3 2 (Corrected second printing, 1998)

ISBN 0-387-94716-7 Springer-Verlag New York Berlin Heidelberg SPIN 10671942

*To Beverly, Robert and Alexandra*

# Preface

The existence of high speed, inexpensive computing has made it easy to look at data in ways that were once impossible. Where once a data analyst was forced to make restrictive assumptions before beginning, the power of the computer now allows great freedom in deciding where an analysis should go. One area that has benefited greatly from this new freedom is that of nonparametric density, distribution, and regression function estimation, or what are generally called smoothing methods. Most people are familiar with some smoothing methods (such as the histogram) but are unlikely to know about more recent developments that could be useful to them.

If a group of experts on statistical smoothing methods are put in a room, two things are likely to happen. First, they will agree that data analysts seriously underappreciate smoothing methods. Smoothing methods use computing power to give analysts the ability to highlight unusual structure very effectively, by taking advantage of people's abilities to draw conclusions from well-designed graphics. Data analysts should take advantage of this, they will argue.

Then, they will strongly disagree about which smoothing methods should be disseminated to the public. These conflicts, which often hinge on subtle technical points, send a garbled message, since nonexperts naturally think that if the experts can't agree, the field must be too underdeveloped to be of practical use in analyzing real data. Besides being counterproductive, these arguments are often pointless, since while some methods are better at some tasks than others, no method is best on all counts. The data analyst must always address issues of conceptual and computational simplicity versus complexity.

In this book, I have tried to sort through some of these controversies and uncertainties, while always keeping an eye towards practical applications. Some of the methods discussed here are old and well understood, while others are promising but underdeveloped. In all cases, I have been guided by the idea of highlighting what seems to work, rather than by the elegance (or even existence) of statistical theory.

This book is, first and foremost, for the data analyst. By "data analyst" I mean the scientist who analyzes real data. This person has a good knowledge of basic statistical theory and methodology, and probably knows certain areas of statistics very well, but might be unaware of the benefits that smoothing methods could bring to his or her problems. Such a person should benefit from the decidedly applied focus of the book, as arguments generally proceed from actual data problems rather than statistical theory.

A second audience for this book is statisticians who are interested in studying the area of smoothing methods, perhaps with the intention of undertaking research in the field. For these people, the "Background material" section in each chapter should be helpful. The section is bibliographic, giving references for the methods described in the chapter, but it also fills in some gaps and mentions related approaches and results not discussed in the main text. The extensive reference list (with over 750 references) also allows researchers to follow up on original sources for more technical details on different methods.

This book also can be used as a text for a senior undergraduate or graduate-level course in smoothing. If the course is at an applied level, the book can be used alone, but a more theoretical course probably requires the use of supplementary material, such as some of the original research papers. Each chapter includes exercises with a heavily computational focus (indeed, some are quite time consuming computationally) based on the data sets used in the book. Appendix A gives details on the data sets and how to obtain them electronically.

I believe that anyone interested in smoothing methods benefits from applying the methods to real data, and I have included sources of code for methods in a "Computational issues" section in each chapter. New code often becomes available, and commercial packages change the functionality that they provide, but I still hope that these sources are useful. Many software packages include the capability to write macros, which means that analysts can write their own code (or perhaps someone else already has). I apologize for any omissions or errors in the descriptions of packages and code. No endorsement or warranty, express or implied, should be drawn from the listing and/or description of the software given in this book. The available code is to be used at one's own risk. See Appendix B for more details on computational issues.

In recent years, several good books on different aspects of smoothing have appeared, and I would be remiss if I did not acknowledge my debt to them. These include, in particular, Silverman (1986), Eubank (1988), Härdle (1990, 1991), Hastie and Tibshirani (1990), Wahba (1990), Scott (1992), Green and Silverman (1994), and Wand and Jones (1995). I own all these books, but I believe that this book is different from them.

The coverage in this book is very broad, including simple and complex univariate and multivariate density estimation, nonparametric regression estimation, categorical data smoothing, and applications of smoothing to other areas of statistics. There are strong theoretical connections between all these methods, which I have tried to exploit, while still only briefly examining technical details in places. Density estimation (besides its importance in its own right) provides a simple framework within which smoothing issues can be considered, which then builds the necessary structure for regression and categorical data smoothing and allows the latter topics to be covered in less detail. Even so, the chapter on nonparametric regression is the longest in the book, reinforcing the central nature of regression modeling in data analysis.

Despite the broad coverage, I have had to omit certain topics because of space considerations. Most notably, I do not describe methods for censored data, estimation of curves with sharp edges and jumps, the smoothing of time series in the frequency domain (smoothed spectral estimation), and wavelet estimators. I hope that the material here provides the necessary background so that readers can pick up the essence of that material on their own.

This book originated as notes for a doctoral seminar course at New York University, and I would like to thank the students in that course, David Barg, Hongshik Kim, Koaru Koyano, Nomi Prins, Abe Schwarz, Karen Shane, Yongseok Sohn, and Gang Yu, for helping to sharpen my thoughts on smoothing methods. I have benefited greatly from many stimulating conversations about smoothing methods through the years with Mark Handcock, Cliff Hurvich, Paul Janssen, Chris Jones, Steve Marron, David Scott, Berwin Turlach, Frederic Udina, and Matt Wand. Samprit Chatterjee, Ali Hadi, Mark Handcock, Cliff Hurvich, Chris Jones, Bernard Silverman, Frederic Udina, and Matt Wand graciously read and gave comments on (close to) final drafts of the manuscript. Cliff Frohlich, Chong Gu, Clive Loader, Gary Oehlert, David Scott, and Matt Wand shared code and data sets with me. Marc Scott provided invaluable assistance in installing and debugging computer code for different methods. I sincerely thank all these people for their help. Finally, I would like to thank my two editors at Springer, Martin Gilchrist and John Kimmel, for shepherding this book through the publication process.

A World Wide Web (WWW) archive at the URL address

`http://www.stern.nyu.edu/SOR/SmoothMeth`

is devoted to this book, and I invite readers to examine it using a WWW browser, such as `Netscape` or `Mosaic`. I am eager to hear from readers about any aspect of the book. I can be reached via electronic mail at the Internet address `jsimonoff@stern.nyu.edu`.

East Meadow, N.Y. Jeffrey S. Simonoff

# Contents

Chapter 1

# Introduction

## 1.1 Smoothing Methods: a Nonparametric/Parametric Compromise

One thing that sets statisticians apart from other scientists is the general public's relative ignorance about what the field of statistics actually is. People have at least a general idea of what chemistry or biology is — but what is it exactly that statisticians *do*?

One answer to that question is as follows: statistics is the science that deals with the collection, summarization, presentation, and interpretation of data. Data are the key, of course — the stuff from which we gain insights and make inferences (or, to paraphrase Sherlock Holmes, the clay from which we make our bricks).

Consider Table 1.1. This data set represents the three-month certificate of deposit (CD) rates for 69 Long Island banks and thrifts, as given in the August 23, 1989, issue of *Newsday*. This table presents a valid data collection but clearly is quite inadequate for summarizing or interpreting the data. Indeed, it is difficult to glean any information past a feeling for the range (roughly 7.5% – 8.8%) and a "typical" value (perhaps around 8.3%).

**Table 1.1.** Three-month CD rates for Long Island banks and thrifts.

| | | | | | | | | | |
|---|---|---|---|---|---|---|---|---|---|
| 7.56 | 7.57 | 7.71 | 7.82 | 7.82 | 7.90 | 8.00 | 8.00 | 8.00 | 8.00 |
| 8.00 | 8.00 | 8.00 | 8.05 | 8.05 | 8.06 | 8.11 | 8.17 | 8.30 | 8.33 |
| 8.33 | 8.40 | 8.50 | 8.51 | 8.55 | 8.57 | 8.65 | 8.65 | 8.71 | |
| 7.51 | 7.75 | 7.90 | 8.00 | 8.00 | 8.00 | 8.15 | 8.20 | 8.25 | 8.25 |
| 8.30 | 8.30 | 8.33 | 8.34 | 8.35 | 8.35 | 8.36 | 8.40 | 8.40 | 8.40 |
| 8.40 | 8.40 | 8.40 | 8.45 | 8.49 | 8.49 | 8.49 | 8.50 | 8.50 | 8.50 |
| 8.50 | 8.50 | 8.50 | 8.50 | 8.50 | 8.50 | 8.52 | 8.70 | 8.75 | 8.78 |

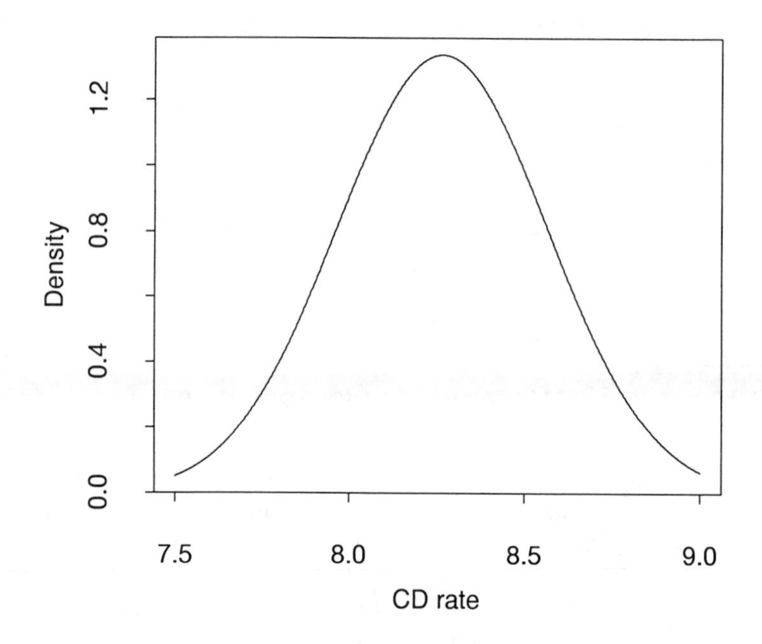


**Fig. 1.1.** Fitted Gaussian density estimate for Long Island CD rate data.

The problem is that no assumptions have been made about the underlying process that generated these data (loosely speaking, the analysis is purely nonparametric, in the sense that no formal structure is imposed on the data). Therefore, no true summary is possible. The classical approach to this difficulty is to assume a parametric model for the underlying process, specifying a particular form for the underlying density. Then, appropriate summary statistics can be calculated, and a fitted density can be presented. For example, a data analyst might hypothesize a Gaussian form for the density $f$. Calculation of the sample mean ($\overline{X} = 8.26$) and standard deviation ($s = .299$) then determines a specific estimate, which is given in Fig. 1.1. This curve provides a wealth of information about the pattern of CD rates, including typical rates, the likelihood of finding certain rates at a randomly selected institution, and so on.

Unfortunately, the strength of parametric modeling is also its weakness. By linking inference to a specific model, great gains in efficiency are possible, but *only if the assumed model is (at least approximately) true.* If the assumed model is not the correct one, inferences can be worse than useless, leading to grossly misleading interpretations of the data.

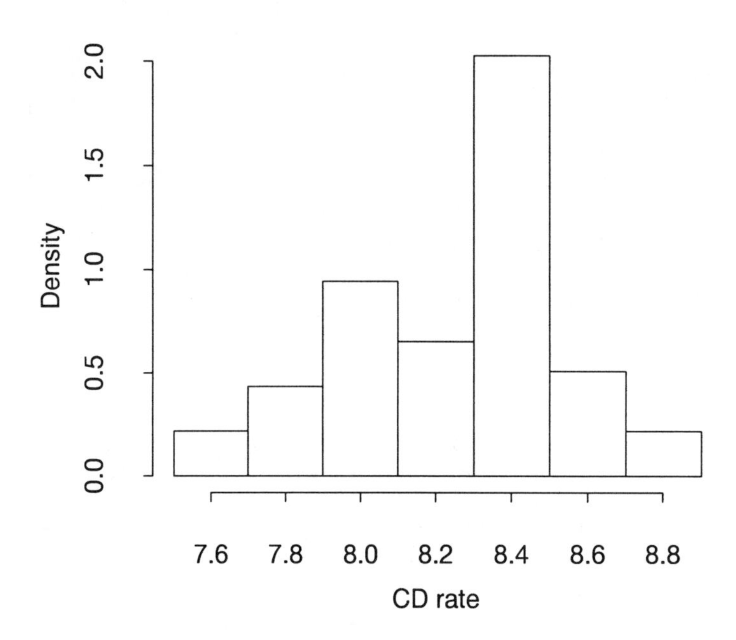


**Fig. 1.2.** Histogram for Long Island CD rate data.

Smoothing methods provide a bridge between making no assumptions on formal structure (a purely nonparametric approach) and making very strong assumptions (a parametric approach). By making the relatively weak assumption that whatever the true density of CD rates might be, it is a smooth curve, it is possible to let the data tell the analyst what the pattern truly is. Figure 1.2 gives a histogram for these data, based on equally sized bins (discussion of histograms and their variants is the focus of Chapter 2). The picture is very different from the parametric curve of Fig. 1.1. The density appears to be bimodal, with a primary mode around 8.5% and a secondary mode around 8.0% (the possibility that the observed bimodality could be due to the specific construction of this histogram should be addressed, and such issues are discussed in Chapter 2).

The form of this histogram could suggest to the data analyst that there are two well-defined subgroups in the data. This is, in fact, the case — the 69 savings institutions include 29 commercial banks and 40 thrift (Savings and Loan) institutions (the CD rates for the commercial banks correspond to the first three rows of Table 1.1, while those for the thrifts appear in the last four rows). These subgroups can be acknowledged parametrically by fitting separate Gaussian densities for the two groups (with means 8.15

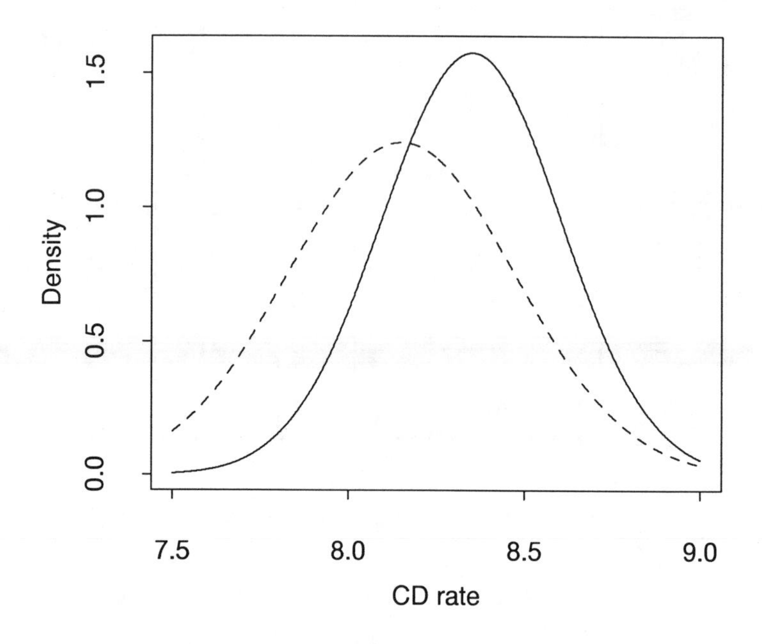


**Fig. 1.3.** Fitted Gaussian density estimates for Long Island CD rate data: commercial banks (dashed line) and thrifts (solid line).

and 8.35, respectively, and standard deviations .32 and .25, respectively).

Figure 1.3 gives the resultant fitted densities (the dashed line refers to commercial banks, while the solid line refers to thrifts). It is apparent that recognizing the distinction between commercial banks and thrifts helps to account for the bimodal structure in the histogram. There are several plausible hypotheses to explain this pattern. The Savings and Loan bailout scandal was just becoming big news at this time, and it is possible that many thrifts felt they had to offer higher rates to attract nervous investors. Another possibility is that these institutions were trying to encourage an influx of deposits so as to ward off bankruptcy. Still, Fig. 1.3 is less than satisfactory, as the modes are not as distinct as they are in Fig. 1.2 (indeed, the mixture density that combines these two Gaussian densities is unimodal).

Figure 1.4 provides still more insight into the data process. It gives two kernel density estimates for these data, corresponding to the commercial banks (dashed line) and the thrifts (solid line). These estimates can be thought of as smoothed histograms with very small bins centered at many

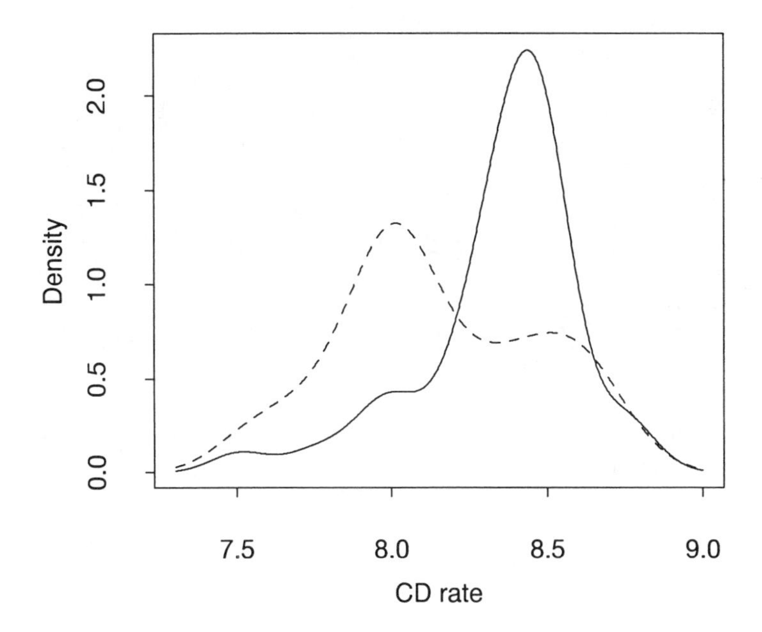


**Fig. 1.4.** Kernel density estimates for Long Island CD rate data.

different CD rate values. The underlying structure in the data is now even clearer. While the distinction between commercial banks and thrifts is a key aspect of the data, there is still more going on. The commercial bank CD rates are bimodal, with a primary mode at 8.0% and a secondary mode around 8.5%, while the distribution for the thrifts has a pronounced left tail and a mode around 8.5%. These modes account for the form of the histogram in Fig. 1.2.

Note also the apparent desirability of "round" numbers for CD rates of both types, as bumps or modes are apparent at 7.5%, 8.0%, and 8.5%. The construction and properties of density estimators of the type constructed in Fig. 1.4 will be discussed in Chapter 3.

It is clear that any model that did not take the subgroups into account would be doomed to failure for this data set. The subgroups do not exactly correspond to the observed modes in the original histogram (since the left mode comes from both subgroups), which shows that there is not necessarily a one-to-one correspondence between modes and subgroups. Still, the observed structure in the histogram was instrumental in recognizing the best way to approach the analysis of these data.

The ability of smoothing methods to identify potentially unexpected

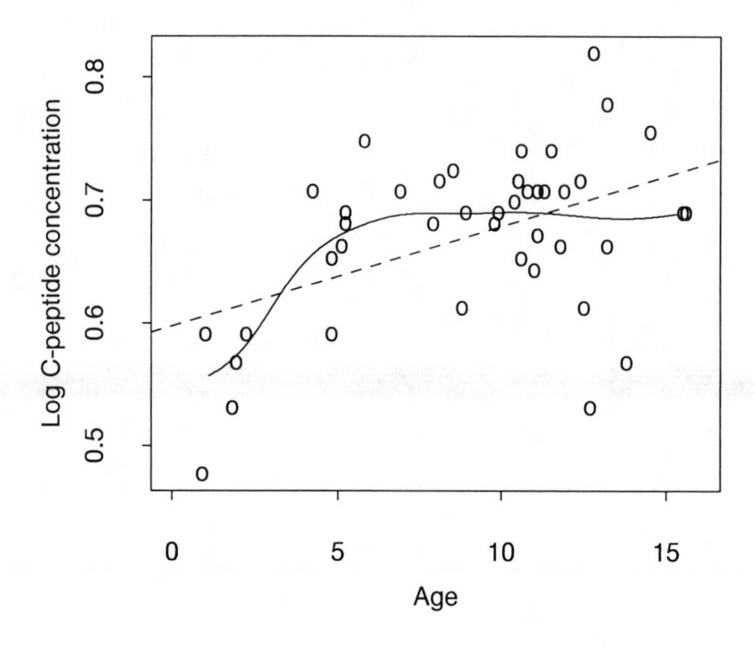


**Fig. 1.5.** Scatter plot of log C-peptide versus age with linear least squares (dashed line) and Nadaraya–Watson kernel (solid line) estimates superimposed.

structure extends to more complicated data analysis problems as well. The scatter plot given in Fig. 1.5 comes from a study of the factors affecting patterns of insulin-dependent diabetes mellitus in children. It relates the relationship of the logarithm of C-peptide concentration to age for 43 children. The superimposed dashed line is the ordinary least squares linear regression line, which does not adequately summarize the apparent relationship in the data. The solid line is a so-called Nadaraya–Watson kernel estimate (a smooth, nonparametric representation of the regression relationship), which shows that log C-peptide increases steadily with age to about age 7, where it levels off (with perhaps a slight rise around age 14). Again, the smoothed estimate highlights structure in a nonparametric fashion. Regression smoothers of this type will be discussed in Chapter 5.

It might be supposed that the benefits of smoothing occur only when analyzing relatively small data sets, but this is not the case. It also can happen that a data set can be so large that it overwhelms an otherwise useful graphical display.

Consider, for example, Fig. 1.6. This scatter plot refers to a data set examining the geographic pattern of sulfate wet deposition ("acid rain").

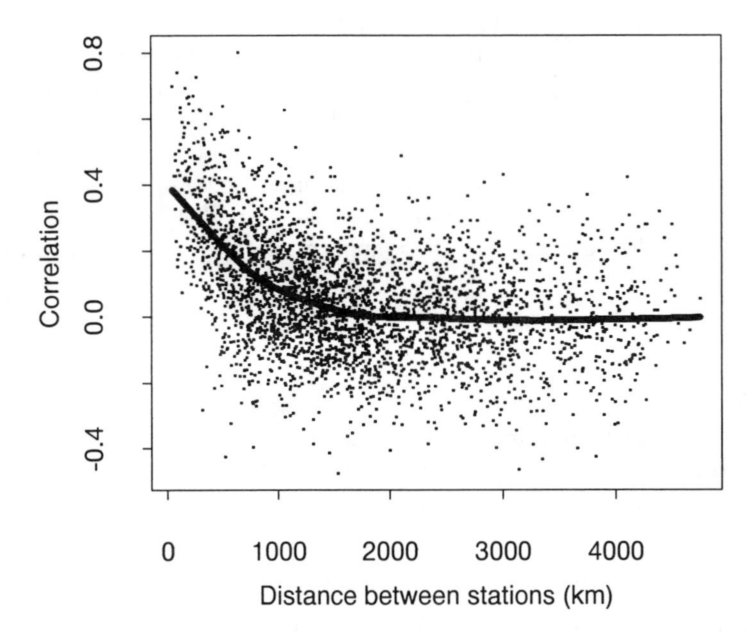


**Fig. 1.6.** Scatter plot of correlation of adjusted wet deposition levels versus distance with lowess curve superimposed.

The plot relates the distance (in kilometers) between measuring stations and the correlation of adjusted deposition levels (adjusted for monthly and time trend effects) for the 3321 pairings of 82 stations. It is of interest to understand and model this relationship, in order to estimate sulfate concentration and trend (and provide information on the accuracy of such estimates) on a regional level.

Unfortunately, the sheer volume of points on the plot makes it difficult to tell what structure is present in the relationship between the two variables past an overall trend. Superimposed on the plot is a *lowess curve*, a robust nonparametric scatter plot smoother discussed further in Chapter 5. It is clear from this curve that there is a nonlinear relationship between distance and correlation, with the association between deposition patterns dropping rapidly until the stations are roughly 2000 km apart, and then leveling off at zero.

## 1.2 Uses of Smoothing Methods

The three preceding examples illustrate the two important ways that smoothing methods can aid in data analysis: by being able to extract more information from the data than is possible purely nonparametrically, as long as the (weak) assumption of smoothness is reasonable; and by being able to free oneself from the "parametric straitjacket" of rigid distributional assumptions, thereby providing analyses that are both flexible and robust.

There are many applications of smoothing that use one, or both, of these strengths to aid in analysis and inference. An illustrative (but nonexhaustive) list follows; many of these applications will be discussed further in later chapters.

### A. Exploratory data analysis

The importance of *looking* at the data in any exploratory analysis cannot be overemphasized. Smoothing methods provide a way of doing that efficiently — often even the simplest graphical smoothing methods will highlight important structure clearly.

### B. Model building

Related to exploratory data analysis is the concept of model building. It should be recognized that choosing the appropriate model as the basis of analysis is an iterative process. Box (1980) stated this point quite explicitly: "Known facts (data) suggest a tentative model, implicit or explicit, which in turn suggests a particular examination and analysis of data and/or the need to acquire further data; analysis may then suggest a modified model that may require further practical illumination and so on." Box termed this the *criticism* stage of model building, and smoothing methods can, and should, be an integral part of it.

The earlier example regarding CD rates illustrates this point. A first hypothesized model for the rates might be Gaussian, yielding Fig. 1.1. Examination of a histogram (Fig. 1.2) suggests that this model is inadequate and that a model that takes account of any subgroups in the data could be more effective. Outside knowledge of the data process then leads to the idea of using the type of banking institution as those subgroups (without seeing the histogram, the existence of these subgroups might easily be ignored). This might suggest a model based on a mixture of two Gaussians, with possibly different variances, as is presented in Fig. 1.3. The kernel estimates in Fig. 1.4 imply that this model is still inadequate, however, and that further refinement is necessary. In this way, both the data and outside knowledge combine to progressively improve understanding of the underlying random process.

### C. Goodness-of-fit

Smoothed curves can be used to test formally the adequacy of fit of a hypothesized model. It is apparent from Figs. 1.2 and 1.4 that a Gaussian model is inadequate for the CD rate data; tests based on the difference between those curves and the Gaussian curve (Fig. 1.1) can be defined to assess this lack of fit formally. Similarly, the difference between the solid and dashed lines in Fig. 1.5 defines a test of the goodness-of-fit of a linear model to the diabetes data presented there. Tests constructed this way can be more powerful than those based on the empirical distribution alone and more robust than those based on a specific distributional form for the errors in a regression model. Alternatively, smoothed density estimates and regression curves can be used to construct confidence intervals and regions for true densities and regression functions, with similar avoidance of restrictive parametric assumptions.

### D. Parametric estimation

Density and regression estimates can be used in parametric inference as well. Suppose the mixture of two Gaussians tied to the type of bank institution that was graphically presented in Fig. 1.3 was hypothesized for the CD rate data (considering the long tails and bimodality apparent in the density estimates of Fig. 1.4, this might be a poor choice, however). An alternative to the usual maximum likelihood estimates would be to fit the two Gaussian densities that are "closest" to the curves in Fig. 1.4, defining closeness by some suitable distance metric. Estimators of this type are often fully efficient compared to maximum likelihood but are more robust, since a density estimate is much less sensitive to an unusual observation (outlier) than are maximum likelihood estimates like the sample mean and variance.

### E. Modification of standard methodology

Standard methodologies can be modified using smoothed density estimates by simply substituting the density estimate for either the empirical or parametric density function in the appropriate place. For example, discriminant analysis is usually based on assuming a multivariate normal distribution for each subgroup in the data, with either common (linear discriminant analysis) or different (quadratic discriminant analysis) covariance matrices. Then, observations are classified to the most probable group based on the normal densities. This procedure can be made nonparametric (and robust to violations of normality) by substituting smoothed density estimates for the normal density and classifying observations accordingly.

The bootstrap is another method where improvement via smoothing is possible. The ordinary bootstrap is based on repeated sampling from the data using the empirical distribution. This can result in bootstrap samples dominated by unusual observations, or repeat values, not typical of the true underlying density. Simulating from a smoothed version of the underlying density avoids these difficulties, leading to potentially more accurate estimates of standard error and confidence regions for statistics.

## 1.3 Outline of the Chapters

The purpose of this book is to provide a general discussion of smoothing methods in statistics, with particular emphasis on the actual application of such methods to real data problems. There is a good deal of common structure in all these methods, which will be emphasized throughout the discussion.

Chapter 2 introduces the ideas of data smoothing through the simplest of all density estimators, the histogram. The required tradeoff of bias versus variance is noted, as well as its connection to the choice of bin widths for the estimate. More efficient variants of the histogram idea, such as the frequency polygon, are then examined, and their strengths and weaknesses are described.

Smoother univariate density estimation methods are the focus of Chapter 3. The kernel estimator is discussed first, including an examination of the properties of the estimator and discussion of various methods that have been suggested to regulate the amount of smoothing of the estimator. Improvements and extensions of the estimator are also examined. In addition, alternative estimators are examined and compared with kernel estimators.

Generalization to multivariate density estimation is the subject of Chapter 4. High-dimensional spaces are difficult to deal with, and methods designed to overcome these problems are described.

Chapter 5 treats nonparametric regression techniques. The three most widely used methods — kernel, local polynomial, and spline estimators — are discussed and compared.

Application of smoothing to contingency tables is the subject of Chapter 6. Smoothing can lead to great gains in accuracy over the usual frequency estimates when the table is sparse. Variations and improvements of different estimators are examined and compared, and connections with nonparametric regression and density estimation are described.

Chapter 7 presents applications of smoothed estimates of the type described in Section 1.2. This includes a more detailed discussion of the uses of smoothing in discriminant analysis, goodness-of-fit, parametric estimation, and the bootstrap.

## Background material

### Section 1.1

The diabetes data come from a study discussed in Sockett *et al.* (1987). Hastie and Tibshirani (1990, p. 10) presented scatter plots with a polynomial fit superimposed, as well as eight different nonparametric smooth curves, each of which has an appearance similar to that of the curve in Fig. 1.5.

Oehlert (1993) discussed extensively the ecological and statistical aspects of the analysis of sulfate wet deposition data. He proposed using a multivariate time series model that incorporates both temporal and spatial correlation, and smoothing observed values based on physical distance between their associated locations.

### Section 1.2

Tukey (1977) emphasized the importance of looking at the data in any statistical analysis. Other authors besides Box (1980) that have noted the iterative nature of model building include Tukey (1977, p. v), Mosteller and Tukey (1977, Chapter 6), Cook and Weisberg (1982, pp. 7–8) and Cleveland (1993, p. 122). Other references for this section will appear in later chapters.

## Computational issues

This section describes the availability of computer software to use the methods discussed in each of the following chapters. The software includes both commercial and free (or shareware) resources. An important resource is the `statlib` archive at Carnegie–Mellon University; information on using `statlib` can be obtained by sending the message `send index` to the electronic mail address `statlib@lib.stat.cmu.edu`. In addition, the authors of many papers will provide code of some sort upon request; if a paper specifically encourages such dissemination of code, the section notes this. On-line directories of generally available resources, such as `archie` (for anonymous `ftp` access), `veronica` (for `gopher` access), and `Lycos` (for World Wide Web access), are invaluable in keeping up with the ever-expanding pool of information available over the Internet. Material that is available via one access method is often also available via other methods (so, for example, code available via anonymous `ftp` is often also available via `gopher` or a World Wide Web browser, such as `Mosaic` or `Netscape`).

## Exercises

**Exercise 1.1.** Consider the histogram in Fig. 1.2. Using that histogram, construct an alternative density estimate by drawing straight lines that connect the points defined by the middle of each bin interval and the height of that bin. Do you feel that either estimate is a better representation of the true underlying density of CD rates than the other? If so, why?

**Exercise 1.2.** Based on the kernel estimates given in Fig. 1.4, try to estimate the density of the CD rate data using a mixture of normal densities. How many such normals, with how many fitted parameters, would be needed to fit all the CD rates? Just the commercial banks? Just the thrifts?

**Exercise 1.3.** Fit a quartic polynomial regression model (that is, with terms up to power 4) to the diabetes data using any linear regression software. How does the fitted line compare to the kernel curve? Do you think a lower order polynomial might fit about as well as the quartic polynomial?

**Exercise 1.4.** Use a nonlinear regression package to fit the model

$$\text{Correlation} = \alpha \times \exp(-\beta \times \text{Distance}) + \epsilon$$

to the sulfate wet deposition data of Fig. 1.6. How does this fit compare to the lowess curve?

# Chapter 2
# Simple Univariate Density Estimation

## 2.1 The Histogram

### 2.1.1 Motivation for the histogram

A fundamental concept in the analysis of univariate data is the *probability density function*. Let $X$ be a random variable that has probability density function $f(x)$. The density function describes the distribution of $X$ and allows probabilities to be determined using the relation

$$P(a < X < b) = \int_a^b f(u)du.$$

A motivation for the construction of a nonparametric estimate of the density function can be found using the definition of the density function. Recall that

$$f(x) \equiv \frac{d}{dx}F(x) \equiv \lim_{h\to 0} \frac{F(x+h) - F(x)}{h}, \tag{2.1}$$

where $F(x)$ is the cumulative distribution function of the random variable $X$. Let $\{x_1, \ldots, x_n\}$ represent a random sample of size $n$ from the density $f$. A natural finite-sample analog of (2.1) is to divide the line into a set of $K$ equisized bins with small bin width $h$ and to replace $F(x)$ with the empirical cumulative distribution function

$$\hat{F}(x) = \frac{\#\{x_i \leq x\}}{n}.$$

This leads to the histogram estimate of the density within a given bin:

$$\hat{f}(x) = \frac{(\#\{x_i \leq b_{j+1}\} - \#\{x_i \leq b_j\})/n}{h}, \qquad x \in (b_j, b_{j+1}],$$

where $(b_j, b_{j+1}]$ defines the boundaries of the $j$th bin, or

$$\hat{f}(x) = \frac{n_j}{nh}, \qquad x \in (b_j, b_{j+1}], \tag{2.2}$$

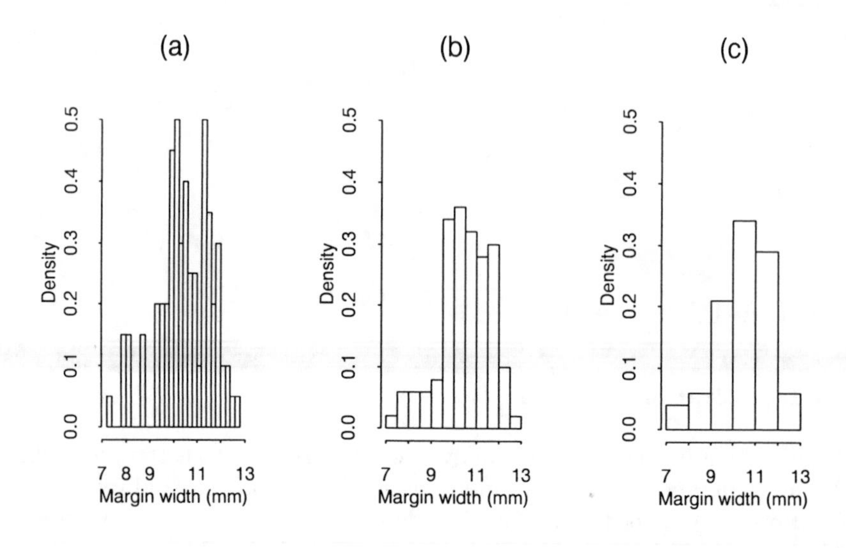


**Fig. 2.1.** Histograms of forged Swiss bank note data based on (a) 28 bins, (b) 12 bins, (c) 6 bins.

where $n_j$ is the number of observations in the $j$th bin and $h = b_{j+1} - b_j$. The histogram (and other density estimators) also are often used to summarize the distribution of observed values even if they are not a random sample from an (unknown) density (although then the usual theory is not relevant). The histogram estimator is undoubtedly the most commonly used univariate density estimator, and there are several good reasons for this. The estimator has the advantages of ease and simplicity of construction, simplicity of interpretation (including for the statistically unsophisticated), and lack of requirement of advanced graphical tools. Its popularity makes it important to understand its strengths and weaknesses and how to overcome those weaknesses.

### 2.1.2 Properties of the histogram

It is clear from (2.2) that the properties of the histogram estimator depend on the bin width $h$ (or, equivalently, the number of bins). The histograms given in Fig. 2.1 refer to observation of the width of the bottom margin (in millimeters) for 100 forged Swiss bank notes. Multimodality is of interest here, since forged bank notes are typically printed in batches (thus, a multimodal distribution would be indicative of possibly forged currency). The first histogram is based on 28 bins; the histogram seems undersmoothed

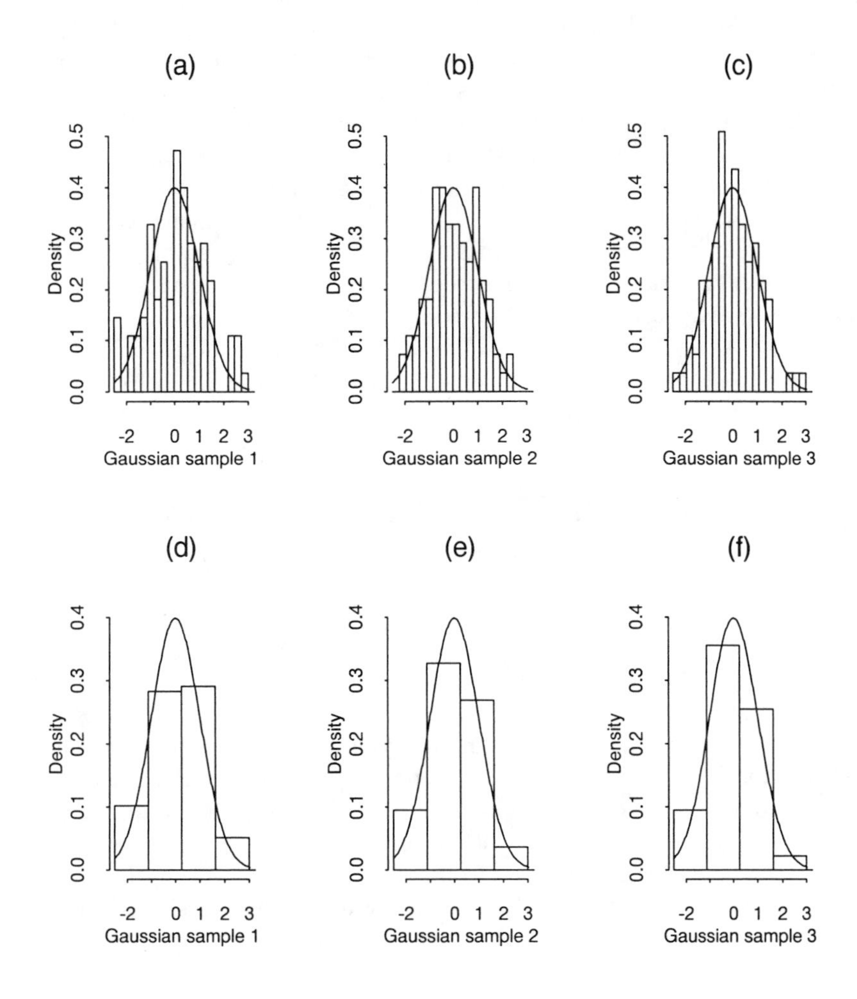


**Fig. 2.2.** Histograms of three randomly generated Gaussian samples: (a) – (c) using 20 bins; (d) – (f) using 4 bins.

(too bumpy), and there is evidence of modes around 8 mm (perhaps), 10 mm, and 11.5 mm. Figure 2.1(b), based on 12 bins, indicates two modes at around 10 and 11.5 mm (although the second mode is somewhat questionable). The third histogram, based on 6 bins, suggests merely one mode, with a possibly skewed density. Thus, the bin width choice is crucial in the construction and interpretation of the histograms.

The patterns in Fig. 2.1 can be viewed as reflecting a fundamental (and unavoidable) tradeoff in all smoothing methods, that of bias versus variance. Figure 2.2 presents histograms that illustrate this tradeoff. Plots (a), (b), and (c) are histograms with 20 bins each from 3 randomly generated samples of size 100 from a standard normal distribution, while plots (d), (e), and (f) are histograms with 4 bins each for the same 3 samples. The true Gaussian density is superimposed on each plot.

Plots (a) – (c) demonstrate the difficulties with undersmoothing. Although the bin heights generally follow the true density, they vary dramatically from plot to plot. That is, the estimator has small bias but large variability. In contrast, plots (d) – (f) highlight the problems with oversmoothing. In these plots, the bin heights are reasonably stable from plot to plot, but they don't follow the true density very well. That is, the estimator has small variability but large bias.

What is needed is some way to evaluate $\hat{f}(x)$ as an estimator of $f(x)$. One way to evaluate $\hat{f}(x)$ is via some measure of its difference from $f(x)$. The measure that is simplest to handle mathematically is squared error, $\mathrm{SE}(x) = \left[\hat{f}(x) - f(x)\right]^2$, and its expected value (mean squared error), $\mathrm{MSE}(x) = E_f\left[\hat{f}(x) - f(x)\right]^2$. Since global accuracy over the entire interval of support is usually of most importance, the integrated squared error,

$$\mathrm{ISE} = \int_{-\infty}^{\infty} \left[\hat{f}(u) - f(u)\right]^2 du,$$

and its expected value, mean integrated squared error (MISE), are also of interest.

Since the value of the histogram estimator in any bin follows a binomial distribution (scaled by a constant), it is possible to calculate the exact MSE of $\hat{f}(x)$, but it is simpler to examine the asymptotic MSE (as the sample size $n \to \infty$). In order for the estimator to be consistent, the bins must get narrower, with the number of observations per bin getting larger, as $n \to \infty$; that is, $h \to 0$ with $nh \to \infty$. If the underlying density is smooth enough ($f'(x)$ is absolutely continuous and square integrable), then

$$\begin{aligned}\mathrm{Bias}\left[\hat{f}(x)\right] &= E_f\left[\hat{f}(x)\right] - f(x) \\ &= \frac{1}{2} f'(x)\left[h - 2(x - b_j)\right] + O\left(h^2\right), \qquad x \in (b_j, b_{j+1}], \end{aligned} \tag{2.3}$$

while the variance is

$$\mathrm{Var}\left[\hat{f}(x)\right] = \frac{f(x)}{nh} + O\left(n^{-1}\right).$$

Combining the squared bias and variance yields the mean squared error,

$$\begin{aligned}\mathrm{MSE}\left[\hat{f}(x)\right] &= \mathrm{Var}\left[\hat{f}(x)\right] + \mathrm{Bias}^2\left[\hat{f}(x)\right] \\ &= \frac{f(x)}{nh} + \frac{f'(x)^2}{4}\left[h - 2(x - b_j)\right]^2 \\ &\quad + O\left(n^{-1}\right) + O\left(h^3\right). \end{aligned} \tag{2.4}$$

Finally, integrating over each bin, and then summing bin by bin, gives

$$\text{MISE} = \frac{1}{nh} + \frac{h^2 R(f')}{12} + O\left(n^{-1}\right) + O\left(h^3\right), \tag{2.5}$$

where $R(\phi)$ represents $\int \phi(u)^2\, du$. The notation AMISE will be used to represent asymptotic MISE (the leading terms in the expansion of MISE).

The tradeoff of bias versus variance previously noted can be seen mathematically in (2.5). The bin width $h$ is directly related to the integrated squared bias $h^2 R(f')/12$ and inversely related to the integrated variance $(nh)^{-1}$. That is, narrower bins give an estimator that is less biased (as $h \to 0$, $\hat{f}$ approaches a set of spikes at each observation and has zero bias) but more variable; making the bins wider increases the number of observations per bin, reducing variance, but increasing the bias. The histograms in Figs. 2.1 and 2.2 illustrate the practical implications of this tradeoff.

The minimization of MISE requires explicitly balancing bias and variance through the choice of the bin width $h$. The minimizer of AMISE is easily determined as

$$h_0 = \left[\frac{6}{R(f')}\right]^{1/3} n^{-1/3}, \tag{2.6}$$

which results in the minimum AMISE,

$$\text{AMISE}_0 = \left[\frac{9R(f')}{16}\right]^{1/3} n^{-2/3}. \tag{2.7}$$

Equations (2.6) and (2.7) show that the roughness of the underlying density, as measured by $R(f')$, determines the optimal level of smoothing and the accuracy of the histogram estimate. Densities with few bumps (smaller $R(f')$) are easier to estimate and require a wider bin, while bumpy densities (larger $R(f')$) are more difficult to estimate and require a smaller bin width.

### 2.1.3 Choosing the bin width

Equation (2.6) provides an unambiguous rule (in terms of AMISE) for choosing the bin width $h$ of a fixed bin-width histogram. Unfortunately, this rule involves the density $f$, which is precisely what is being estimated, so it is of limited usefulness.

The most straightforward approach to choosing $h$ is to pick a particular density $f$ and simply substitute into (2.6) to get a value of $h$. Not surprisingly, the typical choice is a Gaussian density. It can be shown that the minimizer of AMISE then has the form

$$h_0 = 3.491 \sigma n^{-1/3}. \tag{2.8}$$

Practical application of the rule requires an estimate of $\sigma$. The usual estimates, such as the sample standard deviation or interquartile range, can be inadequate for nonnormal data. Multiple modes inflate these measures, meaning that a Gaussian-based histogram based on them will be oversmoothed.

What is needed is a measure of scale with a local focus — that is, one that is dependent on the variability of the observations within homogeneous subgroups (modes), rather than for the entire sample (across several modes). Janssen *et al.* (1995) developed such a measure, which leads to a more effective histogram based on (2.8) for multimodal densities.

The obvious drawback in choosing the bin width based on (2.8) is that there is no theoretical justification for this approach if the underlying density is not Gaussian. A method that does not require assuming the nature of the density would be more generally applicable. Consider again the ISE measure:

$$\begin{aligned}\text{ISE} &= \int [f(u) - \hat{f}(u)]^2\,du \\ &= \int f(u)^2\,du + \int \hat{f}(u)^2\,du - 2\int \hat{f}(u)f(u)du \\ &\equiv R(f) + R(\hat{f}) - 2\int \hat{f}(u)f(u)du. \end{aligned} \tag{2.9}$$

$R(f)$ does not depend on $\hat{f}$, so it is irrelevant for the purposes of choosing $h$. The last term of (2.9) is evidently $-2E[\hat{f}(X)]$ (where the expectation is with respect to the point of evaluation, not over $\mathbf{x}$), which must be estimated. The principle of *cross-validation* provides one way to do that. Dropping the $i$th case from the random sample $\{x_1, \ldots, x_n\}$ leads to an estimate $\hat{f}_{-i}(x_i)$ that has expected value $E[\hat{f}(X)]$ (based on the full sample). That is,

$$E[\hat{f}_{-i}(x_i)] = E[\hat{f}(X)],$$

or

$$E\left[\frac{1}{n}\sum_{i=1}^{n} \hat{f}_{-i}(x_i)\right] = E[\hat{f}(X)] \equiv \int \hat{f}(u)f(u)du.$$

Thus, a data-dependent way to choose $h$ that minimizes an unbiased estimator of ISE is to minimize the cross-validation criterion

$$\text{CV} = R(\hat{f}) - \frac{2}{n}\sum_{i=1}^{n} \hat{f}_{-i}(x_i). \tag{2.10}$$

An equivalent form of (2.10) for the histogram estimator is

$$\text{CV} = \frac{1}{h}\left[\frac{2}{n-1} - \frac{n+1}{n-1}\sum_{j=1}^{K}\left(\frac{n_j}{n}\right)^2\right]. \tag{2.11}$$

The potential strengths and weaknesses of these two approaches can be seen in Fig. 2.3. The given histograms use the Gaussian-based rule (first

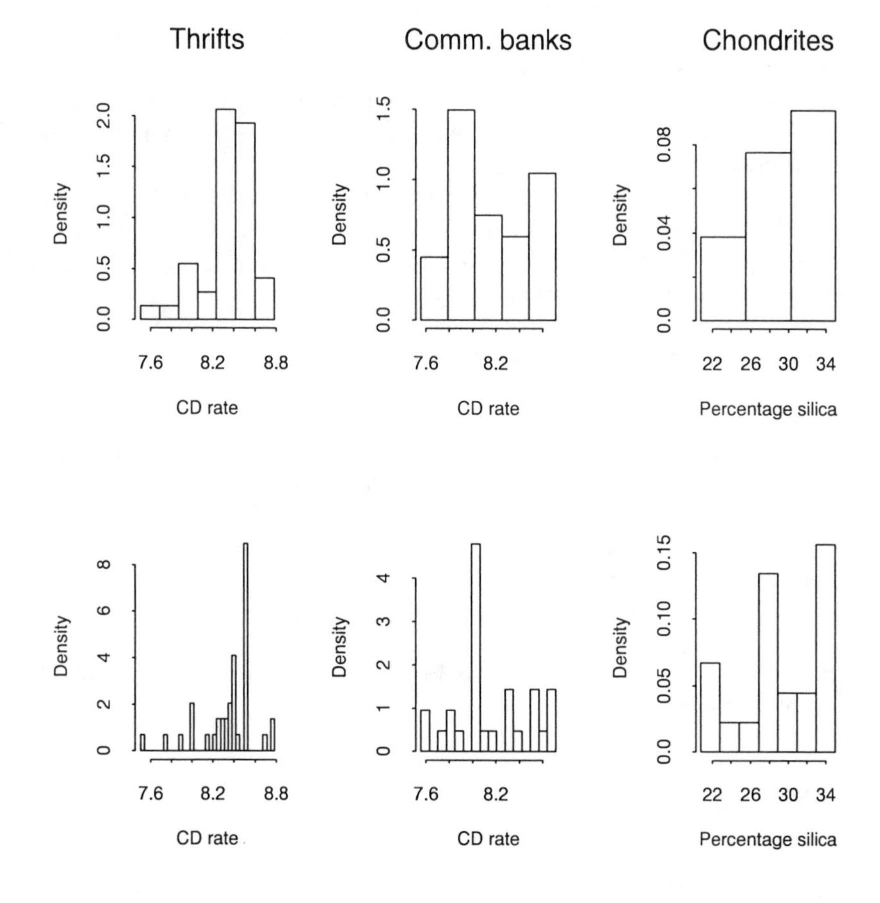


**Fig. 2.3.** Histograms of CD rates for Long Island thrifts, CD rates for Long Island commercial banks, and percentage silica in chondrite meteors, respectively, choosing bin width based on the Gaussian-based rule (top) and cross-validation (bottom).

row) or cross-validation (second row), for three data sets: CD rates for Long Island thrifts (first column), CD rates for Long Island commercial banks (second column), which were discussed in Chapter 1, and the percentage of silica in 22 chondrite meteors, for which a trimodal density has been suggested in many previous analyses.

An encouraging aspect of these plots is that the simple Gaussian-based histograms, despite being based on the density having a roughly Gaussian shape, show the bimodal structure of both CD rate distributions, with modes at 8% and 8.5%. Unfortunately, the histogram for the chondrite data is oversmoothed, suggesting an inability of the scale estimate to focus on the underlying subgroups in the data.

The cross-validation-based histograms also can be problematic. Both of the CD rate histograms are obviously undersmoothed, exhibiting the excess roughness that comes from too many bins. This pattern of undersmoothing is not unusual for density estimates based on cross-validation. On the other hand, this method does bring out the trimodal nature of the chondrite data very nicely.

One other point about a data-dependent choice of the level of smoothing of the estimate should be made. It could be argued that an automatic choice is not even worth investigating, since the choice should be made by the analyst subjectively anyway. Certainly, no data-dependent rule should be followed blindly, and subjective impressions of the proper amount of smoothing for a given set of data are important.

That does not diminish the importance of data-dependent choices, however. In some applications, graphical summaries need to be provided automatically (for screening of large amounts of data, for example), which requires automatic smoothness determination. Statistical investigation of the general properties of different methods via Monte Carlo simulation also requires such automatic choice, as does the use of resampling (bootstrap) methods to assess their properties for a given data set. Finally, even subjective choice requires some sort of "benchmark" to use as a basis from which to (subjectively) vary. For all these reasons, effective automatic determination of the amount of smoothing for any smoothing method is highly desirable if the method is to be applied in practice.

## 2.2 The Frequency Polygon

### 2.2.1 Properties of the frequency polygon

Whatever the usefulness of histograms might be in the presentation of data, it is apparent that they do not provide an adequate description of a smooth density function, because of their inherent piecewise constant nature. More accurate estimators should necessarily be smoother. A simple way to make a histogram appear smoother, by avoiding the discontinuities at the bin edges, is to connect mid-bin values by straight lines. The resultant *frequency polygon* estimator is continuous, but its derivative is undefined at the mid-bin points.

Let $\{b_1, \ldots, b_{K+1}\}$ again represent bin edges of bins with width $h$, with $\{n_1, \ldots, n_K\}$ being the number of observations falling in the bins, and define $\{c_0, \ldots, c_{K+1}\}$ to be the midpoints of the bin intervals (that is, $c_j = (b_j + b_{j+1})/2$, $j = 1, \ldots, K$, with $c_0 = b_1 - h/2$ and $c_{K+1} = b_{K+1} + h/2$). The frequency polygon is then defined as

$$\hat{f}(x) = \frac{1}{nh^2}\left[n_j c_{j+1} - n_{j+1} c_j + (n_{j+1} - n_j)x\right], \quad x \in [c_j, c_{j+1}], \qquad (2.12)$$

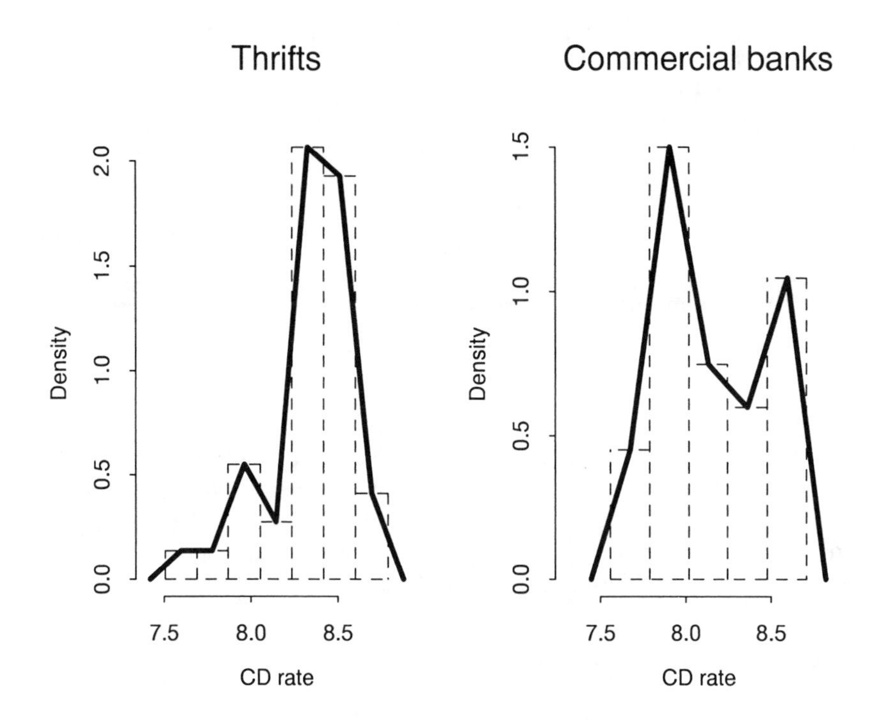


**Fig. 2.4.** Frequency polygon construction based on histograms of CD rates for Long Island thrifts and CD rates for Long Island commercial banks.

where $n_0 = n_{K+1} \equiv 0$. Figure 2.4 gives examples of the construction of frequency polygons based on histograms given previously in Fig. 2.3 for Long Island CD rates. The resultant density estimate gives a more aesthetically pleasing representation of the density than the histogram, while retaining its simplicity and ease of interpretation.

Asymptotic analysis shows that this aesthetic improvement can translate into improved accuracy of the estimator. Assume that $f''$ is absolutely continuous, and $R(f)$, $R(f')$, $R(f'')$, and $R(f''')$ are all finite (note that additional smoothness of $f$ is being assumed here, compared with what is required for histogram estimation). Proceeding in a manner analogous to that for histograms (basing properties on two adjacent bin counts, rather than only one) gives the form of the MISE:

$$\text{MISE} = \frac{2}{3nh} + \frac{49h^4 R(f'')}{2880} + O\left(n^{-1}\right) + O\left(h^6\right). \tag{2.13}$$

As was true for histograms (see Eq. (2.5)), the first (variance) term varies inversely with bin size, while the second (bias) term varies directly with it. The remarkable change, however, is that the squared bias term is now $O(h^4)$, rather than $O(h^2)$; evaluating the histogram only at the mid-bin value, where the bias is $O(h^2)$, rather than $O(h)$, reduces the bias by an order of magnitude.

The implications of this bias reduction are immediate. Minimization of the leading terms of (2.13) provides the asymptotically optimal bin width

$$h_0 = 2\left[\frac{15}{49R(f'')}\right]^{1/5} n^{-1/5}, \tag{2.14}$$

which gives the minimal asymptotic MISE

$$\text{AMISE}_0 = \frac{5}{12}\left[\frac{49R(f'')}{15}\right]^{1/5} n^{-4/5}. \tag{2.15}$$

Thus, the simple strategy of imposing straight-line interpolation onto a histogram estimate results in a density estimator with an improved convergence rate. Comparison of (2.14) and (2.6) shows that the optimal bin width of a frequency polygon is different from that of a histogram ($O(n^{-1/5})$ versus $O(n^{-1/3})$), and will asymptotically be larger.

### 2.2.2 Choosing the bin width

The simplest way to choose the bin width $h$ for a frequency polygon is to substitute a particular form of $f$ into (2.14). The resultant Gaussian reference rule is

$$h_0 = 2.15\sigma n^{-1/5}, \tag{2.16}$$

with an appropriate estimate of scale substituted for $\sigma$.

The frequency polygon given in Fig. 2.5 uses this Gaussian-based construction. The estimator refers to the distribution of velocities (in kilometers per second) of 82 galaxies relative to the Milky Way. If galaxies cluster together into so-called superclusters, the distribution of velocities would be multimodal, with each mode representing a cluster as it moves away at its own speed. The frequency polygon supports the supercluster hypothesis, possessing modes at around 10,000, 19,000, 23,000, 27,000, and 33,000 km per second.

## 2.3 Varying the Bin Width

The histogram (2.2) and frequency polygon (2.12) estimators discussed thus far are based on a fixed (constant) bin width for all cells. This is not, however, the best approach (at least theoretically). The bin width $h$ controls the

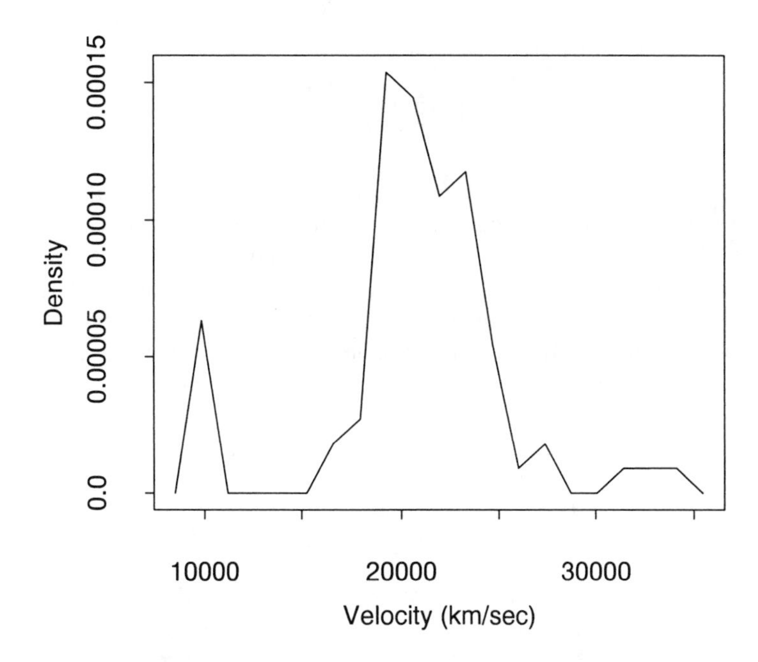


**Fig. 2.5.** Gaussian-based frequency polygon of velocities of galaxies data.

bias – variance tradeoff at any point, and the best tradeoff varies depending on the local properties of the true density.

Equation (2.4) quantifies this tradeoff in terms of MSE for the histogram, as it implies that the local properties of the density at any point $x$ determine the accuracy of a histogram at $x$. In particular, the bin width should be larger in regions of high density, to reduce the first (variance) term, while it should be inversely related to $|f'(x)|$, in order to minimize the second (bias) term. The latter pattern is intuitive, since it implies that more detail will be available in the histogram estimate in areas where the density is changing rapidly. Thus, a histogram with locally varying bin widths should be more accurate than one with fixed bin widths.

Generalizing the definition of a histogram to locally varying bin widths is straightforward. Differentiation of the empirical cumulative distribution function motivates the appropriate estimate:

$$\hat{f}(x) = \frac{n_j}{n(b_{j+1} - b_j)}, \qquad x \in (b_j, b_{j+1}]. \tag{2.17}$$

Thus, the problem becomes one of choosing the bin edges $\{b_1, \ldots, b_{K+1}\}$.

A similar pattern emerges for the frequency polygon. The asymptotic MISE of the estimator over the bin containing $x$ has the form

$$\text{AMISE}(x) = \frac{2f(x)}{3nh} + \frac{49}{2880} h^4 f''(x)^2, \tag{2.18}$$

implying that the optimal local bin width satisfies

$$h_x = 2 \left[ \frac{15 f(x)}{49 f''(x)^2} \right]^{1/5} n^{-1/5}.$$

Comparison of (2.18) and (2.6) shows that the locally optimal bin width for a frequency polygon is different from that for a histogram, with $f''(x)$ determining the bias, rather than $f'(x)$. The behavior of these two measures will often coincide in the tails (with both being small, implying larger bin widths), but otherwise can be quite different.

The frequency polygon can be generalized to allow varying bin widths by simply allowing $b_{j+1} - b_j$ to vary, giving the general form

$$\begin{aligned} \hat{f}(x) = & \frac{1}{n(c_{j+1} - c_j)} \left[ \frac{n_j c_{j+1}}{b_{j+1} - b_j} - \frac{n_{j+1} c_j}{b_{j+2} - b_{j+1}} \right. \\ & \left. + \left( \frac{n_{j+1}}{b_{j+2} - b_{j+1}} - \frac{n_j}{b_{j+1} - b_j} \right) x \right], \qquad x \in [c_j, c_{j+1}]. \end{aligned} \tag{2.19}$$

A simple way to construct locally varying bin-width histograms or frequency polygons that often works well in practice is by transforming the data to a different scale and then smoothing the transformed data. This can sometimes remove characteristics of a density that can cause trouble for a fixed bin-width estimator. Then, the final estimate is formed by simply transforming the constructed bin edges $\{b_j\}$ back to the original scale and using (2.17) or (2.19) (this assumes, of course, that the transformation is monotonic).

Figure 2.6 illustrates how this method can work. The data are the 1993 salaries of the 118 Major League baseball players who were eligible for salary arbitration before the 1993 season (these are players who have between two and six years of Major League service and whose contracts have expired). Figure 2.6(a) is a fixed bin-width frequency polygon of the data in the original dollar scale, based on a bin width of \$300,000. The long right tail of the distribution is apparent, with the highest mode at about \$1 million, but the roughness in the tail makes it difficult to decide if any other structure might be present in the data.

Salaries are often modeled using the lognormal distribution, and the long right tail of the distribution suggests that a logarithmic transformation can reveal new structure by deemphasizing the long tail. The frequency polygon in Fig. 2.6(b) is a fixed bin-width frequency polygon (with bin width .1) in the $\log_{10}$(salary) scale. After exponentiating back to the original scale, the locally varying bin-width frequency polygon given in Fig. 2.6(c) is the result. While this plot still shows the long tail and primary mode at \$1 million, it also exhibits minor modes at \$250,000, \$500,000, and \$2.5 million. This is an intuitive result, as these values correspond to appealing

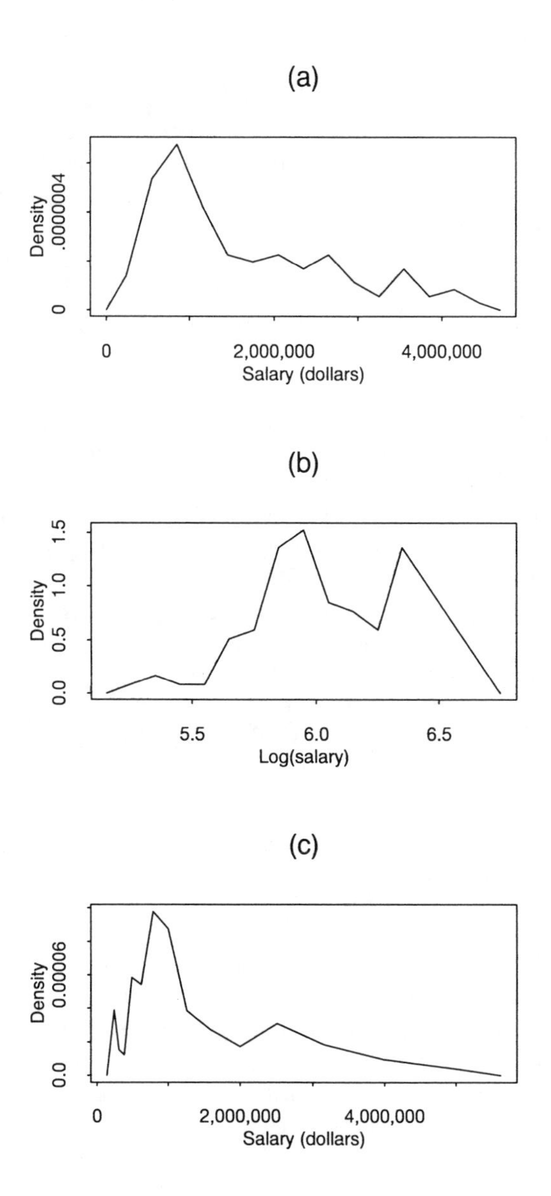


**Fig. 2.6.** Frequency polygons of baseball player salary data. (a) Fixed bin-width frequency polygon in original scale. (b) Fixed bin-width frequency polygon in logarithmic scale. (c) Locally varying bin-width frequency polygon based on exponentiation from logarithmic scale.

"round" numbers and reflect the inherent inhomogeneity in the data (while some players who are eligible for arbitration have already been successful in the Major Leagues and earn salaries in the millions, a notable percentage are utility players and earn considerably less).

Other (monotone) transformations can be useful for particular applications. The histograms in Fig. 2.7 are based on values of shooting percentage for the 292 National Hockey League players who played at least 60 games during the 1991–1992 season and scored at least one goal (the shooting percentage is the proportion of shots by a player on the opposition goal that are actually goals). Figure 2.7(a) is a 13-cell, fixed bin-width histogram in the original scale. Most players have a shooting percentage between 2% and 15%, but the structure in that area is very hard to see.

Since the data values are proportions here, a reasonable target scale is the logistic, $\ell = \log[p/(1-p)]$, where $p$ is the observed shooting percentage. After constructing a fixed bin-width histogram in that scale, it is then transformed back using the inverse transformation $p = \exp(\ell)/[\exp(\ell)+1]$. Figure 2.7(b) is the resultant variable bin-width histogram based on 13 cells. It now appears that the distribution of shooting percentages is trimodal. The reason for this pattern can be seen in Fig. 2.7(c). This plot gives two fixed bin-width histograms (bin width .01) for shooting percentage, separated by whether the player was a forward (shaded bars) or a defenseman (unshaded bars). The distribution of shooting percentages for defensemen is multimodal, with modes corresponding to defensively oriented defensemen and offensively oriented ones. After combining these players with the forwards (who are generally more offensively oriented, since they score most of the goals), the trimodal pattern in Fig. 2.7(b) emerges.

## 2.4 The Effectiveness of Simple Density Estimators

The examples given thus far show that simple density estimators like the histogram and frequency polygon can be very informative in highlighting interesting structure in a univariate data set. Still, the question remains whether such estimators are adequate for general use or should be replaced with better (and more complex) methods.

From the point of view of AMISE, the choice between the histogram and frequency polygon estimators is easy. Equations (2.7) and (2.15) imply that, as the sample size increases, the frequency polygon dominates the histogram (in terms of MISE) if the bin width is chosen appropriately. Monte Carlo simulations support this for small samples also, whether choosing the bin width to minimize the actual ISE, or in a data-dependent way, as the frequency polygon consistently has smaller average ISE than the histogram in the situations examined. Since there is virtually no additional effort in calculating or presenting a frequency polygon compared to a histogram,

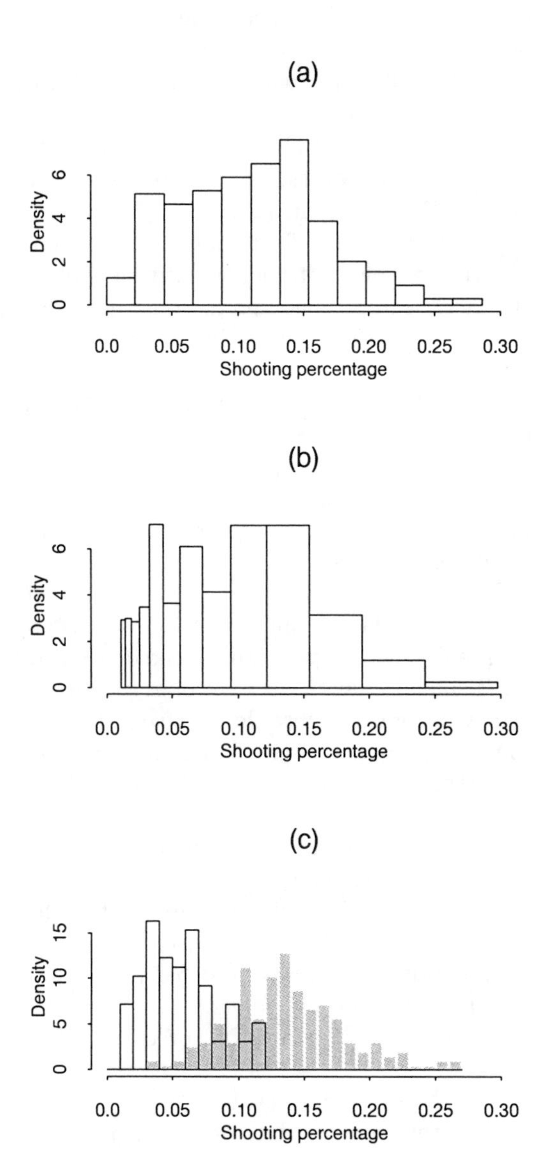


**Fig. 2.7.** Histograms of hockey player shooting percentage data. (a) Fixed bin-width histogram. (b) Locally varying bin-width histogram based on inverse transformation from logistic scale. (c) Fixed bin-width histograms separated by whether player was a defenseman (unshaded bars) or a forward (shaded bars).

frequency polygons should be used routinely instead of histograms, and histograms should be replaced as standard output of statistical packages.

Simulations also provide guidance about how to choose the bin width in practice. More complicated approaches, such as cross-validation (and its variants), do not seem to be worth the effort, since the Gaussian reference rules (2.8) and (2.16) beat them (with respect to ISE), even if the true underlying density is far from Gaussian.

A related question is that of choosing the anchor of a histogram or frequency polygon ($b_1$ in the notation of Sections 2.1.1 and 2.2.1, respectively). The form of MISE given in (2.5) and (2.13), respectively, shows that the choice has a lower-order effect than does the choice of $h$ asymptotically, but it is still possible that anchor position could be important in small samples. In fact, Monte Carlo simulations indicate that the choice of anchor position has a (perhaps surprisingly) small effect on the ISE of the resultant histograms or frequency polygons, as long as the underlying bins do not cross a discontinuity point (for example, the point $x = 0$ for an exponential density).

Do these results mean that a simple Gaussian-based estimator, with arbitrarily chosen anchor, is the final density estimator of choice? Unfortunately, no. Consider the six histograms given in Fig. 2.8. These histograms are all estimates for one data set ($n = 50$) generated from the bimodal normal mixture distribution $.5\,N(0,1) + .5\,N(3,1)$, all based on the same bin width $h = 1$, but with different anchor positions. It is apparent that the histograms give quite different impressions of the density, with from one to three modes indicated, and different degrees of asymmetry being suggested (obviously, frequency polygons also would have very different appearances). Despite their different appearances, all of these histograms have ISE in the range $[.0153, .0169]$ (that is, the least accurate of these histograms has ISE only 10% higher than the most accurate). The ISE measure does not adequately reflect the performance of the estimate in the way that matters to the data analyst — that is, the ability to identify interesting structure in the data accurately. Monte Carlo simulations confirm this result, in that seemingly accurate frequency polygon rules (in terms of ISE) can be very poor at resolving modes in the underlying density. The stability of the appearance of a histogram or frequency polygon also depends on the precision to which the data are reported and how that precision relates to the chosen bin width.

The implication of these results is that the simple density estimation methods described in this chapter don't take advantage of the power of smoothing methods well enough. Apparently, reaping the practical benefits of smoothing requires better smoothing methods, which are necessarily more complex. Such methods are the subject of the succeeding chapters.

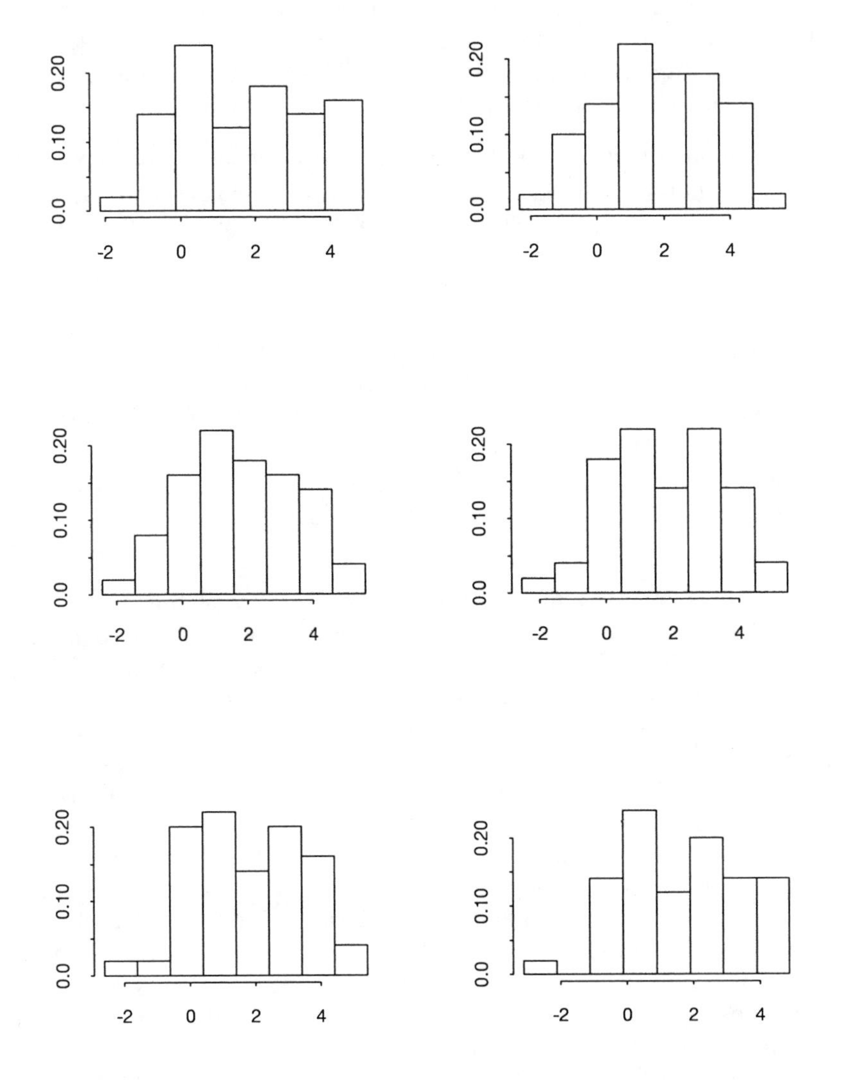


**Fig. 2.8.** Histograms of bimodal mixture data using different anchor positions. All histograms have bin width equal to 1.

## Background material

### Section 2.1

**2.1.1.** The histogram estimator can be motivated using ordinary (likelihood-based) theory, as it is the unique maximum likelihood estimator over the set of estimates of $f$ that are piecewise constant on the set of bins $\{b_j\}$ (de Montricher, Tapia, and Thompson, 1975). This is true for both equal and unequal bin-width histograms.

**2.1.2.** The observations of bottom margin width for Swiss bank note data come from a larger data set given in Flury and Riedwyl (1988).

Standard notation concerning the convergence properties of sequences is as follows:

(1) *Deterministic sequences:* Let $x_n$ and $y_n$ be two real-valued deterministic (nonrandom) sequences. Then, as $n \to \infty$,
  (a) $x_n = O(y_n)$ if and only if $\limsup_{n\to\infty} |x_n/y_n| < \infty$,
  (b) $x_n = o(y_n)$ if and only if $\lim_{n\to\infty} |x_n/y_n| = 0$.

(2) *Random sequences:* Let $X_n$ and $Y_n$ be two real-valued random sequences. Then, as $n \to \infty$,
  (a) $X_n = O_p(Y_n)$ if and only if for all $\epsilon > 0$, there exist $\delta$ and $N$ such that $P(|X_n/Y_n| > \delta) < \epsilon$, for all $n > N$,
  (b) $X_n = o_p(Y_n)$ if and only if for all $\epsilon > 0$, $\lim_{n\to\infty} P(|X_n/Y_n| > \epsilon) = 0$.

Many authors have proposed rules to determine the number of bins to use in the construction of a histogram. In a weak sense, the choice of $h$ is not crucial, since $\hat{f}(x)$ is a consistent estimator of $f(x)$ as long as $n \to \infty$ with $h \to 0$ and $nh \to \infty$ (Abou-Jaoude, 1976a,b), but this does not reflect the very different appearances histograms have for finite samples when based on different bin widths. Suggested bin-selection methods include rules based on the logarithm of the sample size (Sturges, 1926; Larson, 1975, p. 15; Doane, 1976; Becker, Chambers, and Wilks, 1988; Dixon, 1988), the square root of the sample size (Duda and Hart, 1973; Davies and Goldsmith, 1980, p. 11; Numerical Algorithms Group, 1986), and the sample size itself (Ishikawa, 1986, p. 8). Scott (1979) and Freedman and Diaconis (1981) established the optimal rate, in terms of minimizing AMISE. The optimal $O(n^{-2/3})$ rate for AMISE is uncomfortably slow, as Boyd and Steele (1978) showed that the fastest possible rate is $O(n^{-1})$. This is the rate achieved in the parametric situation when using a $\sqrt{n}$-consistent estimator, so this deficiency can be viewed as the price one pays (when using the histogram) for removing parametric assumptions.

Equation (2.5) provides a way to approximate the effect of choosing the bin width $h$ suboptimally. Choosing the bin width $h^* = ch_0$, substituting into (2.5), and comparing with the value given in (2.7) shows that $\text{AMISE}^* = (c^3 + 2)/(3c) \times \text{AMISE}_0$. The AMISE is fairly insensitive to errors in the choice of $h$ as high as 30% or so. Proportional errors above the optimal $h$ are somewhat less deleterious than errors below the optimal $h$; for example, there is a 13% increase in AMISE associated with choosing $h$ 40% too high, as opposed to a 23% increase when choosing $h$ 40% too low. Thus, oversmoothing a bit hurts accuracy (as measured by squared error) less than a similar amount of undersmoothing.

**2.1.3.** Lugosi and Nobel (1996) gave sufficient conditions for the consistency of (possibly multivariate) histograms based on data-dependent partitioning rules. Scott (1979) proposed applying the Gaussian-based rule (2.8) using the sample standard deviation $s$ as the estimate of $\sigma$. Freedman and Diaconis (1981) noted that $s$ is not appropriate for non-Gaussian densities and suggested using the interquartile range, with the rule $h = 2(\text{IQR})n^{-1/3}$. Silverman (1986, p. 47) combined these two ideas by proposing the rule $h = 3.49\hat{\sigma}n^{-1/3}$, where $\hat{\sigma} = \min(s, \text{IQR}/1.34)$.

It is possible to justify the Gaussian-based rule without the requirement of underlying normality. Terrell and Scott (1985) noted that while determination of the optimal number of bins from (2.6) requires exact knowledge of $f$, it is possible to estimate a lower bound for $R(f')$, based on the data, thereby giving a lower bound for the optimal number of bins. They termed this *oversmoothing*, since any wider bin width cannot be the optimal choice. They determined the density that minimizes $R(f')$ among bounded densities and then showed that using the range of the data as a scale estimate implies that the optimal number of bins must be at least $(2n)^{1/3}$. This corresponds to a bin-width requirement $h \leq 3.55\sigma n^{-1/3}$, which is very similar to the Gaussian-based rule. Terrell (1990) investigated the oversmoothing idea further by exploring measures of scale other than the range, leading to other rules (still similar to the Gaussian one).

Rudemo (1982) and Bowman (1984) originally suggested the use of cross-validation for choosing the bin width of a histogram. Stone (1985a) provided theoretical justification for the CV rule (2.11), showing that this rule is asymptotically optimal, in the following sense: Let $\hat{h}_0$ be the true minimizer of ISE, and assume that there exists a nonempty subset of $\mathbb{R}$ on which $f'$ is continuous and nonzero (actually, the required condition is slightly weaker than this). Then,

$$\lim_{n\to\infty} \frac{\text{ISE}(\hat{h}_{\text{CV}})}{\text{ISE}(\hat{h}_0)} = 1 \tag{2.20}$$

with probability 1. Burman (1985) established a similar result assuming only boundedness of $f$, but required stronger conditions on $h$. This asymptotic optimality of the CV choice does not address the variability of CV as an estimator of ISE, which is unfortunately high. Daly (1988) suggested an alternative criterion that is based on estimating MISE.

An attempt to address this high variability is the *biased cross-validation* (BCV) method of Scott and Terrell (1987), which sacrifices the unbiasedness of CV as an estimator of MISE $- R(f)$ in order to reduce variability around MISE $- R(f)$. The BCV criterion is based on substituting an estimate for $R(f')$ into the AMISE given by the leading terms of (2.5). For this reason, it can be viewed as an example of a *plug-in method* for choosing the bin width; such methods will be more important in succeeding chapters. The BCV criterion has the form

$$\text{BCV} = \frac{5}{6nh} + \frac{1}{12n^2h}\sum_{j=0}^{K}(n_{j+1} - n_j)^2$$

(defining $n_0 = n_{K+1} = 0$). Scott and Terrell showed that in most cases Var[BCV($h$)] is considerably smaller than Var[CV($h$)], which translates into correspondingly reduced variation in the minimizers $\hat{h}_{\text{CV}}$ and $\hat{h}_{\text{BCV}}$, respectively. Both $\hat{h}_{\text{CV}}$ and $\hat{h}_{\text{BCV}}$ have the slow relative convergence rate of

$$\frac{\hat{h}}{\hat{h}_0} = 1 + O_p(n^{-1/6}),$$

but Hall and Marron (1987a,b) showed that this is, in fact, the optimal rate. Unfortunately, $\hat{h}_{\text{BCV}}$ tends to oversmooth; indeed, for smaller samples the BCV criterion sometimes does not have any local minima at all.

Scott (1992, pp. 77–80) hypothesized possible causes of the tendencies for CV to undersmooth (the existence of (near) ties in the data) and BCV to oversmooth (the global minimum of BCV is $h = \infty$, and local minima can be correspondingly large). Marron (1992a) proposed that the practical implementation of the cross-validation rules should be that $\hat{h}_{\text{CV}}$ is the *largest* value of $h$ that is a local minimum of CV($h$), while $\hat{h}_{\text{BCV}}$ is the *smallest* value of $h$ that is a local minimum of BCV($h$) (presumably less than the oversmoothed choice). It should be noted, however, that this recommendation was made regarding the more accurate density estimators discussed in the next chapter, where it would be expected that the CV and BCV functions would have fewer local minima.

The chondrite meteor data are originally from Ahrens (1965). The trimodal nature of the underlying density has been noted by several authors, including Good and Gaskins (1980) and Simonoff (1983).

### Section 2.2

**2.2.1.** The frequency polygon is the unique maximum likelihood estimator (MLE) over the set of functions that are piecewise linear between a set of equispaced grid points, but it is not the MLE if the grid points are unequally spaced (de Montricher, Tapia, and Thompson, 1975).

Scott (1985a) provided an asymptotic analysis of the accuracy (in terms of MISE) of the fixed bin width frequency polygon, leading to Eqs. (2.13) – (2.15). Samiuddin, Jones, and El-Sayyad (1993) showed that the gain in AMISE of the frequency polygon over the histogram is due to the evaluation of the histogram at only the mid-bin value, where the bias is $O(h^2)$, rather than $O(h)$.

There are other simple histogram-based estimators that achieve this reduced bias; see, for example, Minnotte (1996). Jones *et al.* (1998) suggested connecting bin edge, rather than mid-bin values, with the height at the edge being the average of contiguous bin heights. This *bin edge frequency polygon* (EFP) has the form

$$\hat{f}_{\mathrm{EFP}}(x) = n^{-1}\left\{\frac{n_I + \frac{1}{2}(n_{I+1} + n_{I-1})}{2h} + \frac{(x - m_I)(n_{I+1} - n_{I-1})}{2h^2}\right\}, \tag{2.21}$$

where $x$ falls in bin $I$ and $m_I$ is the midpoint of the bin. The optimal $\hat{f}_{\mathrm{EFP}}$ has 11% smaller AMISE than the optimal frequency polygon. The Gaussian reference rule is to take $h = 1.50\sigma n^{-1/5}$.

Equation (2.13) can be used in the same way as Eq. (2.5) to approximate the effect of choosing the bin width $h$ suboptimally. If a bin width $h^* = ch_0$ is chosen, substituting into (2.13) shows that $\mathrm{AMISE}^* = (c^5 + 4)/(5c) \times \mathrm{AMISE}_0$.

**2.2.2.** Terrell and Scott (1985) and Terrell (1990) investigated oversmoothing in the frequency polygon context. Minimization of $R(f'')$, rather than $R(f')$, determines the bin width. Using the range of the data as a scale estimate implies that the optimal number of bins must be at least $(147n/2)^{1/5}$. For fixed $\sigma^2$, the corresponding rule is $h \le 2.33\sigma n^{-1/5}$. The similarity of this to (2.16) reinforces the impression that the Gaussian density is quite smooth.

Roeder (1990) gave data on velocities of galaxies relative to the Milky Way and found the same five modes as are evident in the frequency polygon using a mixture of Gaussian densities.

Cross-validation, as in (2.10), can be used to choose the bin width of a frequency polygon. For a fixed bin-width estimator, if $x \in [b_j, b_{j+1}]$, then $\hat{f}_{-i}(x_i)$ has the form

$$\hat{f}_{-i}(x_i) = \begin{cases} \frac{1}{(n-1)h}\left[n_{j-1} + \frac{n_j - n_{j-1} - 1}{h}(x_i - c_{j-1})\right], & x_i \in [b_j, c_j], \\ \frac{1}{(n-1)h}\left[n_{j+1} + \frac{n_{j+1} - n_j + 1}{h}(x_i - c_{j+1})\right], & x_i \in [c_j, b_{j+1}] \\ . \end{cases}$$

Substituting these values (for each $x_i$) into (2.10) gives the CV criterion.

The BCV criterion also can be defined for a fixed bin-width frequency polygon. Here an estimate of $R(f'')$ is required, based on the available frequency polygon. Scott and Terrell (1987) proposed an estimate based on second differences:

$$\hat{R}(f'') = \frac{1}{n^2 h^5}\sum(n_{j+1} - 2n_j + n_{j-1})^2 - \frac{6}{nh^5}.$$

Plugging into (2.13) then gives

$$\text{BCV} = \frac{271}{480nh} + \frac{49}{2880n^2h}\sum(n_{j+1} - 2n_j + n_{j-1})^2$$

(Scott, 1992, p. 101).

## Section 2.3

Scott (1992, Sect. 4.1.3) derived the optimal bin width of a frequency polygon at a point $x$ based on the asymptotic MSE.

Early attempts at constructing variable bin-width histograms were based on sample percentiles; that is, setting $n_j$ equal to a fixed constant (say, $k$). See, for example, Rodriguez and Van Ryzin (1985), Van Ryzin (1973) and Wegman (1969, 1970a,b, 1975). Unfortunately, as was demonstrated by Scott (1982), choosing bin widths in this fashion results in histograms with MISE many times larger than that of the best fixed bin-width histogram, for a wide range of densities (see also Scott, 1992, Sect. 3.2.8.4). The problem is that these rules imply bin widths that are inversely related to $f(x)$, rather than $|f'(x)|$ (as they should be), leading to inflated bias.

Kogure (1987) derived the optimal MISE for the variable bin-width histogram. This optimal value is typically 15%–30% smaller than the optimal value for a fixed bin-width histogram (Scott, 1992, p. 68). Kogure showed that this optimal rate can be achieved using a recursive partitioning method, where the interval of interest is divided into equisized bins, each of which might then be subdivided further.

The data on Major League Baseball salaries come from the February 21, 1993, issue of *Newsday* (Newsday, 1993a). The data on shooting percentages for National Hockey League players were extracted from National Hockey League (1992).

## Section 2.4

The Monte Carlo simulations of Simonoff and Hurvich (1993) indicated the superiority (based on ISE) of simple Gaussian-based rules for choosing the bin width of histograms and frequency polygons over the cross-validated choice. Their results also established the much smaller average ISE of frequency polygons compared with histograms in the situations studied for even very small samples.

Many authors have focused on the apparent effect of the choice of anchor position on the appearance of the resultant histogram as one of the biggest drawbacks of the histogram estimator (see, for example, Silverman, 1986, Sect. 2.2; Härdle, 1991, Sect. 1.4; Izenman, 1991; Härdle and Scott, 1992; Scott, 1992, Sect. 4.3; Samiuddin, Jones, and El-Sayyad, 1993). Simonoff (1995a) provided Monte Carlo evidence for the insensitivity of average ISE to the choice of anchor position for both histograms and frequency polygons. That paper also examined the mode resolution ability of the estimators (a property much more related to its appearance) and showed the potential inability of histograms and frequency polygons to identify true underlying modes correctly. That is, a functional norm like ISE does not necessarily reflect what a data analyst would mean by effective estimation of the density.

Simonoff and Udina (1996) proposed an index of the stability of the appearance of histograms for a given bin width if the anchor position is changed, based on the statistical roughness ($R(f')$) of the histogram. They found that underlying densities with more structure (modes and long tails) lead to greater instability in the appearance of histograms. They also found that it is very important to choose the bin width to be consistent with the precision of the data, in order to avoid unstable histograms. That is, if the data are reported to the nearest integer (for example), using a non-integral bin width greatly increases the chances of producing histograms whose appearance changes dramatically if the anchor position is changed. In comparisons with other bin-based methods, they found that the bin edge frequency polygon (2.21) is generally less sensitive to shifts in anchor position than the histogram and ordinary frequency polygon.

ISE, as a measure of the quality of a density estimate, has the advantage of mathematical tractability and the availability of the intuitive variance versus bias tradeoff. Still, there is nothing inherently more reasonable about the use of ISE compared with other possible criteria (there is no least squares/Gaussian distribution/maximum likelihood justification, as is often the case in parametric modeling). The ISE criterion can be criticized on the grounds that it puts too little emphasis on the tails of the density, because of the squaring of $\hat{f}(x) - f(x)$.

A natural alternative to the use of squared error is to use integrated absolute error to evaluate the accuracy of a density estimate:

$$\text{IAE} = \int |\hat{f}(u) - f(u)| du,$$

with expected value mean integrated absolute error

$$\text{MIAE} = E\left[\int |\hat{f}(u) - f(u)| du\right].$$

Devroye and Györfi (1985) and Devroye (1987) provided a thorough discussion of density estimation and the absolute error criterion. They derived bounds on the asymptotic MIAE for fixed bin-width histograms and showed that MIAE $= O(n^{-1/3})$, as would be expected from the $O(n^{-2/3})$ rate of MISE.

The upper bound on MIAE provides a criterion to minimize to determine an optimal bin width. The resultant minimizer is

$$h_0 = \left\{ \frac{\left[\int f(u)^{\frac{1}{2}} du\right]^2}{8n\pi \left[\int |f'(u)| du\right]^2} \right\}^{-1/3},$$

or $h_0 = 2.72\sigma n^{-1/3}$ for Gaussian data. This is considerably smaller than the AMISE-optimal bin width, but Scott (1992, Sect. 3.6.1) reported that Monte Carlo simulations with Gaussian data show that the minimizer of MIAE $h_0 \approx 3.37\sigma n^{-1/3}$, which is close to the AMISE-optimal bin width given in (2.8) (a pattern that persists for more accurate density estimation methods). See also Hall and Wand (1988a). Thus, there appears to be little gain in using the more mathematically complicated absolute error.

The likelihood principle states that inference should proceed based on the likelihood function, and measures like ISE and IAE are artificial. The Kullback–Leibler information provides a likelihood-based distance measure for densities:

$$\begin{aligned} \mathrm{KL}(f, \hat{f}) &= \int f(u) \log[f(u)/\hat{f}(u)]du \qquad (2.22) \\ &= \int f(u) \log[f(u)]du - \int f(u) \log[\hat{f}(u)]du. \end{aligned}$$

The first term is not a function of $\hat{f}$, so minimizing KL is equivalent to maximizing $\int f(u) \log[\hat{f}(u)]du$.

KL is not defined if $\hat{f} = 0$; that is, a strategy to choose the bin size to minimize KL requires that the minimum bin count must be 1. Hall (1990a) pointed out that the minimization of KL can violate the principle of trading off bias versus variance, depending on the tail properties of $f$. For any histogram estimator (with at least two bins), there is always a nonzero probability that a bin will be empty, so $E(\mathrm{KL}) = \infty$ and cannot be used as a target criterion.

The same cross-validation argument as was used for squared error can be used to form a data-based estimate of KL as well. Since the goal is to maximize $\int f(u) \log[\hat{f}(u)]du$, the *likelihood cross-validation* approach is to maximize

$$\mathrm{KLCV} = \sum_{i=1}^{n} \log\left[\hat{f}_{-i}(x_i)\right]$$

(Habbema, Hermans, and van den Broek, 1974). Chow, Geman, and Wu (1983) established consistency of $\hat{f}$ based on KLCV if $f$ is bounded with compact support, but for unbounded densities, there are problems in the tails. Hall (1990a) provided careful asymptotic analysis of KL and KLCV, showing that often when KLCV leads to a consistent estimator, it still gives bin widths that are not $O(n^{-1/3})$.

An alternative likelihood-based approach is by using Akaike's Information Criterion, or *AIC* (Akaike, 1973). *AIC* is based on maximizing a penalized likelihood function, where the penalty function is typically the number of unknown parameters. For histograms, the quantity to be maximized is

$$AIC = \sum_{j=1}^{K} n_j \log n_j - n \log(nh) - K^*$$

(Taylor, 1987; Atilgan, 1990). Here $K^*$, the number of unknown parameters, can be taken as either the number of nonempty bins or the total number of bins between the smallest and largest data points ($K$). See Atilgan (1990) and Hall (1990a) for a discussion of the properties of histograms constructed based on $AIC$.

Hall and Hannan (1988) proposed a criterion based on stochastic complexity to choose the bin width of a histogram, recommending that it be chosen to maximize

$$\sum_{j=1}^{K}\left(n_j - \frac{1}{2}\right)\log n_j + n\log K.$$

They found that, from a practical point of view, such histograms are very similar to those based on KLCV.

Kanazawa (1988, 1992) used the Hellinger distance

$$\sum_{j=1}^{K}\int_{b_j}^{b_{j+1}}\left[\hat{f}(u)^{\frac{1}{2}} - f(u)^{\frac{1}{2}}\right]^2 du$$

as a criterion for density estimation, and investigated the function $\hat{f}$ that estimates its minimizer among piecewise constant functions. Although the resultant estimator has the appearance of a locally varying bin-width histogram, it is not, in fact, a histogram estimator, in the sense of (2.17). Kanazawa (1988) provided a dynamic programming algorithm to obtain the required estimate. Kanazawa (1993) proved the equivalence of minimizing Hellinger distance and maximizing $AIC$ for fixed bin-width histograms; see also Jones (1995a).

## Computational issues

Virtually any statistical package can produce fixed bin-width histograms, usually with the ability to control the position of the anchor and the bin width. Unfortunately, often the estimate is presented in a nondensity form, with the vertical axis representing the count of observations in a bin, rather than the appropriate density estimate value (this doesn't change the appearance of the estimate, of course, since the density estimate is a constant multiplier of the bin count for every bin). This can make it more difficult for a package to produce a locally varying bin-width histogram, since that density estimate is not a constant multiplier of the bin count for every bin. Correct locally varying bin width histograms can be constructed in MINITAB and S–PLUS.

Frequency polygon construction often must be done by hand using a package that can draw straight lines between specified $[x, \hat{f}(x)]$ values.

Fortran code that determines such values for the Gaussian-based and locally varying bin-width frequency polygons discussed in Simonoff and Hurvich (1993) can be obtained using a World Wide Web (WWW) browser at `http://www.stern.nyu.edu/~jsimonof/frpoly.f`. MINITAB, NCSS, and Systat can produce fixed bin-width frequency polygons, as can the Fortran code of Exponent Graphics (IMSL). MINITAB also gives variable bin-width frequency polygons. Venables and Ripley (1994) gave S–PLUS code to construct frequency polygons automatically.

The collection `haerdle` in the `S` directory of `statlib` contains S–PLUS functions to construct fixed bin-width histograms and frequency polygons based on code discussed in Härdle (1991).

XLISP–STAT code to calculate the anchor position stability index of Simonoff and Udina (1997) is available via anonymous `ftp` at the address `libiya.upf.es` in the directory `/pub/stat/anchor-position`.

## Exercises

**Exercise 2.1.** Write code to implement the cross-validation and biased cross-validation criteria for choosing the bin width of a fixed bin-width histogram (this involves evaluating the CV and BCV criteria over a grid of possible bin-width choices). Apply your code to the forged Swiss bank note data of Fig. 2.1. Do either of these criteria yield a satisfactory estimate? Recall that multimodality is of particular interest for data of this type — what is your impression of the existence of more than one mode for these data? BCV is known to have a tendency to oversmooth, while CV tends to undersmooth — is that the case here?

**Exercise 2.2.** Write computer code to implement the cross-validation and biased cross-validation criteria for choosing the bin width of a fixed bin-width frequency polygon. Apply your code to the forged Swiss bank note data. What are your impressions? How do the results compare with those of Exercise 2.1?

**Exercise 2.3.** The Swiss bank note data presented in Fig. 2.1 are part of a larger set, which includes both real notes and forged notes and also includes measurement of the image diagonal length of the bill. Compare histograms and frequency polygons, using various criteria for choosing the bin width, for the additional variables and cases. Do the real notes exhibit multimodality? Does the pattern for diagonal length differ from that for bottom margin?

**Exercise 2.4.** Write computer code to implement non-ISE-based criteria for choosing the bin width of a fixed bin-width histogram, such as using KLCV, AIC, or stochastic complexity. Apply your code to the Swiss bank note data. Do your impressions change from those of Exercises 2.1 – 2.3?

Theory suggests that histograms using such criteria will have wider bins than those based on ISE-based measures — is that the case here?

**Exercise 2.5.** Data-dependent criteria such as CV, BCV, KLCV, AIC, etc., can be used to choose the anchor position of an estimate in addition to the bin width. Write computer code to allow the anchor position to be chosen using such criteria, and apply it to the CD rate and Swiss bank note data. Does using a data-dependent choice of anchor position seem to lead to a better estimate?

**Exercise 2.6.** Construct the bin edge frequency polygon (2.21), using a Gaussian reference rule to choose the bin width, for the CD rate and Swiss bank note data. Is its appearance very different from that of a Gaussian-reference frequency polygon for each data set? Is that appearance stable if the anchor position is changed?

Chapter 3

# Smoother Univariate Density Estimation

## 3.1 Kernel Density Estimation

### 3.1.1 Motivation for the kernel estimator

The simple density estimators of Chapter 2 are informative, but they suffer from two serious drawbacks: they are not smooth, and they are not sensitive enough to local properties of $f$. It is easy to solve both of these problems.

Consider again the definition of $f(x)$:

$$f(x) \equiv \frac{d}{dx}F(x) \equiv \lim_{h \to 0} \frac{F(x+h) - F(x-h)}{2h} \tag{3.1}$$

(this is equivalent to definition (2.1)). The histogram estimates (3.1) by dividing the line into bins, but a more sensible approach is to estimate the derivative separately at each point $x$. Replacing $F(x)$ with the empirical cumulative distribution function gives

$$\hat{f}(x) = \frac{\#\{x_i \in (x-h, x+h]\}}{2nh}.$$

This can be rewritten as

$$\hat{f}(x) = \frac{1}{nh} \sum_{i=1}^{n} K\left(\frac{x - x_i}{h}\right), \tag{3.2}$$

where

$$K(u) = \begin{cases} \frac{1}{2}, & \text{if } -1 < u \le 1, \\ 0, & \text{otherwise.} \end{cases}$$

The form (3.2) is that of a *kernel density estimator*, with kernel function $K$. Note that this kernel function is a uniform density function on $(-1, 1]$.

Figure 3.1 gives a kernel density estimate of the CD rate data with $h = .14$ and can be compared with the histograms of Fig. 2.3 and the frequency polygons of Fig. 2.4. The "rug" along the bottom of the plot gives the positions of the observations. The estimate is more local in nature but is hardly a reasonable estimate of a smooth density.

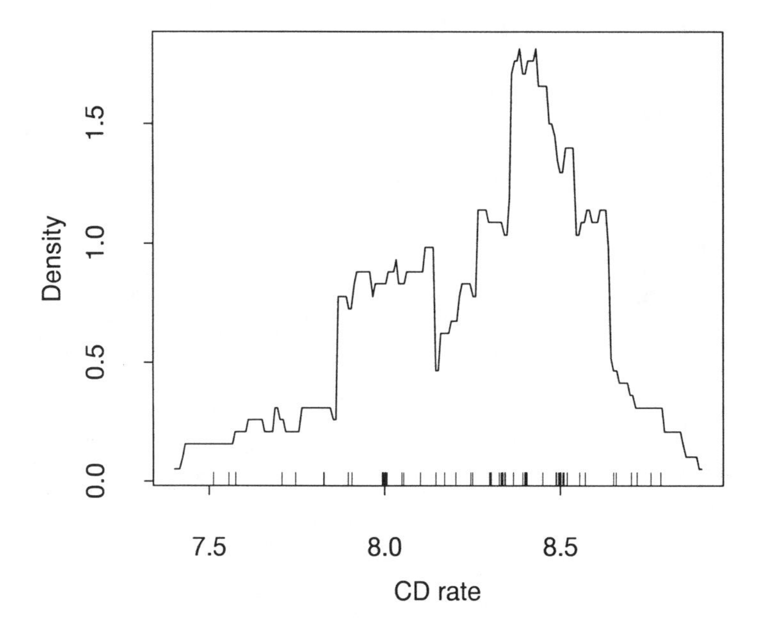


**Fig. 3.1.** Kernel estimate of CD rate data using uniform kernel.

The problem is that the additive form of (3.2) implies that the estimate $\hat{f}$ retains the continuity and differentiability properties of $K$. Since the uniform density is discontinuous, so is the kernel density estimate based on a uniform kernel function. A smoother kernel function will thus lead to a smoother kernel density estimate.

Figure 3.2 presents a kernel estimate for these data using a Gaussian density for $K$, with $h = .08$. The curve is appealingly smooth, and a trimodal form, with modes at 7.5%, 8.0%, and 8.5%, is apparent. The curves along the bottom of the plot illustrate the additive form in (3.2); the density estimate at any point (the solid curve) is an average of the Gaussian densities centered at each observation (the dashed curves). For clarity of presentation, only a few of the dashed curves are given.

### 3.1.2 Properties of the kernel estimator

The degree to which the data are smoothed has a strong effect on the appearance of $\hat{f}(x)$ through the setting of the smoothing parameter (or bandwidth) $h$. The kernel estimates in Fig. 3.3 correspond to that of Fig. 3.2, except that the bandwidth is half as large in plot (a) (that is, $h = .04$),

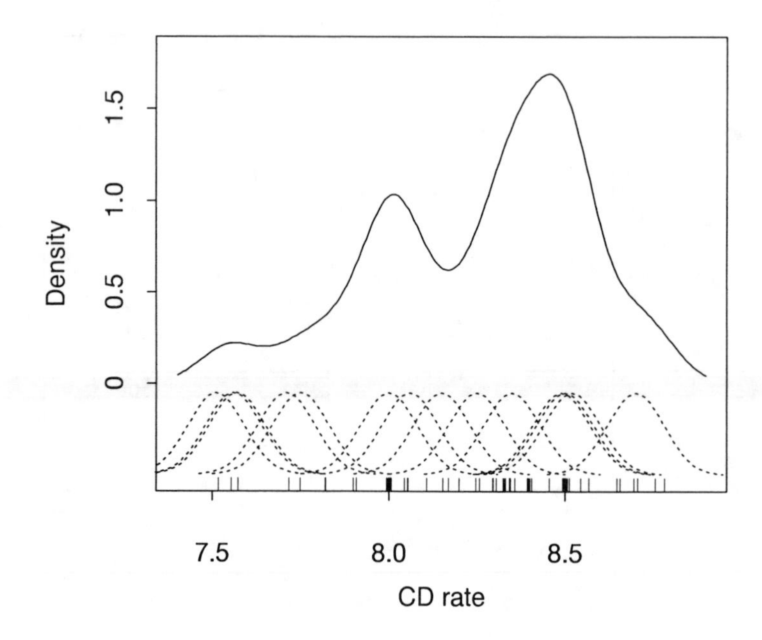


**Fig. 3.2.** Kernel estimate of CD rate data using Gaussian kernel, with some underlying evaluations of the kernel function.

while it is twice as large in plot (b) ($h = .16$). The former estimate is very undersmoothed, with the previously noted three modes joined by three additional modes and a bump. The latter estimate is oversmoothed, with only a slight bulge remaining of two of the modes noted earlier.

Once again, the tradeoff of bias versus variance that results from choosing the amount of smoothing can be quantified through a measure of accuracy of $\hat{f}$, such as MISE. Define $K$ to satisfy the conditions

$$\int K(u)du = 1, \int uK(u)du = 0, \int u^2K(u)du = \sigma_K^2 > 0,$$

and assume that the underlying density is sufficiently smooth ($f''$ being absolutely continuous and $f'''$ being square integrable). If $h \to 0$ with $nh \to \infty$ as $n \to \infty$, then by Taylor Series expansions,

$$\text{Bias}[\hat{f}(x)] = \frac{h^2\sigma_K^2 f''(x)}{2} + O(h^4)$$

and

$$\text{Var}[\hat{f}(x)] = \frac{f(x)R(K)}{nh} + O(n^{-1}).$$

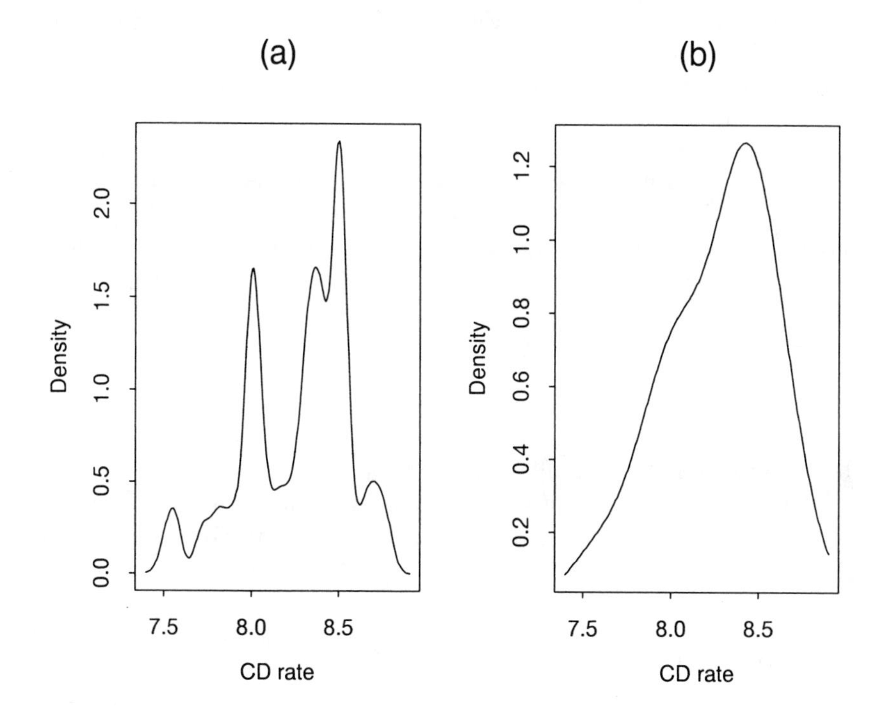


**Fig. 3.3.** Kernel estimates of CD rate data: (a) $h = .04$, (b) $h = .16$.

Combining variance and squared bias gives the mean squared error

$$\mathrm{MSE}[\hat{f}(x)] = \frac{f(x)R(K)}{nh} + \frac{h^4\sigma_K^4[f''(x)]^2}{4} + O(n^{-1}) + O(h^6). \tag{3.3}$$

Integrating over the entire line then gives the asymptotic MISE

$$\mathrm{AMISE} = \frac{R(K)}{nh} + \frac{h^4\sigma_K^4 R(f'')}{4}. \tag{3.4}$$

Note the similarity of (3.4) to the corresponding form of MISE for frequency polygons (2.13). The $O(h^4)$ magnitude of the squared bias term implies the same optimal convergence rate as was noted there; that is, the asymptotically optimal bandwidth satisfies

$$h_0 = \left[\frac{R(K)}{\sigma_K^4 R(f'')}\right]^{1/5} n^{-1/5}, \tag{3.5}$$

implying minimal AMISE

$$\text{AMISE}_0 = \frac{5}{4}[\sigma_K R(K)]^{4/5} R(f'')^{1/5} n^{-4/5}. \tag{3.6}$$

The term $R(f'')$ measures the roughness of the true underlying density and is (of course) out of the control of the data analyst. In general, rougher densities are more difficult to estimate (higher AMISE) and require a smaller bandwidth.

The term $[\sigma_K R(K)]^{4/5}$, being a function only of the kernel function $K$, is under the control of the data analyst, so a natural question is to ask how to choose $K$ in the "best" way; that is, to minimize $[\sigma_K R(K)]^{4/5}$. If the kernel $K$ is restricted to be a proper density function (which means that $\hat{f}$ will be also), the minimizer is a scaled version of a quadratic density,

$$K(u) = \begin{cases} \frac{3}{4}(1 - u^2), & \text{if } |u| \leq 1, \\ 0, & \text{otherwise} \end{cases}$$

(this is often called the Epanechnikov kernel).

The value of $\sigma_K R(K)$ for the Epanechnikov kernel is $3/(5\sqrt{5})$; thus, the ratio $\sigma_K R(K)/[3/(5\sqrt{5})]$ provides a measure of the relative inefficiency of using other kernel functions (this ratio is the multiplicative factor for the equivalent sample size needed to achieve the same AMISE). Table 3.1 gives values for this ratio for several common kernel functions.

**Table 3.1.** Inefficiency of various kernels relative to the Epanechnikov kernel. All kernels, except for the Gaussian, are zero outside the interval $[-1, 1]$.

| *Kernel* | *Form* | *Inefficiency* |
|---|---|---|
| Epanechnikov | $\frac{3}{4}(1 - u^2)$ | 1 |
| Biweight | $\frac{15}{16}(1 - u^2)^2$ | 1.0061 |
| Triweight | $\frac{35}{32}(1 - u^2)^3$ | 1.0135 |
| Gaussian | $(2\pi)^{-1/2} e^{-u^2/2}$ | 1.0513 |
| Uniform | $\frac{1}{2}$ | 1.0758 |

The obvious message from the values in Table 3.1 is that the AMISE is insensitive to the choice of the kernel, so $K$ should be chosen based on other issues, such as ease of computation or properties of $\hat{f}$. In particular, an argument against using the Epanechnikov kernel is that since it is not everywhere differentiable, neither will $\hat{f}$ be everywhere differentiable (despite the assumption that three derivatives of $f$ exist).

### 3.1.3 Choosing the bandwidth

The simplest way to choose the bandwidth $h$ is by choosing a reference density for $f$ and substituting into (3.5). So, for example, if the reference density is Gaussian, and a Gaussian kernel $K$ is used, then

$$h_0 = 1.059\sigma n^{-1/5}. \tag{3.7}$$

Substituting an estimate for $\sigma$ into (3.7) gives a data-based version of this rule.

This Gaussian reference rule assumes use of a Gaussian kernel, but it is straightforward to convert a rule based on one kernel function (such as the Gaussian) to any other kernel, using (3.5). Since $R(K) = (2\sqrt{\pi})^{-1}$ and $\sigma_K = 1$ for the Gaussian kernel, the equivalent asymptotically optimal bandwidth for any density using a different kernel $K^*$ satisfies

$$h_{0,K^*} = c_{K^*} h_{0,G},$$

where

$$c_{K^*} = \left[\frac{2\sqrt{\pi}R(K^*)}{\sigma^4_{K^*}}\right]^{1/5}, \tag{3.8}$$

and $h_{0,G}$ is the optimal bandwidth using a Gaussian kernel. Thus, any rule derived based on a Gaussian kernel (which is often more convenient to work with theoretically) can be converted to one based on any other kernel (which might be more useful computationally) using a simple multiplier. Table 3.2 gives the appropriate multipliers for different kernel functions. So, for example, the asymptotically optimal bandwidth for a Gaussian density when using a biweight kernel is $h_0 = (2.623)(1.059\sigma n^{-1/5}) = 2.778\sigma n^{-1/5}$.

**Table 3.2.** Multiplier (3.8) for converting smoothing parameter based on Gaussian kernel to other kernels.

| *Kernel* | *Multiplier* |
|---|---|
| Epanechnikov | 2.214 |
| Biweight | 2.623 |
| Triweight | 2.978 |
| Uniform | 1.740 |

Depending as it does on the assumption that the true density is Gaussian, the rule (3.7) is of limited value. Cross-validation, as in (2.10), is general enough to be applied to bandwidth choice for kernel density estimation. Unfortunately, the cross-validated choice of bandwidth for the

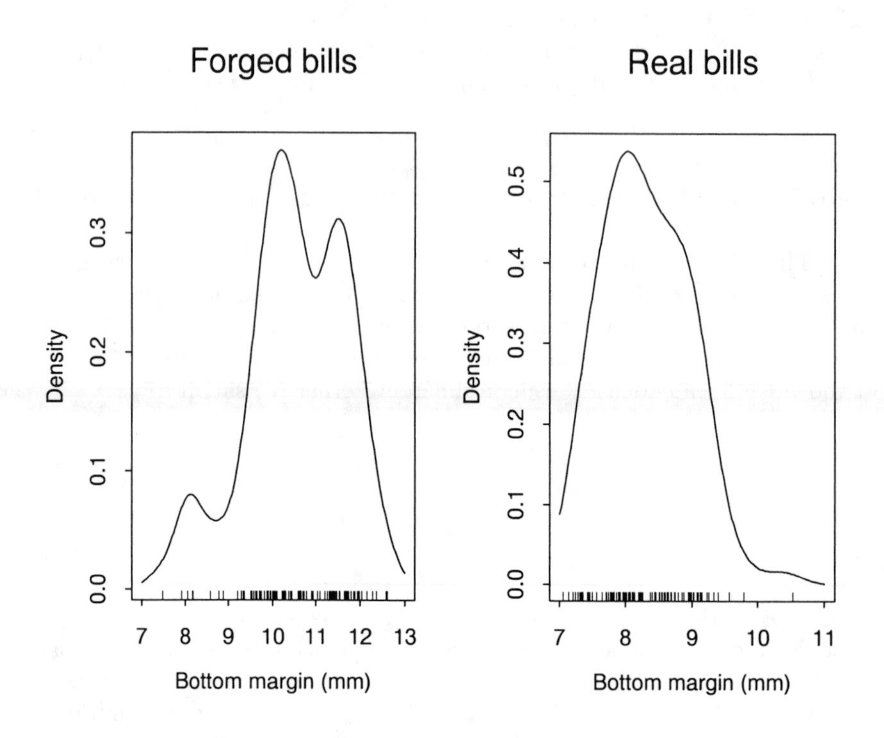


**Fig. 3.4.** Kernel estimates of Swiss bank note data (bottom margin) using Sheather–Jones bandwidth choice. (a) Forged bills. (b) Real bills.

kernel estimator is highly variable, and often undersmooths, yielding spurious bumpiness in the estimate. A different approach is based on the *plug-in principle*, where the asymptotically optimal $h_0$ is estimated by substituting an estimate of $R(f'')$ into (3.5); that is,

$$\hat{h} = \left[\frac{R(K)}{\sigma_K^4 \widehat{R(f'')}}\right]^{1/5} n^{-1/5}. \tag{3.9}$$

Implementation of (3.9) requires choosing the estimate $\widehat{R(f'')}$. The current "state of the art" appears to be the method of Sheather and Jones (1991), which takes $\widehat{R(f'')} = R(\hat{f}'')$. The estimate $\hat{f}$ used here is based on a different bandwidth from the one that is appropriate for estimation of $f$ itself (determined by theoretical considerations), and is estimated from the data.

Figure 3.4 presents kernel estimates for the bottom margins of 100 forged Swiss bank notes and 100 real Swiss bank notes, using the Sheather–

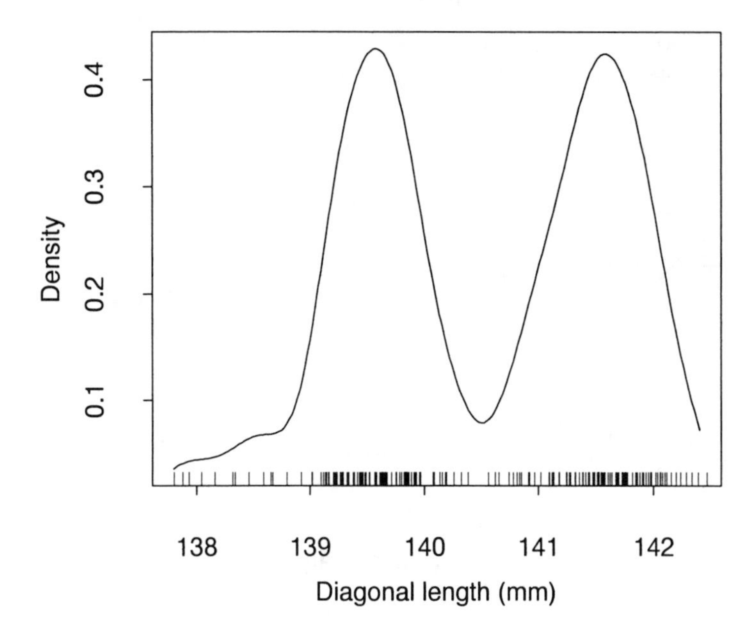


**Fig. 3.5.** Kernel estimate of Swiss bank note data (diagonal length) using Sheather–Jones bandwidth choice.

Jones bandwidth ($\hat{h}_{\rm SJ}$). The margins of the forged bills exhibit a trimodal structure, with modes around 8, 10, and 11.55 mm, which suggests production of separate batches of bills. The real bills exhibit considerably less structure, but there is a hint of bimodality (which, considering government quality control efforts in currency manufacture, is somewhat surprising). If anything, the rug along the bottom of the plot suggests stronger bimodality, meaning that the chosen bandwidth has oversmoothed a bit.

The curve in Fig. 3.5 is a kernel density estimate for the diagonal length of all 200 bills, again using the Sheather–Jones bandwidth. A clear bimodal structure is apparent, where the left mode corresponds to forged bills, while the right mode corresponds to real bills. The small bump around 138.5 mm also corresponds to forged bills.

The impressions from Figs. 3.4 and 3.5 are potentially misleading, in the sense that the given density estimates do not reflect the inherent variability in the estimation process. There are two sources of this variability: that of the estimator from sample to sample drawn from a given population (for a given choice of $h$), and that associated with a data-based choice of $h$.

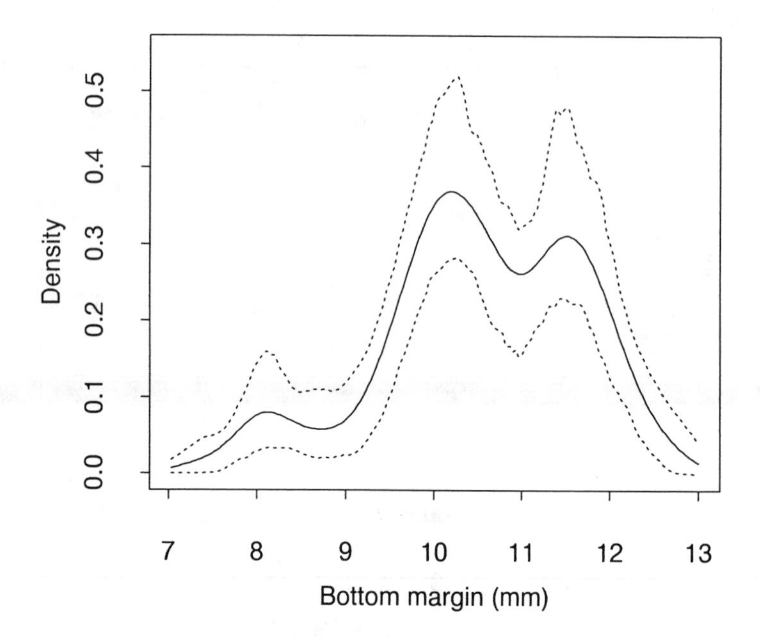


**Fig. 3.6.** Variability plot for kernel estimate of Swiss bank note data (bottom margin).

A *variability plot*, as given in Fig. 3.6, can help give a fuller picture of the degree of certainty associated with a given density estimate. The plot is based on bootstrapping. A sample of size $n$ is drawn with replacement from the given data set. Then, the bandwidth is chosen for the new sample based on the Sheather–Jones method, and a density estimate is determined. This process is then repeated many times, with the values of $\hat{f}$ being recorded at a fixed grid of values of $x$.

The dashed lines in Fig. 3.6 are the upper and lower pointwise 2.5% points from 200 resamples from the bottom margins of forged Swiss bills data. The dashed envelope is **not** a 95% confidence interval for the true $f$, but rather a representation of the inherent variability in the process yielding $\hat{f}$. The variability envelope for these data supports the trimodal impression of the original density estimate, as it is fairly narrow. The plot widens at the peaks and valleys of $\hat{f}$ and narrows where $\hat{f}$ is flat; this is consistent with the mean squared error of $\hat{f}(x)$, which is directly related to $[f''(x)]^2$, as in (3.3).

If dynamic graphics are available, the sensitivity of the fitted estimate to the choice of $h$ can be examined interactively. A natural mechanism is the use of a *slider*, which determines the bandwidth. Markings on the slider corresponding to data-dependent choices such as cross-validation or Sheather–Jones could designate benchmark values, and the data analyst could then observe the effects on $\hat{f}$ when $h$ is changed (the multiplicative effect of $h$ on AMISE suggests that such markings should be on a logarithmic scale). In any event, the estimates corresponding to several choices of $h$ should be examined in order to find the most reasonable $h$ (and $\hat{f}$).

## 3.2 Problems with Kernel Density Estimation

Despite (or perhaps because of) the simplicity and intuitive appeal of the kernel estimator, it is not without faults as a general-purpose density estimation tool. Boundary bias, lack of local adaptivity, and the tendency to flatten out peaks and valleys are all potential difficulties for this estimation method.

### 3.2.1 Boundary bias

Kernel estimation can fail dramatically when the region of definition of the data at hand is not unbounded. Consider Fig. 3.7. The data are the time intervals (in days) between accidents resulting in fatalities in mines in Division 5 of the Great Britain National Coal Board over a 245 day period in 1950. The kernel density estimate, based on a biweight kernel with $h = 10.5$, peaks at around 3 days between accidents, even though roughly 35% of the data points fall in the range $[0, 3)$. That is, the estimate is biased downward near the boundary.

The reason for this is apparent from the original definition of $f(x)$, given in (3.1), which motivates the kernel estimate with uniform kernel:

$$g(x) \equiv \lim_{h \to 0} \frac{F(x+h) - F(x-h)}{2h}.$$

Of course, typically $g(x) = f(x)$, the density function. Say the observed density has a lower boundary at 0, and $f$ is being estimated at some $x$ within the boundary region; that is, $x = ph$, $0 \le p < 1$. Then (3.1) becomes

$$g(x) \equiv \lim_{h \to 0} \frac{F(x+h) - F(0)}{2h},$$

since $x < h$. This implies that

$$\begin{aligned} g(x) &= \left(\frac{x+h}{2h}\right) \lim_{h \to 0} \frac{F(x+h) - F(0)}{x+h} \\ &= \left(\frac{p+1}{2}\right) f\left(\frac{x+h}{2}\right), \end{aligned}$$

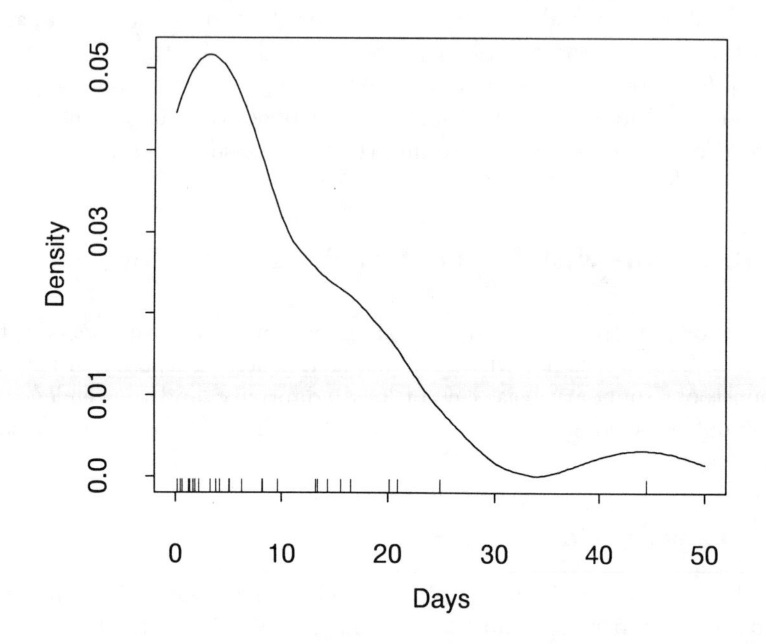


**Fig. 3.7.** Kernel estimate of mine accident data.

since $x = ph$. That is, the usual kernel formulation is estimating a value that is biased downward near the boundary, unless $f(x) = 0$ there.

Taylor Series expansions similar to those discussed in Section 3.1.2 formalize this effect for general kernel density estimation. Consider any kernel $K$ defined on $(-1, 1)$. Let $a_\ell(p) = \int_{-1}^{p} u^\ell K(u)du$ and $b(p) = \int_{-1}^{p} K^2(u)du$. Then

$$E[\hat{f}(x)] = a_0(p)f(x) - ha_1(p)f'(x) + \frac{h^2 a_2(p)f''(x)}{2} + o(h^2) \qquad (3.10)$$

and

$$\text{Var}[\hat{f}(x)] = \frac{f(x)b(p)}{nh} + O(n^{-1}).$$

Away from the boundary ($p \geq 1$), these expressions reduce to the usual formulas, but near the boundary the kernel estimate is not even consistent, unless $f(x) = 0$ (since $a_0(p) < 1$). Even if the kernel is locally renormalized to integrate to 1 (by dividing $\hat{f}(ph)$ by $a_0(p)$), the bias in the boundary region is $O(h)$, rather than the $O(h^2)$ of the interior.

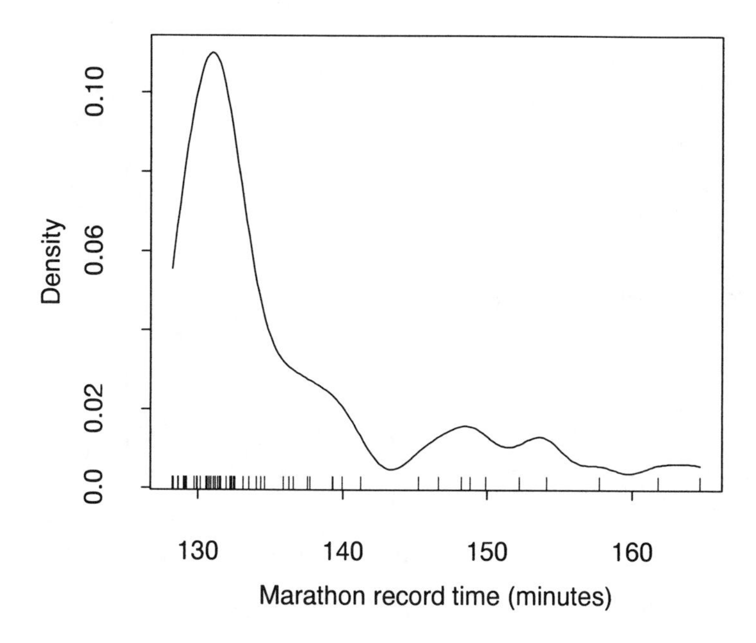


**Fig. 3.8.** Kernel estimate of men's marathon record data.

### 3.2.2 Lack of local variation in smoothing

The ordinary kernel estimator does not allow for different levels of smoothing at different parts of the density, as it is controlled by the single bandwidth $h$. This is obviously not optimal, as Eq. (3.3) shows that the mean squared error of $\hat{f}(x)$ at any point $x$ is directly related to $f(x)/h$ and $h^4[f''(x)]^2$. That is, in order to reduce MSE, $h$ should increase with $f(x)$ (to reduce variance) and should decrease with $|f''(x)|$ (to reduce bias).

From a practical point of view, this lack of adaptivity tends to manifest itself as oversmoothing in regions with high structure (where $|f''|$ is large) and undersmoothing (bumpiness) in the tails (where $|f''|$ is small), with automatic bandwidth selectors (such as plug-in selectors) sometimes performing poorly. Figure 3.8 gives a kernel estimate based on 55 national men's record times for the marathon, based on a Gaussian kernel and $h = 1.435$. The estimate peaks at around 132 minutes (2 hours 12 minutes) and shows interesting structure near the "round" numbers of 135 minutes and 150 minutes ($2\frac{1}{4}$ and $2\frac{1}{2}$ hours, respectively), perhaps reflecting psychological barriers at those values. It is difficult to assess how genuine the structure in the tails is, however, given the apparently spurious bumpiness there.

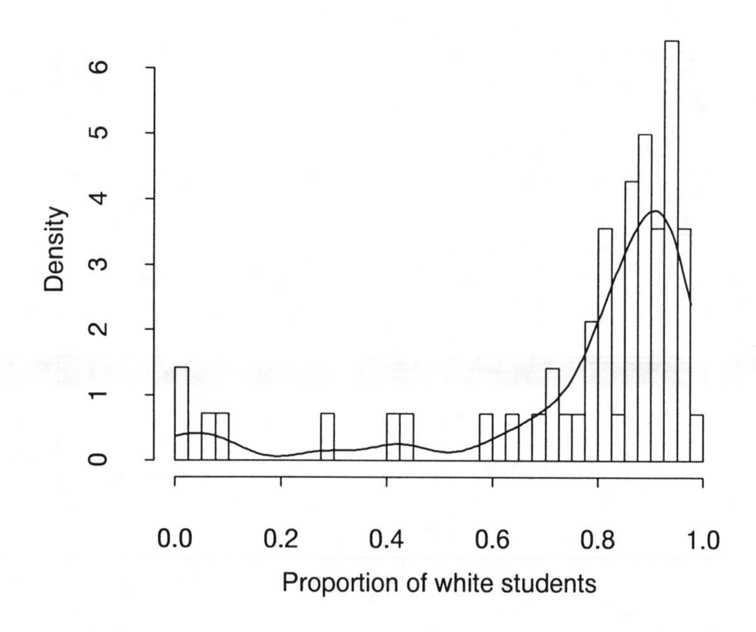


**Fig. 3.9.** Kernel estimate of Nassau County racial distribution data, superimposed on a histogram of the same data.

### 3.2.3 Flattening of peaks and valleys

The bias of the kernel estimator often shows up as a flattening of peaks and troughs of the density. Recall that the first term in the asymptotic bias of $\hat{f}(x)$ has the form $h^2\sigma_K^2 f''(x)/2$. This value is usually largest in absolute value at local maxima and minima, where $f'(x)$ is changing fastest, being negative at peaks and positive at valleys.

Figure 3.9 refers to data reporting the proportion of white students in 56 public school districts in Nassau County (New York) for the 1992–1993 school year. The solid line is a kernel density estimate (using a Gaussian kernel, with $h = .052$), superimposed on a histogram with a narrow bin width (.025). The distribution of proportion of white students is apparently bimodal, with most districts being at least 70% white, while a small number are heavily populated with minority students. The kernel estimate of the right mode appears to be too low (compared with the histogram bin heights), suggesting flattening of the peak there. In addition, spurious bumpiness in the center of the density estimate and boundary bias near the left boundary are also apparent.

## 3.3 Adjustments and Improvements to Kernel Density Estimation

### 3.3.1 Boundary kernels

The boundary bias of the kernel estimator can be corrected by using special kernels termed *boundary kernels.* Boundary kernels are weight functions that are used only within the boundary region (with the usual kernel $K$ used in the interior). Let $c_\ell(p) = \int_{-1}^{p} u^\ell L(u)du$, with $L$ being some kernel function different from, but related to, $K$. Let $\hat{f}_K$ be a kernel estimate based on $K$, while $\hat{f}_L$ is an estimate based on $L$. By (3.10),

$$E[c_1(p)\hat{f}_K(x)] = c_1(p)a_0(p)f(x) - ha_1(p)c_1(p)f'(x) + O(h^2)$$

and

$$E[a_1(p)\hat{f}_L(x)] = a_1(p)c_0(p)f(x) - ha_1(p)c_1(p)f'(x) + O(h^2).$$

This immediately implies that the linear combination of the two kernels

$$B(x) = \frac{c_1(p)K(x) - a_1(p)L(x)}{c_1(p)a_0(p) - a_1(p)c_0(p)} \tag{3.11}$$

satisfies

$$E[\hat{f}_B(x)] = f(x) + O(h^2);$$

that is, the bias near the boundary is restored to the $O(h^2)$ level achieved in the interior.

Many different forms of $L$ are possible, each leading to a different boundary kernel (or family of boundary kernels). One useful form is to take $L(x) = xK(x)$, which results in the boundary kernel

$$B(x) = \frac{[a_2(p) - a_1(p)x]K(x)}{a_0(p)a_2(p) - a_1^2(p)}. \tag{3.12}$$

Figure 3.10 gives a density estimate for the mine accident data of Fig. 3.7 based on this kernel, using the biweight kernel for $K$ and $h = 10.5$. The boundary bias apparent in Fig. 3.7 is now gone, and the characterization of the density as roughly exponential (as would be implied by a Poisson process for the occurrence of the accidents), but with a bulge at around 20 days between accidents, is clear.

The bias correction in Fig. 3.10 comes with a cost. Figure 3.11 gives a 95% variability plot of the boundary kernel estimate for the mine explosion data (taking the bandwidth to be the biweight equivalent to $1.683\hat{h}_{\mathrm{SJ}}$ for each bootstrap resample, as that was the value used in Figs. 3.7 and 3.10). The most striking property of this plot is the great widening of the envelope as the boundary is approached. That is, the reduction in bias is accompanied by a notable increase in variance. The important practical message is that

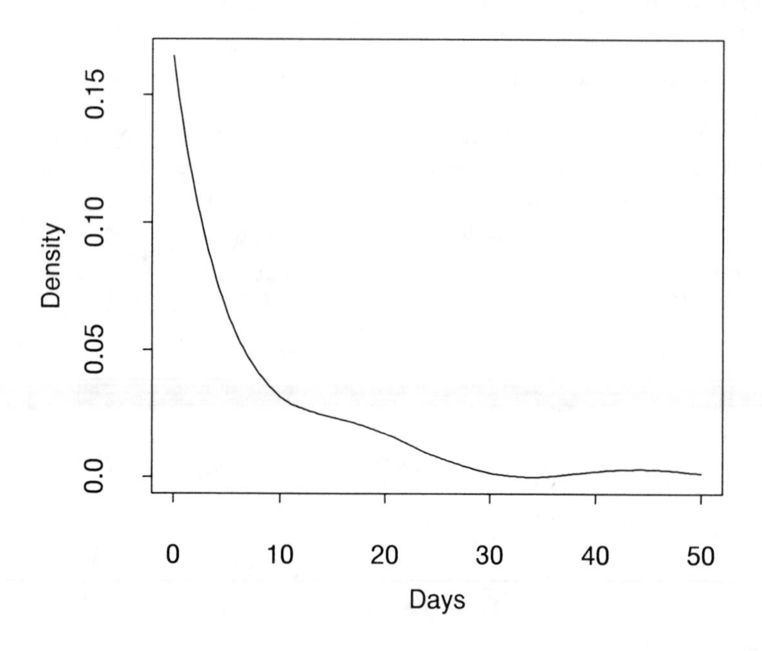


**Fig. 3.10.** Boundary kernel estimate of mine accident data.

while the appearance of $\hat{f}$ near zero is more appealing in Fig. 3.10 than in Fig. 3.7, the actual value of $\hat{f}$ near zero cannot be viewed as known with great accuracy.

The density estimate $\hat{f}_B$ is not necessarily a bona fide density, since it can take on negative values and does not necessarily integrate to 1. A simple solution is to define $\hat{f}$ to equal zero when it is negative and then normalize to force a unit integral. More complicated corrective procedures are also possible.

### 3.3.2 Varying the bandwidth

The usual kernel density estimator (3.2) is susceptible to bumpiness in the tails, since it does not adapt to local variations in smoothness. The estimator can be generalized to allow this, by using broader windows for the contribution of values associated with regions of low density and narrower windows for values associated with regions of high density. The general formula for one such estimator, the *variable-bandwidth kernel estimator*, is

$$\hat{f}(x) = \frac{1}{n} \sum_{i=1}^{n} \frac{1}{h(x_i)} K\left[\frac{x - x_i}{h(x_i)}\right].$$

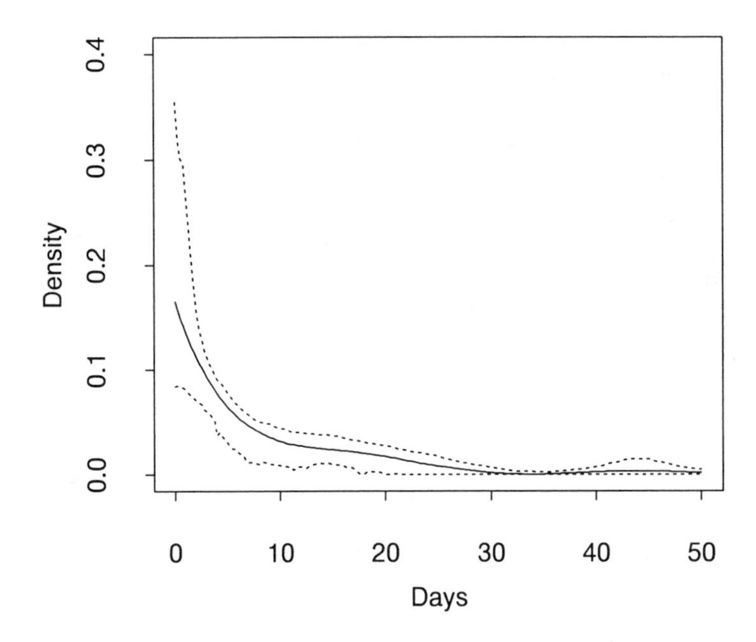


**Fig. 3.11.** Variability plot for boundary kernel estimate of mine accident data.

Since the goal is to smooth less where there is more structure (and more where there is less structure), it is natural to have $h(x_i)$ vary inversely with the underlying density. Consider taking $h(x_i) = h_v \times f(x_i)^{-1/2}$; that is,

$$\hat{f}_v(x) = \frac{1}{nh_v} \sum_{i=1}^{n} f(x_i)^{1/2} K \left[ \frac{(x - x_i) f(x_i)^{1/2}}{h_v} \right]. \tag{3.13}$$

This choice is particularly advantageous, since (subject to some technical details) it results in the bias of $\hat{f}_v(x)$ being $O(h^4)$, rather than the usual $O(h^2)$, while leaving the variance $O(n^{-1}h^{-1})$. Taking $h_v = O(n^{-1/9})$ then gives the improved convergence rate of MSE $= O(n^{-8/9})$.

Some issues need to be addressed before the estimator (3.13) can be applied in practice. The estimator depends on the values $f(x_i)$, which are of course unknown. A "pilot" estimate $\tilde{f}$ is usually used, based on, for example, a fixed-bandwidth kernel estimate (note that taking $\tilde{f}$ to be a uniform density gives the fixed-bandwidth kernel as a special case of the variable-bandwidth estimator).

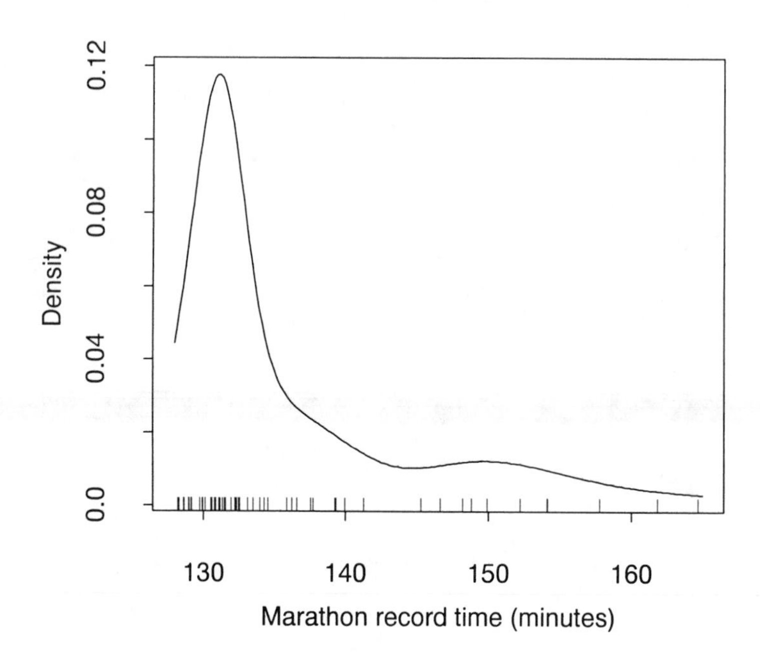


**Fig. 3.12.** Variable kernel estimate of men's marathon record data.

Another difficulty is that the estimator (3.13), as given, does not achieve bias of order $O(h^4)$. The problem is that observations $x_i$ far from the evaluation point $x$ can have a large effect on $\hat{f}_v(x)$. A simple corrective action is to restrict the pilot density $\tilde{f}$ to be bounded away from zero.

Both theory and practical experience show that the precise form of the pilot estimate does not have a strong effect on the final $\hat{f}_v$. The bandwidth $h_v$ does, however, exert a strong influence on the final estimate. It is straightforward to apply variations of the cross-validation criterion (2.10) to the choice of $h_v$, but little work on plug-in type methods for this estimator has been done.

Figure 3.12 gives a variable kernel estimate for the men's marathon data discussed earlier. The pilot estimate is the fixed-bandwidth kernel pictured in Fig. 3.8, and $h_v = .466$ was chosen based on cross-validation. This estimate captures the main features of the density (the peak at about 132 minutes, the bulge at around 135 minutes, and the bump at around 150 minutes) while avoiding the spurious bumpiness apparent in Fig. 3.8.

Figure 3.13 illustrates the different effects that varying the two bandwidths (that of the pilot estimate, and $h_v$) has on $\hat{f}_v$ for the marathon data. In the first column, $h_v$ is fixed at $h_v = .466$, while the pilot bandwidth is

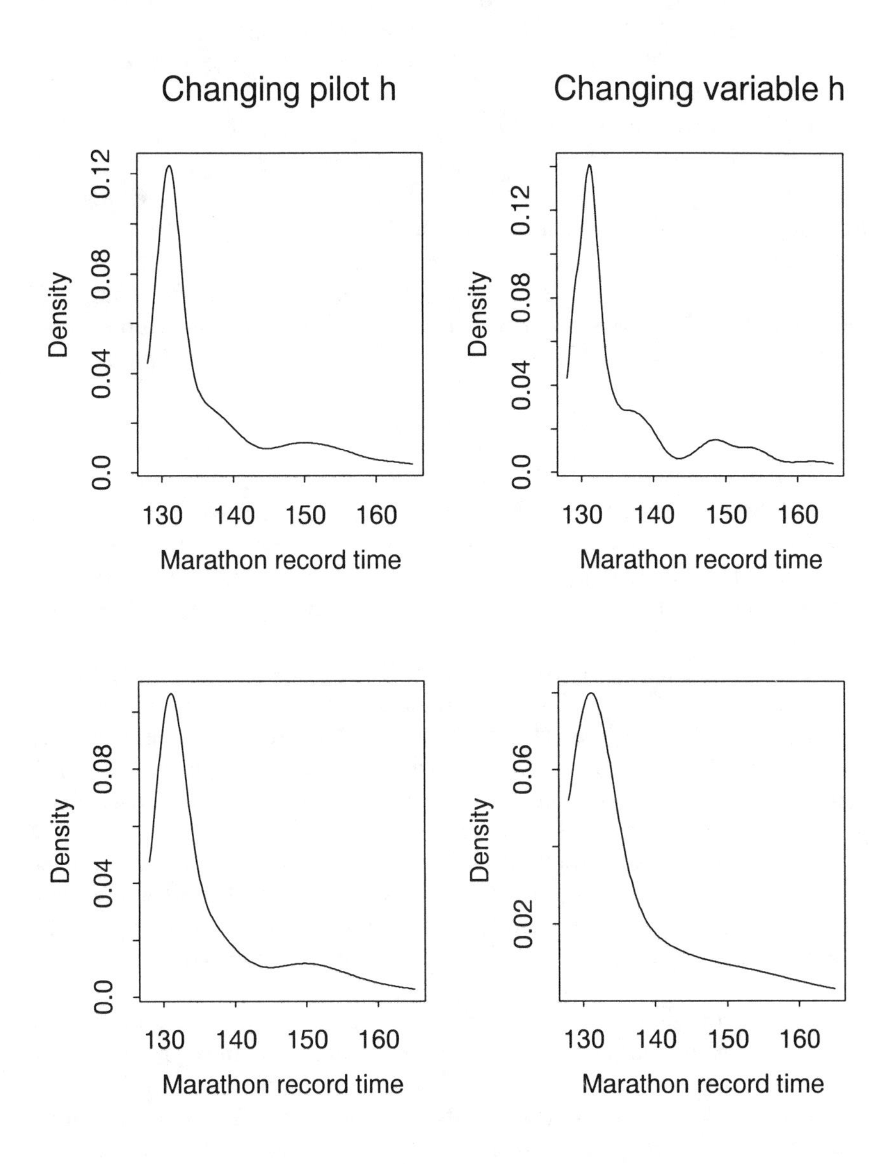


**Fig. 3.13.** Variable kernel estimates of men's marathon record data, changing either the pilot bandwidth (first column) or variable bandwidth $h_v$ (second column).

either halved to $h = .7675$ (first row) or doubled to $h = 2.87$ (second row). These changes do not affect the variable kernel estimate very much, reinforcing the impression that the final estimate is insensitive to the form of the pilot estimate.

In the second column of the figure, the pilot $h$ is fixed at $h = 1.435$, while $h_v$ is either halved to $h_v = .233$ (first row) or doubled to $h_v = .932$ (second row). The variable kernel estimate $\hat{f}_v$ is very sensitive to changes in $h_v$, with too small values leading to undersmoothing and too large values leading to oversmoothing.

A different way to vary $h$ in the kernel estimator to try to improve performance is to choose $h(x)$ as a function of the evaluation point $x$ as discussed in Section 3.2.2; that is, a *local-bandwidth kernel estimator*

$$\hat{f}_L(x) = \frac{1}{nh(x)} \sum_{i=1}^{n} K\left[\frac{x - x_i}{h(x)}\right]. \tag{3.14}$$

At any given value $x$, this is no different from the fixed-bandwidth kernel estimator (3.2) — the only change is that different values of $h$ can be used at different locations $x$. Since (3.14) represents an ordinary fixed-bandwidth kernel estimator at each point $x$, it cannot improve on the usual $O(n^{-4/5})$ convergence rate of MISE.

The most common choice of $h(x)$ is as the $k$th nearest neighbor distance from the data points to $x$ (that is, the distance from $x$ of the $k$th closest data point). The number of nearest neighbors $k$ controls the level of smoothing, with larger values of $k$ corresponding to more smoothing. The use of nearest neighbors results in more smoothing occurring in regions of low density (and less smoothing in regions of high density).

Unfortunately, the resultant nearest neighbor estimate is unsatisfactorily rough and does not provide the desired improved local adaptivity. Figure 3.14 gives nearest neighbor estimates for the men's marathon data based on using distances of 10, 20, 30, and 40 nearest neighbors, respectively (with $K$ being a biweight kernel). The estimate based on 10 nearest neighbors is too rough, but increasing the number of nearest neighbors does not remove all of the jaggedness of the estimate, as in plot (d). The estimate also does not capture the interesting features of the density (as the variable kernel does in Fig. 3.12), with the the bump at around 150 minutes being smoothed over in plots (b)–(d). It is apparent that the variable kernel estimator is a better way to allow local levels of smoothing than the local-bandwidth estimator using nearest neighbors (though other ways of choosing a local bandwidth can do better than nearest neighbors).

### 3.3.3 Higher order kernels

The use of a density function as the kernel function $K$ in a fixed-bandwidth kernel estimator has intuitive advantages, in that the resultant estimate

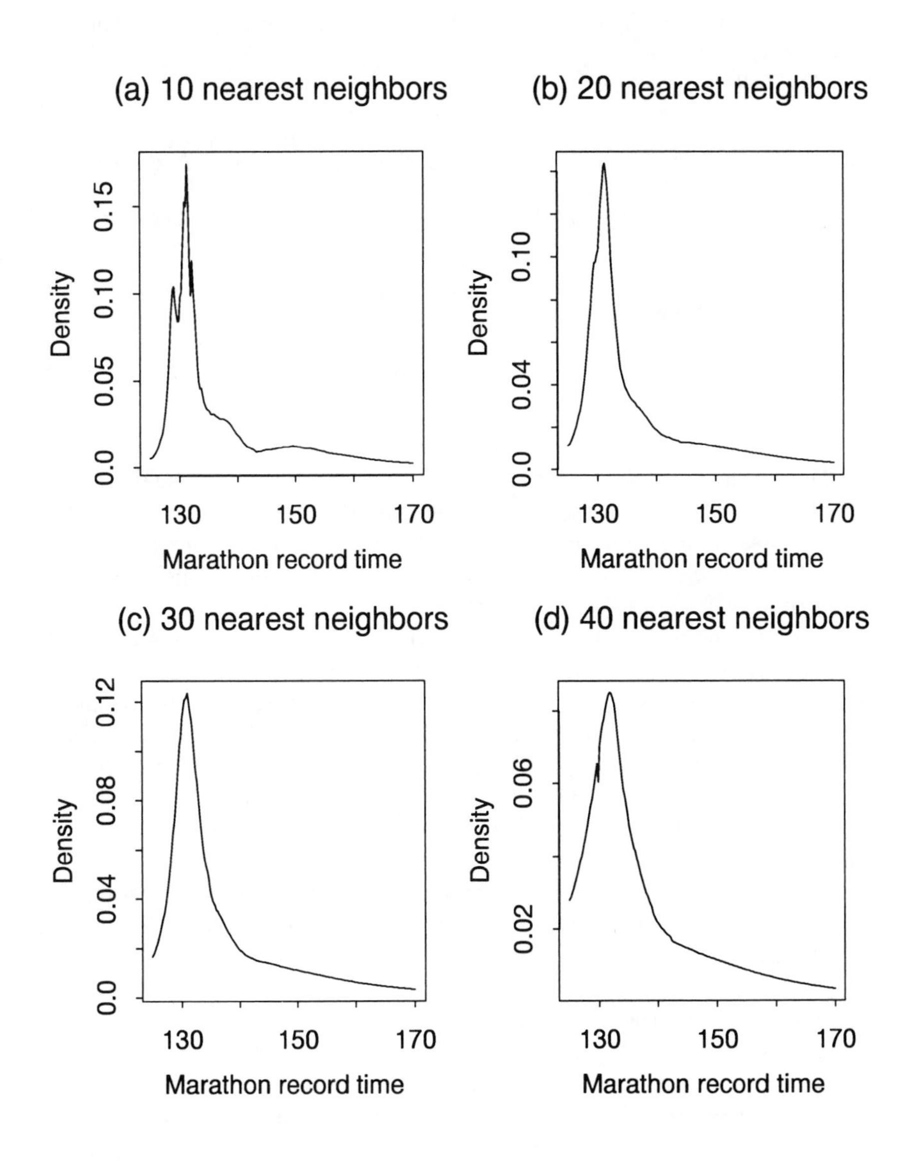


**Fig. 3.14.** Nearest neighbor estimates of men's marathon record data, based on (a) 10, (b) 20, (c) 30, (d) 40 nearest neighbors.

also will be a density (that is, nonnegative and integrating to 1). From a theoretical point of view, however, there is an advantage to allowing the choice of $K$ to be more general.

The full Taylor Series expansion for $E[\hat{f}(x)]$ has the form

$$\begin{aligned} E[\hat{f}(x)] =& f(x) \int K(u)du - hf'(x) \int uK(u)du + \frac{h^2 f''(x)}{2} \int u^2 K(u)du \\ &+ \cdots + (-1)^p \frac{h^p f^{(p)}(x)}{p!} \int u^p K(u)du + \cdots . \end{aligned} \tag{3.15}$$

Let $K$ be a symmetric function satisfying

$$\int K(u)du = 1, \int u^r K(u)du = 0, r = 1, \ldots, p-1, \int u^p K(u)du \neq 0$$

(note that $p$ must be even, and for $p > 2$ the function $K$ must take on negative values). This is termed a *kernel of order* $p$, with the usual nonnegative kernel being of order 2. If $f$ is sufficiently differentiable, the integrated squared bias of an estimator based on a $p$th order kernel is $O(h^{2p})$. Since the integrated variance is still $O(n^{-1}h^{-1})$, using higher order kernels improves the optimal MISE to $O(n^{-2p/(2p+1)})$.

Despite the apparent benefit to using higher order kernels, there are several reasons why they have not had much impact in practice:

(1) Since the kernel function $K$ takes on negative values, the resultant density estimate also can have negative values. Although such negative values can be "clipped" to be zero, this then results in a nondifferentiable estimate, which is particularly counterintuitive given the assumed increased differentiability of $f$ that justifies taking the Taylor Series (3.15) to more terms.

(2) Sample sizes at least in the hundreds are apparently necessary for higher order kernels to outperform nonnegative kernels for simple (Gaussian-type) densities, with far larger samples necessary for densities with more complicated structure.

(3) The choice of the bandwidth is more complex for higher order kernel estimates, since the smoothness of the resultant estimate is not a monotone function of $h$ (that is, both smaller and larger values of $h$ can lead to estimates that appear rougher than ones for $h$ in the middle ground). A practical plug-in estimate of $h$ has not been constructed (one could be constructed using the same principles as for the nonnegative kernel estimates), although cross-validation is easily implemented. In any event, higher order kernel estimates often exhibit spurious bumpiness (and negativity) in the tails.

Other variations of the higher order kernel theme are possible. The order can be taken to the limit $p \to \infty$, resulting in the best possible MISE rate (assuming sufficient smoothness of $f$). One example of such an estimator is the Fourier integral estimator, which is based on the kernel

$$K(u) = \frac{\sin(u)}{\pi u}.$$

This estimate integrates to 1, but the regions corresponding to both the positive and the negative values, respectively, are not integrable with probability 1.

A different way to achieve faster convergence of MISE is to decrease the bias in estimation of $\log f$ (rather than $f$) and then exponentiate the result. Equivalently, the *geometric combination estimator* of $f$ has the form

$$\hat{f}_h(x)^{c^2/(c^2-1)} / \hat{f}_{ch}(x)^{1/(c^2-1)}, \tag{3.16}$$

where $\hat{f}_{ch}(x)$ is a kernel estimator with bandwidth $ch$. Taking $c = 2$, for example, gives the estimator

$$\hat{f}_h(x)^{4/3} / \hat{f}_{2h}(x)^{1/3}. \tag{3.17}$$

The resultant estimator is positive everywhere but does not integrate to 1.

### 3.3.4 Transformation-based estimation

A natural approach to estimate hard-to-handle (complex) densities is to transform the data to a more appropriate scale, estimate the transformed density, and then transform back. Figures 2.6 and 2.7 showed how this can work when using histograms or frequency polygons, and similar results are possible using kernel estimates. In particular, transformation-based estimators can remove spurious bumpiness in the tails (by smoothing more there) and reduce boundary bias (by eliminating the boundary in the transformed scale).

Let $f_X(x)$ be the density in the original scale (which is the density being estimated), while $f_Y(y)$ is the density of the transformed random variable $Y = g(X)$, with $g(\cdot)$ being a monotonic (increasing) transformation. Then, a simple change of variable implies that

$$f_X(x) = f_Y[g(x)]g'(x).$$

If $f_Y$ is estimated using a kernel estimator, the transformation-based estimator of $f_X$ is

$$\hat{f}_X(x) = \frac{g'(x)}{nh_Y} \sum_{i=1}^{n} K\left[\frac{g(x) - g(x_i)}{h_Y}\right],$$

where $h_Y$ is chosen based on the $Y$ scale. So, for example, for long right-tailed (positive) data, a logarithmic transformation might be used, yielding the estimator

$$\hat{f}_X(x) = \frac{1}{nxh_Y} \sum_{i=1}^{n} K\left[\frac{\log(x) - \log(x_i)}{h_Y}\right].$$

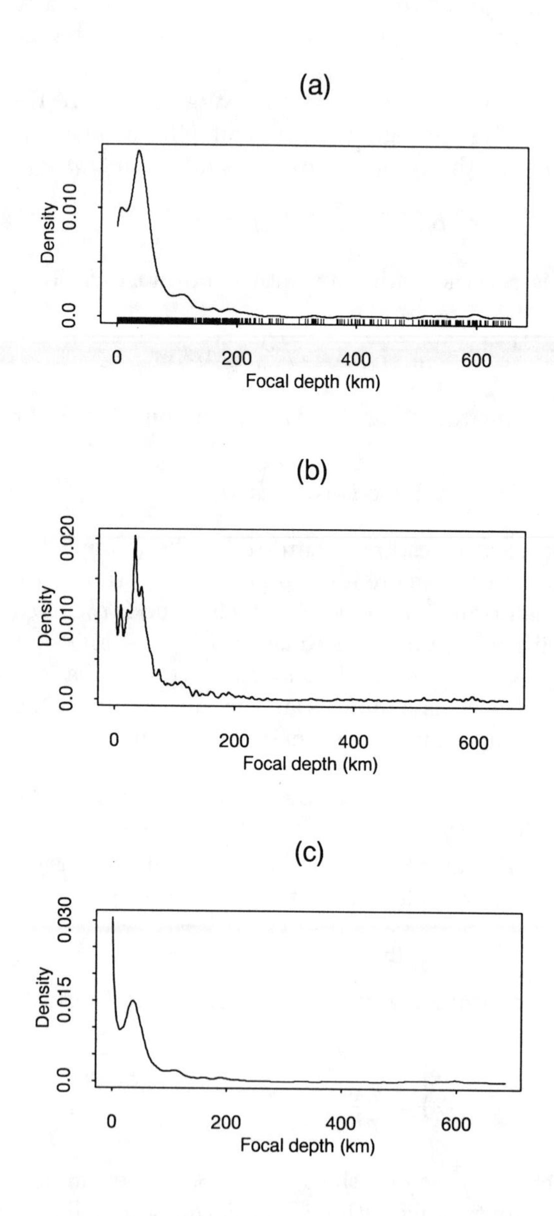


**Fig. 3.15.** Density estimates of earthquake depth data. (a) Kernel estimate with $h = 5.5$. (b) Kernel estimate with $h = 2$. (c) Boundary kernel estimate with $h = 5.5$.

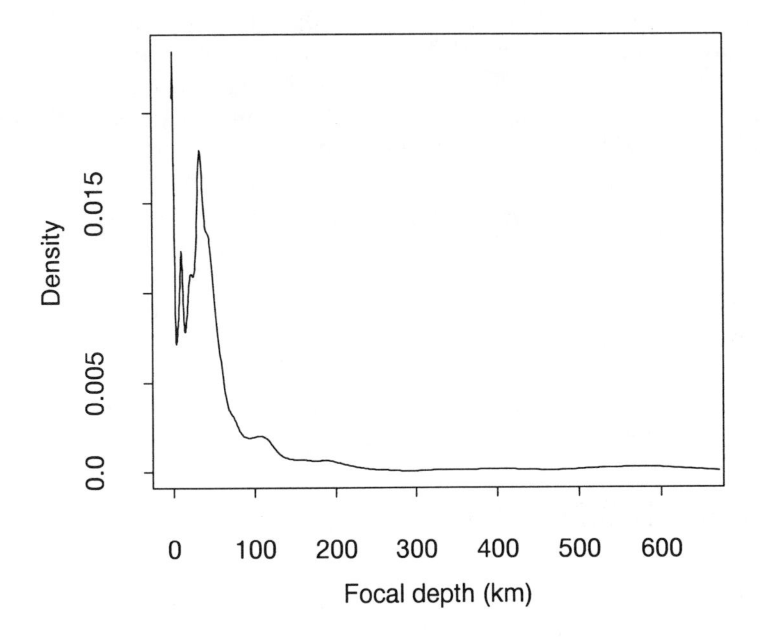


**Fig. 3.16.** Transformation-based kernel estimate of earthquake depth data.

The properties of the log function mean that the data are smoothed much less for small values of $x$ than for large values.

Consider Fig. 3.15, which gives estimates of the density for the focal depth in kilometers of 2178 earthquakes with body wave magnitude at least 5.8 on the Richter scale occurring between January 1964 and February 1986. Figure 3.15(a) is a kernel estimate (based on a Gaussian kernel with $h = 5.5$) of the density, which exhibits modes at 10, 33, 100, and 200 km but (given its decrease towards zero for a sample with many small values, including 145 zero values) seems to suffer from boundary bias. Figure 3.15(b), a kernel estimate with $h = 2$, reinforces this point, as it shows a peak at zero depth (but it is disturbingly rough, including in the long right tail). A boundary kernel estimate (Fig. 3.15(c)), smoothed as in Fig. 3.15(a), shows the peak at zero depth but smooths over the possible mode at 10 km.

Figure 3.16 gives a transformation-based estimate for these data. The depths were transformed using a Johnson family I transformation,

$$y = 99 \log[1 + (x - \overline{X})/99],$$

where $\overline{X} = 74.36$ km, smoothed based on a Sheather–Jones bandwidth in that scale, and then back-transformed. The resultant estimate captures the previously noted structure (modes at 0, 10, 33, 100, and 200 km), while avoiding roughness and spurious bumpiness.

It is possible to automate this procedure by choosing the appropriate transformation in a data-based way, although it is unlikely that any parametric family of transformations would be rich enough to account for all possible densities. For example, long right-tailed data probably would benefit from a shifted power transformation,

$$g(x) = \begin{cases} (x+\lambda_1)^{\lambda_2}\text{sign}(\lambda_2), & \text{if } \lambda_2 \neq 0, \\ \log(x+\lambda_1), & \text{if } \lambda_2 = 0, \end{cases}$$

with $\lambda_1 > -\min(x)$. Other transformations could be more appropriate for fat-tailed (kurtotic) data, or correction of boundary bias. Alternatively, the transformation could be estimated nonparametrically from the data based on a pilot estimate of the density $f_X(x)$.

## 3.4 Local Likelihood Estimation

The density estimation methods discussed so far are quite nonparametric, in that the estimate is not affected by any predisposition towards a particular parametric family. In some circumstances, it could be that a particular family is appealing as a "first guess" for the local behavior of the density, which could then be incorporated into the estimator.

A flexible way to do this is by using *local likelihood* estimation. The idea is to estimate the parameter vector $\boldsymbol{\theta}$ of a parametric family $f(x, \boldsymbol{\theta})$ locally for each $x$ and then estimate $f$ as $\hat{f}_\ell(x) = f(x, \hat{\boldsymbol{\theta}})$.

In order to preserve likelihood-based estimation while still operating locally, a local log-likelihood function is defined as

$$L(x,\boldsymbol{\theta}) = \sum_{i=1}^{n} \frac{1}{nh} K\left(\frac{x-x_i}{h}\right) \log f(x_i,\boldsymbol{\theta}) - \int \frac{1}{h} K\left(\frac{x-u}{h}\right) f(u,\boldsymbol{\theta}) du. \tag{3.18}$$

When $h$ is large, this is close to a constant times the ordinary log-likelihood function, and the maximizers of $L(x, \boldsymbol{\theta})$ are the ordinary maximum likelihood estimators, giving a fully (global) parametric fit. For smaller $h$, the maximizers provide a local fit instead.

The local log-likelihood (3.18) also provides a link to ordinary kernel estimation. If the parametric family is a (log) constant $[f(t, \theta) = \exp(\theta)]$, then (3.18) has the form

$$L(x,\theta) = \theta \sum_{i=1}^{n} \frac{1}{nh} K\left(\frac{x-x_i}{h}\right) - e^{\theta} \int \frac{1}{h} K\left(\frac{x-u}{h}\right) du.$$

The maximizer of this is $\hat{f} = \exp(\hat{\theta})$ and (if the range of $x$ is unbounded) is the ordinary kernel estimator (for bounded range, it is the locally renormalized kernel estimator discussed in Section 3.2.1).

More flexible choices of $f(\cdot, \boldsymbol{\theta})$ result in better local behavior. In particular, the distribution of $\hat{f}_\ell$ converges to a normal distribution with bias

$$\text{Bias}[\hat{f}_\ell(x)] = \frac{h^2\sigma_K^2 b(x)}{2} + O(h^4 + n^{-1})$$

and variance

$$\text{Var}[\hat{f}_\ell(x)] = \frac{f(x)R(K)}{nh} + O(n^{-1}).$$

These are similar to the values for the kernel estimator (substituting $b(x)$ for $f''(x)$), with $b(x)$ a measure of the difference between $f(x)$ and $f(x, \boldsymbol{\theta})$ whose exact form is related to $p$, the dimension of $\boldsymbol{\theta}$. In particular, if $p = 2$, $b(x) = f''(x) - f''(x, \boldsymbol{\theta}_0)$, where $f(x, \boldsymbol{\theta}_0)$ is the locally closest member of the parametric family.

Further, while a scalar $\theta$ ($p = 1$) yields an estimator that suffers from boundary bias (as happens for the kernel estimator, a special case of scalar $\theta$), for bivariate $\boldsymbol{\theta}$ ($p = 2$) the asymptotic bias of the distribution of the local likelihood estimator is identical to that of the boundary kernel (3.12). That is, any local likelihood estimator with bivariate $\boldsymbol{\theta}$ automatically corrects for boundary bias without explicitly defining a boundary kernel. Taking $f(x, \boldsymbol{\theta}) = \theta_0 + \theta_1(t - x)$ as the local model around $x$ (a linear model) gives a bias representation identical to that of a kernel, while $f(x, \boldsymbol{\theta}) = \theta_0 \exp[\theta_1(t - x)]$ (a local log-linear model) applies a correction factor to the bias.

The dimension of the parameter $\boldsymbol{\theta}$ continues to determine the bias properties of the local likelihood estimator for dimensions greater than 2. If $p = 3$ or 4, the $O(h^2)$ term in the bias representation in the interior drops out, making it $O(h^4)$. An even dimension ($p = 4$) again results in a simplification to the bias term and automatic boundary bias correction from $O(h^3)$ to $O(h^4)$. So, for example, a local log cubic polynomial estimator roughly corresponds to a boundary-corrected, fourth order kernel estimator (in the $\log f$ scale).

Local likelihood estimation does suffer from an interesting difficulty. For $K$ bounded and fixed $h$, there is positive probability that no observations will fall within the support of $K$ for any $x$. This means that $\text{MSE}[\hat{f}_\ell(x)]$ is not defined for any $x$. One way (but not the only way) to allow $h$ to vary locally to avoid this is to base it on a nearest neighbor distance (so that there are always sufficient observations close to $x$ within the support of $K$). By incorporating the nearest neighbor distance indirectly (through the local likelihood), the extreme roughness of the ordinary nearest neighbor estimate (as in Fig. 3.14) can often be avoided.

Figure 3.17 gives a local likelihood estimate for the racial distribution data (from Fig. 3.9) using a local log-quadratic function, with $h(x)$ a nearest

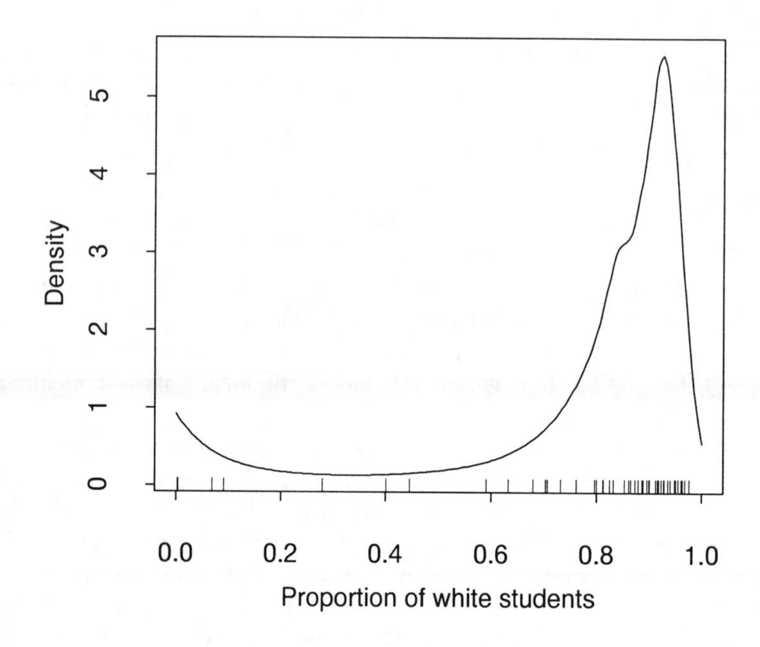


**Fig. 3.17.** Local log-quadratic density estimate of racial distribution data, using 70% nearest neighbor span.

neighbor span covering 70% of the observations closest to $x$. This estimate corrects the deficiencies of the kernel estimate (Fig. 3.9) — the peak at $x = .9$ is not flattened, the spurious bumps around $x = .4$ are not present, and there is no evidence of boundary bias near $x = 0$. The resultant estimate is quite smooth, showing that the local likelihood approach smooths out the roughness of the nearest neighbor distance.

The automatic boundary bias correction achieved by the local likelihood estimator is illustrated in Fig. 3.18. This is a variability plot for a local log-quadratic density estimate of the mine accident data, with $h(x)$ having a nearest neighbor span of 65% of the observations. The estimate is similar to that of the boundary kernel (Fig. 3.10), and the variability envelope exhibits the expected widening at $x = 0$ and around $x = 20$.

There is also, however, a puzzling "jag" in the upper limit of the envelope at around 5 days between accidents. This is a direct consequence of the use of nearest neighbor distance and shows that the roughness associated with nearest neighbors is not completely avoided when using local likelihood estimation.

Figure 3.19 is a local log-quadratic density estimate of these data with

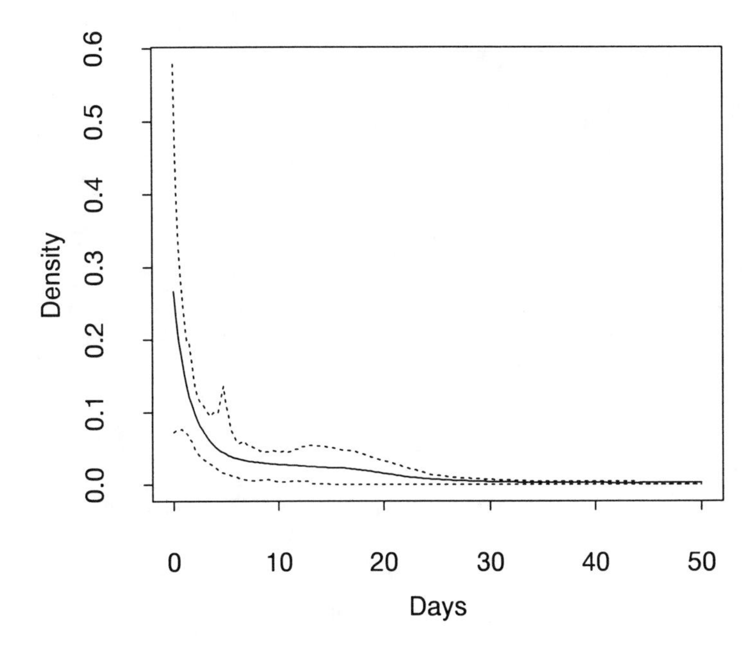


**Fig. 3.18.** Variability plot of local log-quadratic density estimate of mine accident data, using 65% nearest neighbor span.

a nearest neighbor span of 64%, rather than 65%, of the observations. The estimate is much less satisfactory, as the problems in the region of 5 to 10 days between accidents are apparent. The local log-quadratic fit with fixed $h$ does not exhibit this pattern for these data (but is also not locally adaptive).

## 3.5 Roughness Penalty and Spline-Based Methods

The nonparametric maximum likelihood estimator (MLE) of $f$ is not a viable density estimator; the maximizer of the log-likelihood

$$\ell(f) = n^{-1} \sum_{i=1}^{n} \log f(x_i)$$

is a set of Dirac spikes at the observations $\{x_i\}$. Local likelihood estimation avoids this problem by operating locally rather than globally. A different approach is to modify the log-likelihood to discourage the roughness of the MLE by maximizing a penalized (log) likelihood

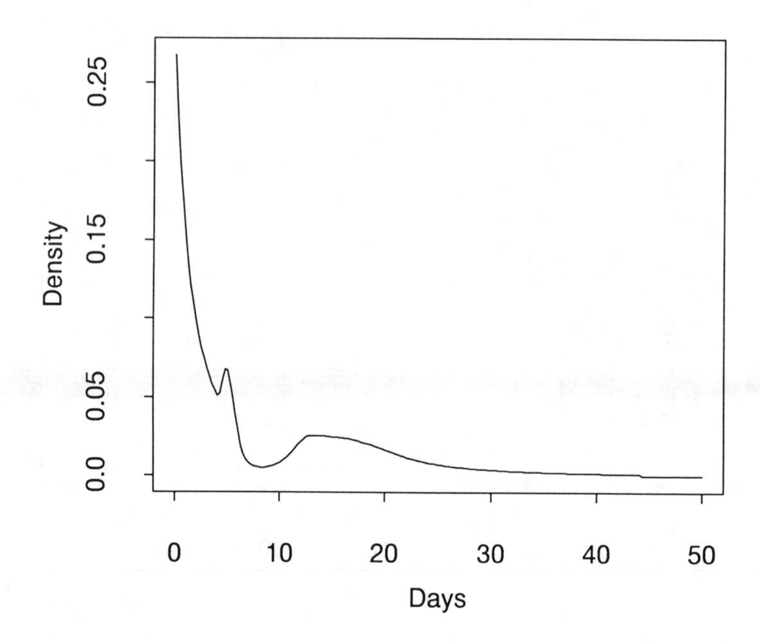


**Fig. 3.19.** Local log-quadratic density estimate of mine accident data, using 64% nearest neighbor span.

$$L(f) = n^{-1} \sum_{i=1}^{n} \log f(x_i) - \Phi(f) \tag{3.19}$$

subject to $\int f = 1$, where $\Phi(f) \geq 0$ is a roughness penalty that decreases as $f$ gets smoother. The resultant *maximum penalized likelihood estimator* (MPLE) provides a tradeoff between fidelity to the data (from the log-likelihood) and smoothness (from the roughness penalty).

Different choices of $\Phi$ yield different estimators. The penalty could be based on first derivatives, such as $R[(\log f)'] = \int [f'(u)/f(u)]^2 \, du$ or $R[(f^{1/2})']$, or second derivatives, such as $R[(\log f)'']$ or $R[(f^{1/2})'']$. The advantage of basing the penalty on $\log f$ or $f^{1/2}$ is that negativity can be avoided. For example, take $g = \log f$ and $\Phi(g) = \alpha R(g''), \alpha \geq 0$. Maximizing (3.19) is then equivalent to minimizing

$$L(g) = -n^{-1} \sum_{i=1}^{n} g(x_i) + \alpha \int [g''(u)]^2 \, du + \int e^{g(u)} du \tag{3.20}$$

(the last term effectively enforces the constraint $\int f = 1$). Then, $\hat{f} = \exp(\hat{g})$.

As $\alpha \to 0$, the MPLE approaches the nonparametric MLE of Dirac spikes, while as $\alpha \to \infty$, the MPLE becomes the MLE within a parametric family that depends on $\Phi$. For example, if $f$ is defined on the nonnegative numbers, and $\Phi(g) = \alpha R(g'')$, the limiting family is exponential, while if $\Phi(g) = \alpha R(g''')$, the limiting family is Gaussian. If $f$ is bounded away from zero, and $R[f^{(2m)}]$ is finite, then a roughness penalty based on $m$ derivatives gives an estimator whose MSE converges to zero at the rate $O(n^{-4m/(4m+1)})$.

Penalized likelihood estimators are often called spline estimators, since many such estimators take the form of polynomial splines with knots at the order statistics. A polynomial spline is a function that is a piecewise polynomial of degree $r$ on any subinterval defined by adjacent knots, has $r-1$ continuous derivatives, and has an $r$th derivative that is a step function with jumps at the knots. Asymptotically, the MPLE is approximately a local-bandwidth kernel estimator, as in (3.14). If $\Phi(g) = \alpha R(g'')$, for example,

$$\hat{f}(x) \approx \frac{f(x)^{1/4}}{n\alpha^{1/4}} \sum_{i=1}^{n} K\left[\frac{(x-x_i)f(x)^{1/4}}{\alpha^{1/4}}\right],$$

with

$$K(u) = \frac{1}{2} e^{-|u|/\sqrt{2}} \sin\left(\frac{|u|}{\sqrt{2}} + \frac{\pi}{4}\right),$$

away from the boundary. This is a fourth order kernel, which accounts for the $O(n^{-8/9})$ convergence rate of the MSE (since the roughness penalty is based on $m = 2$ derivatives). In addition, since the MPLE is defined through exponentiating an estimate of the log-density, it is nonnegative.

The exact MPLE is difficult to calculate, which has motivated several approximations and spline-based estimators. One such approach is to define the estimator directly as a cubic spline fit to the logarithm of the density $g$, with knot placement determining the smoothness of the estimate. The *logspline density estimate* then takes $\hat{f} = \exp(\hat{g})$.

Computational details become key in practical application of logspline density estimation. For example, restrictions must be placed on the estimate below the first knot and above the last knot, such as being linear in the log scale. If such exponential behavior is inappropriate, some correction (such as transformation) must be made.

Knot placement is crucial in logspline density estimation. A natural set of potential knots is the set of order statistics (as penalized likelihood estimation would suggest), with the first and last knots being the minimum and maximum of the sample values, respectively. Ideally, the number of knots would be chosen in a data-dependent way. One possibility is to start with the entire set of order statistics being the initial set of knots and then delete knots in a stepwise manner if they do not improve the fit (as measured by some criterion, such as $AIC$). Previously deleted knots also could be added back in if that helps. Once the knots are determined, the spline is estimated by maximum likelihood.

The importance of the placement, as well as the number, of knots for the logspline density estimate is apparent in Fig. 3.20. The three pairs of density estimates (for the mine accident, racial distribution, and earthquake depth data, respectively) refer to estimates where the number and placement of knots were chosen automatically by stepwise deletion of knots, and where the number was chosen manually (with placement in a roughly symmetric pattern). The automatic estimates for the mine accident and racial distribution data are oversmoothed, being based on only four and three knots, respectively; the manual versions are each based on six knots and recover most of the structure. The automatic estimate for the earthquake depth data (based on 14 knots) is a dismal failure, while the manual version (again based on 14 (different) knots) recovers the structure for depths less than 50 km well (although it is less successful above that).

## 3.6 Comparison of Univariate Density Estimators

The most important lesson to be drawn from this chapter is that the appealing simplicity of the estimators described in Chapter 2 is probably not a good enough reason to use them. The estimators discussed in this chapter are superior at recovering interesting structure and should be used instead. Taking the trouble to construct a variability plot envelope is also a good idea, in order to get a sense of how the variability of the estimator changes over the range of the data.

Having said that, which is the estimator of choice? The easy answer is the kernel estimator, which is highly intuitive, well understood, and amenable to routine and automatic application. But, this answer is not right either, as boundary bias and the lack of local adaptivity limit the applicability of the estimator (using boundary kernels can fix the former problem, but not the latter).

Currently no method can claim to be "best," which might be a fundamental characteristic of the problem. Candidates for a good general approach are

(1) Kernel-type estimators such as the local- and variable-bandwidth kernels and transformation-based kernel, which achieve desirable local adaptivity and retain the intuitive appeal of the kernel, but also can share problems such as boundary bias. Many routine problems can be handled using transformations, which allow the use of well-established automatic bandwidth selectors (although in the transformed, rather than original, scale).

(2) Estimators based on the log-density, such as local likelihood and logspline estimators, which achieve automatic boundary correction and

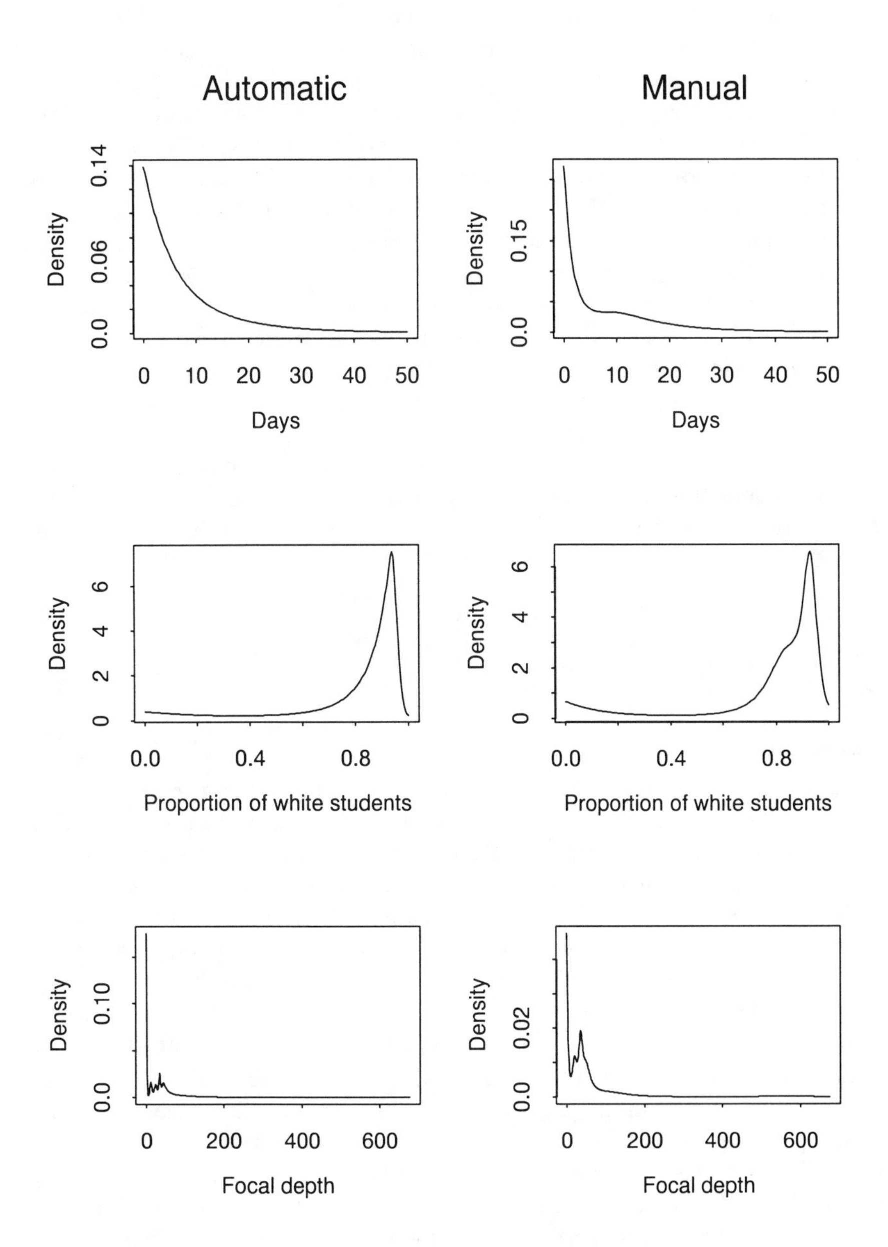


**Fig. 3.20.** Logspline density estimates of mine accident data (first row), racial distribution data (second row), and earthquake depth data (third row), based on whether knots are chosen automatically using stepwise deletion (left column) or manually (right column).

local adaptivity, but can be more difficult to study theoretically. Another appealing possibility that is closely related to the local likelihood estimator will be described in Section 6.4.

One of the most controversial issues in the density estimation literature is that of automatic smoothing parameter selection. If an impression of the broad features in the density is desired, stable bandwidth methods such as plug-in selectors have an advantage, even though they might oversmooth and miss some effects. If it was desired to notice all structure, no matter how small, less biased methods like cross-validation have an advantage, although they might undersmooth and yield many false signals. It is always a good idea to look at a wide range of smoothing parameters to minimize the chances of missing anything important and to get a sense of whether locally varying the smoothing parameter would be useful.

Still, a strong determining aspect in whether any one method becomes the "gold standard" is success in determining the degree of smoothing automatically, a criterion that would work against the spline-based methods. In any event, the similarity of many of the plots in this chapter suggests that most "smart" estimators are likely to give similar impressions of the data.

## Background material

### Section 3.1

**3.1.2.** Fix and Hodges (1951) first proposed the kernel estimator, using a uniform kernel; see Silverman and Jones (1989) for discussion (and publication) of this unpublished paper. Akaike (1954) studied consistency properties of the estimator based on a uniform kernel using the properties of the binomial distribution. Whittle (1958) examined some finite sample properties of the estimator, but the results cannot be used generally in a practical way. Rosenblatt (1956) and Parzen (1962) first examined the asymptotic properties of the kernel estimator in terms of MSE. Parzen (1962), Bertrand-Retali (1978) and Devroye and Györfi (1985) examined the consistency properties of $\hat{f}$, establishing consistency under very weak (or no) conditions on $f$. Wand and Devroye (1993) examined a measure of the difficulty in estimating a density based on absolute error.

Bartlett (1963) and Epanechnikov (1969) (and Hodges and Lehmann, 1956, in a different context) proved that the quadratic kernel is asymptotically optimal when the kernel function is restricted to be a proper density. Trosset (1993) derived the form of the corresponding kernel for finite samples. Cline (1988) showed that any estimator based on an asymmetric kernel can be beaten (in the sense of smaller MISE) by an estimator based on a symmetric kernel, so (in this sense) asymmetric kernels are inadmissible (any asymmetric square integrable kernel can be improved by reflecting about zero, as in $K^*(u) = [K(u) + K(-u)]/2$). Parzen (1962) calculated $R(K)$ for several kernel functions and noted the connection with AMISE.

The formula (3.2), which involves the sum of $n$ values for each evaluation point $x$, is not computationally feasible for very large sample sizes. Similarly, repeated calculation of $\hat{f}$, such as would be required for Monte Carlo simulations or bootstrapping, could become impossible if it was done using (3.2).

For this reason, computationally more efficient forms of $\hat{f}$ have become generally available in recent years. Most are based on binning the data and then smoothing the resultant histogram form. Silverman (1982a) took advantage of binning and the Fast Fourier Transform in computer code based on using the Gaussian kernel.

Scott (1985b) motivated this idea as a way to eliminate the dependence of histograms on the anchor position, by estimating $f$ using an average of histograms with different anchor positions. Let $M$ be the number of shifted histograms that constitute the average. The resultant average shifted histogram (ASH) estimator, as $M$ gets larger, has the form (3.2) with a triangular kernel function

$$K(u) = \begin{cases} 1 - |u|, & \text{if } |u| \leq 1, \\ 0, & \text{otherwise.} \end{cases}$$

The squared error properties of the ASH are similar to those of the kernel estimator with triangular kernel, with added terms that are a function of $M$:

$$\text{AMISE} = \frac{2}{3nh}\left(1 + \frac{1}{2M^2}\right) + \frac{h^2 R(f')}{12M^2} + \frac{h^4 R(f'')}{144}\left(1 - \frac{2}{M^2} + \frac{3}{5M^4}\right).$$

Note that the additional terms become negligible for $M$ as small as 10. See also Jones and Lotwick (1983), Scott and Sheather (1985), Jones (1989), and Hall and Wand (1996).

This estimator can be generalized to approximate any kernel estimator with any kernel function by using weighted averages of histograms rather than a simple average. This weighted averaging of rounded points, or WARPing, is described in detail in Härdle (1991) and Härdle and Scott (1992).

Several authors have proposed kernel-type estimators (of a form termed generalized kernel estimators — see Földes and Révész, 1974, and Walter and Blum, 1979) based on smoothing a histogram without focusing specifically on computational savings. Examples include the estimators of Vitale (1975), Gawronski and Stadtmüller (1980, 1981) and Stadtmüller (1983).

A simple improvement to the WARPing estimate is to use as the estimate a piecewise linear interpolant of the WARPing estimates at the histogram bin centers. The similarity of this estimate to a frequency polygon led Scott (1985b) to refer to this as the frequency polygon-average shifted histogram (FP-ASH) estimate. The resultant estimator appears much smoother to the eye, which translates into improved AMISE,

$$\text{AMISE} = \frac{2}{3nh} + \frac{h^4 R(f'')}{144}\left(1 + \frac{1}{M^2} + \frac{9}{20M^4}\right),$$

which is virtually identical to that of the triangular kernel estimator for $M$ as small as 3. Jones (1989) also investigated this interpolated density estimator.

The standard MISE results for kernel estimators are based on assuming that the observations constitute a random sample from the density $f$; that is, they are independent and identically distributed. If instead the observations are a sample from a strictly stationary process, they are still identically distributed (so estimation of their marginal density $f$ makes sense) but are no longer independent.

Several authors have studied the effect of dependence on the asymptotic properties of the kernel estimator under various models for dependence, including Markov and autoregressive processes, and mixing-type conditions; see, for example, Roussas (1969, 1988), Rosenblatt (1970, 1971), Nguyen (1979), Masry (1983), Hart (1984), Tran (1989, 1990), Györfi *et al.* (1990), and Bosq (1995). These authors found that, if the dependence is not too strong, the AMISE of the kernel estimator is identical to that when the observations are independent. In particular, Hall and Hart (1990a) showed that for linear processes

$$\text{MISE}(\hat{f}_h) \approx \text{MISE}(\hat{f}_h^0) + \text{Var}(\overline{X})R(f'),$$

where $\hat{f}_h^0$ is the kernel estimate of the density function obtained from a random sample (the extra term comes from the asymptotic variance, as the bias is unaffected by dependence). Under short-range dependence, $\text{Var}(\overline{X}) = O(n^{-1})$, which is smaller than $\text{MISE}(\hat{f}_h^0) = O(n^{-4/5})$, so asymptotically the effects of dependence disappear. Under long-range dependence, however, the rate of MISE is dominated by the slower convergence of $\text{Var}(\overline{X})$ to zero and does not even depend on the smoothing parameter (see Beran, 1994, for a discussion of such long memory processes). See also Csörgő and Mielniczuk (1995a).

Despite the lower order effect of short-range dependence on AMISE, for finite samples any kind of dependence can affect the efficiency of $\hat{f}$ in a significant way; see Hart (1984) and Wand (1992a) for discussions relating to first-order autoregressive and linear processes, respectively.

**3.1.3.** Many of the results discussed in Chapter 2 about bin-width determination for histograms and frequency polygons carry over to kernel estimation. For example, Terrell and Scott (1985) and Terrell (1990) addressed oversmoothing in this context, determining that the oversmoothed choice when using a Gaussian kernel is $h = 1.144\sigma n^{-1/5}$. This is close to the optimal value for a Gaussian density (3.7), reinforcing the notion that the Gaussian density is quite smooth.

Kraft, Lepage, and van Eeden (1983) and Dodge (1986) studied the performance of the minimizer of AMISE for finite samples via Monte Carlo simulation. They found that the asymptotically optimal choice works well for small samples (as small as $n = 10$) for shorter-tailed densities, such as the Gaussian and logistic, but is much less effective for longer-tailed densities, such as the double exponential, gamma, and Cauchy (where sample sizes of at least 50 to 100 are necessary for the asymptotics to take over).

Rudemo (1982) and Bowman (1984) proposed using the minimizer $\hat{h}_{CV}$ of the cross-validation function for kernel density bandwidth choice. Let $h_{opt}$ be the value of $h$ that minimizes MISE, while $\hat{h}_{opt}$ is the value that minimizes ISE (for a given sample). Hall (1983a) showed that under smoothness conditions on $f$ and $K$, $\hat{h}_{CV}/h_{opt} \to 1$ in probability as $n \to \infty$. Stone (1984) strengthened this result: if $K$ is symmetric, compact, and Hölder continuous, and $f$ is bounded, then $\hat{h}_{CV}$ is asymptotically optimal, in the sense that

$$\lim_{n\to\infty} \frac{\text{MISE}(\hat{h}_{CV})}{\text{MISE}(h_{opt})} = 1,$$

with probability 1. Burman (1985) used a different approach to obtain similar results.

Diggle and Marron (1988) used the equivalence of $\hat{h}_{CV}$ based on a uniform kernel and an empirical Bayes smoothing parameter method of Diggle (1985a) for intensity function estimation to motivate a strong connection between density and intensity estimation (in the intensity estimation framework, observations are treated as a realization on an interval $[0, T]$ of a nonstationary Poisson process with intensity function $\lambda(x)$).

The value $h_{opt}$ only reflects average performance over all possible data sets from a given population, rather than performance for the observed data set. For this reason, from a conceptual point of view, the value $\hat{h}_{opt}$ is a more reasonable target, since it reflects the accuracy of $\hat{f}$ for the data set at hand.

Unfortunately, $\hat{h}_{opt}$ is an exceedingly difficult target to try to hit. Hall and Marron (1987a) showed that the best possible convergence rate of any data-dependent $\hat{h}$ to $\hat{h}_{opt}$ is

$$\frac{\hat{h}}{\hat{h}_{opt}} = 1 + O_p(n^{-1/10}). \tag{3.21}$$

This is the rate achieved by $\hat{h}_{CV}$ (Hall and Marron, 1987b) and also the rate if $h_{opt}$ itself were available (Hall and Marron, 1987a). Hall and Johnstone (1992) showed that $h_{opt}$ (and any data-based choice satisfying $\hat{h}/h_{opt} = 1 + o_p(n^{-3/10})$) does improve on $\hat{h}_{CV}$ as an estimate of $\hat{h}_{opt}$, however, in the sense of smaller asymptotic variance (although it does not reach the minimal variance). See Mammen (1990), Hall and Marron (1991a), Jones (1991), Jones and Kappenman (1991), and Grund, Hall, and Marron (1994) for further discussion of the issues in estimating $h_{\text{opt}}$ versus $\hat{h}_{\text{opt}}$. Gu (1995a) argues in favor of using ISE, rather than MISE, as a target, further proposing that $R(\hat{f}'')$ is more meaningful than $\hat{h}$ is across sampling replicates.

The slow rate in (3.21) has been the motivation for much research that has focused on the estimation of $h_{\text{opt}}$, rather than $\hat{h}_{\text{opt}}$. If only two bounded derivatives of $f$ exist, a rate of $n^{-1/10}$ to $h_{\text{opt}}$ is optimal, but assuming additional smoothness implies that the optimal rate of convergence is as fast as $\hat{h}/h_{\text{opt}} = 1 + O_p(n^{-1/2})$ (Hall and Marron, 1991a). Note, however, that the kernel estimator itself only assumes two derivatives, so these "improved" methods assume an underlying smoothness that makes the density estimator itself suboptimal, a point noted by Terrell (1992) and Loader (1995), among others.

The minimizer of CV, $\hat{h}_{\text{CV}}$, achieves the very slow rate $\hat{h}_{\text{CV}}/h_{\text{opt}} = 1 + O_p(n^{-1/10})$ (Hall and Marron, 1987b; Scott and Terrell, 1987). This slow rate translates into high variability of $\hat{h}_{\text{CV}}$, as has been noted in many Monte Carlo simulation studies (Bowman, 1985; Scott and Terrell, 1987; Park and Marron, 1990; Chiu, 1991a; Jones, Marron, and Park, 1991; Marron, 1991; Simonoff and Hurvich, 1993; Cao, Cuevas, and González-Manteiga, 1994; Jones, Marron, and Sheather, 1996). These studies have also revealed that the CV function often exhibits multiple local minima; Hall and Marron (1991b) provide theoretical support for this observation.

The value $\hat{h}_{\text{CV}}$, besides being highly variable, often undersmooths in practice, in that it leads to spurious bumpiness in the estimated density ($\hat{h}_{\text{CV}}$ is an unbiased estimator of $h_{\text{opt}}$, so in that MISE-based sense it is not necessarily undersmoothing); see Scott and Terrell (1987), Chiu (1991a), Park and Turlach (1992), and Sheather (1992). Silverman (1986, pp. 51–52), Chiu (1991b), and Hall and Marron (1991b) provided theoretical arguments to explain this observed behavior.

Woodroofe (1970) and Nadaraya (1974) proposed the idea of plug-in bandwidth selection without investigating issues of practical implementation of the idea. An early plug-in approach to bandwidth selection was biased cross-validation (Scott and Terrell, 1987), but this bandwidth choice shares the slow $n^{-1/10}$ convergence rate to $h_{\text{opt}}$ of $\hat{h}_{\text{CV}}$ (Jones and Kappenman, 1991, described other such methods). Later methods have improved on this rate, giving $\hat{h}/h_{\text{opt}} = 1 + O_p(n^{-\alpha})$, with $\alpha = 4/13$ (Park and Marron, 1990), $\alpha = 5/14$ (Sheather and Jones, 1991; Hall, Marron, and Park, 1992; Cao, Cuevas, and González-Manteiga, 1994; Engel, Herrmann, and Gasser, 1994), and $\alpha = 1/2$ (Chiu, 1991a, 1992; Hall *et al.*, 1991; Jones, Marron, and Park, 1991; Hall, Marron, and Park, 1992; Marron, 1992b; Kim, Park, and Marron, 1994). See also Fan and Marron (1992) and Park and Marron (1992) for further theoretical discussion of these methods.

A uniformly best bandwidth selection method for the fixed-bandwidth kernel estimator has not been found, which (given the complexity of the problem and the limitations of the estimator) is probably not surprising. The results in Park and Turlach (1992), Sheather (1992), Cao, Cuevas, and González-Manteiga (1994), and Jones, Marron, and Sheather (1996) suggest that the best methods are those of Park and Marron (1990), Sheather and Jones (1991), Chiu (1992), Hall, Marron, and Park (1992), Cao, Cuevas, and González-Manteiga (1994), and Engel, Herrmann, and Gasser (1994), with the Sheather–Jones method being the single best choice, if one method had to be chosen.

Wand and Jones (1995, Chapter 3 and Appendix D) discussed many of the issues raised in this section in more detail.

Loader (1995) argued forcefully against the use of plug-in selectors, and $\hat{h}_{\mathrm{SJ}}$ in particular. He noted that these selectors require restrictive smoothing assumptions on the true density, and he claimed that selectors based on $AIC$ or cross-validation that make similar assumptions are generally at least as good as plug-in selectors. He noted that plug-in selectors also tend to oversmooth in situations where the density exhibits widely varying $|f''(x)|$, choosing an $\hat{h}$ larger than $h_{\mathrm{opt}}$ (estimators designed for densities of this type are discussed in Section 3.3.2). This oversmoothing tends to reduce variance, but at the cost of increased bias in $\hat{f}$.

The plug-in approach to bandwidth selection is just addressing estimation of $R(f'')$, as in (3.9). Thus, a good bandwidth selection method corresponds to an accurate $\widehat{R(f'')}$, which also can be used in other applications. For example, a plug-in estimate of the optimal (minimal AMISE) bin width for a constant bin-width frequency polygon is

$$\hat{h} = 2\left[\frac{15}{49\widehat{R(f'')}}\right]^{1/5} n^{-1/5},$$

based on (2.14). It is likely that if a complex bin-width selection method (which itself uses kernel estimation techniques) is used, then a better (kernel) density estimator would be used as well, but a plug-in-based method provides a useful "best case" data-based choice of bin width for comparative purposes, or if a simpler estimator was desired. Simonoff (1995a) and Jones *et al.* (1998) examined the use of this plug-in rule as applied to such estimators. The functional $R(f')$, which is needed for optimal histogram bin-width choice (see Eq. 2.6), can sometimes be estimated in a similar way, since for many densities $R(f') = -\int f''(u)f(u)du$; see Wand (1997).

These various bandwidth selection methods can be made more computationally efficient by using binning (as WARPing does for the kernel estimate itself). González-Manteiga, Sánchez-Sellero, and Wand (1996) examined the effects of such binning on cross-validation and plug-in methods. They found that for many univariate densities, binning over a grid of a few hundred values provides an accurate approximation to the actual target bandwidth measure.

All these bandwidth choices are motivated through the use of squared error, which can be criticized as an arbitrary (and unrealistic) loss function. Devroye and Györfi (1985) examined the use of absolute error, IAE $= \int |\hat{f}(u) - f(u)|du$. Hall and Wand (1988a,b) showed that the minimizers of MISE and MIAE are very close for symmetric, light-tailed densities, implying that the methods based on MISE perform similarly with respect to MIAE as well. See also Dodge (1986), Schucany (1989), and Cao, Cuevas, and González-Manteiga (1994).

Habbema, Hermans, and van den Broek (1974) and Duin (1976) proposed likelihood cross-validation, which is based on minimizing an estimate of Kullback–Leibler distance. Chow, Geman, and Wu (1983) established consistency of the resultant bandwidth choice if both $f$ and $K$ are bounded, although Hall (1982) showed that it can be suboptimal even if it is consistent. Gregory and Schuster (1979) and Schuster and Gregory (1981) noted that the likelihood cross-validation bandwidth can be inconsistent for unbounded densities. Hall (1987) sharpened these results theoretically, proving that the tail properties of $f$ and $K$ have a profound influence on the properties of the likelihood cross-validation bandwidth (see also Broniatowski, Deheuvels, and Devroye, 1989). Marron (1985a) discussed a modification of the likelihood cross-validation rule that results in better performance. Marron (1987) compared the asymptotic behavior of least squares and likelihood cross-validation in terms of the distances each rule minimizes.

It can be argued that none of these distances is a sensible target for density estimation, since they do not reflect what a data analyst would consider the quality of a density estimate. In particular, squared (or absolute) error heavily penalizes a misplaced mode, even if the qualitative shape of the density estimate is close to that of the true density (see Kooperberg and Stone, 1991, for an example of this phenomenon).

Park and Turlach (1992) investigated this issue by evaluating various bandwidth selectors using the number of modes in the resultant density estimate, but this does not address the size and shape of those estimated modes. Cuevas and González-Manteiga (1991) and Mammen (1995) focused on characteristic points of the density, such as inflection points. Marron and Tsybakov (1995) proposed the use of "visual error criteria" to try to assess the qualitative effectiveness of smoothing, based on both vertical and horizontal distances between $\{x, f(x)\}$ and $\{x, \hat{f}(x)\}$. Marron (1998) examined these measures further, concluding that the bandwidth selector of Sheather and Jones (1991) is still a good all-around choice.

If the observed data constitute a stationary time series rather than a random sample, then the optimal bandwidth changes. Roughly speaking, dependence implies less information about $f$, and hence an effectively smaller sample size than $n$. This suggests the use of a larger bandwidth than results based on independence would imply.

Hart and Vieu (1990) proposed a modification to cross-validation, wherein a sequence of $\ell_n + 1$ consecutive observations are dropped out, rather than one observation at a time. They showed that under an $\alpha$-mixing condition, as long as $\ell_n$ does not increase too quickly, the cross-validated bandwidth is asymptotically optimal. Although the usual (leave-out-one) version satisfies this condition, they found that for sufficiently strong dependence, taking $\ell_n > 0$ can improve ISE performance.

Cao, Quintela del Río, and Vilar Fernández (1993) examined the behavior of the cross-validation and biased cross-validation bandwidths, as well as those of Jones, Marron, and Park (1991) and Sheather and Jones (1991) (and other proposals), under dependence using Monte Carlo simulations. They found that the Jones–Marron–Park and Sheather–Jones bandwidths were fairly insensitive to the degree of dependence and were the best behaved. They also noted, however, that all the bandwidths tended to decrease with increasing autocorrelation, rather than increase as they should. Quintela del Río and Vilar Fernández (1992) adapted local cross-validation to dependent data.

Efron and Tibshirani (1993) gave an overview of the bootstrap approach to statistical inference. Jones and Rice (1992) described a way of summarizing a set of smooth curves using principal component analysis, which could be used as an alternative to a variability plot.

Tierney (1990) emphasized the usefulness of sliders to assess the sensitivity of estimation procedures to user-chosen parameters, with particular emphasis on kernel density estimation in Section 10.1.3. Marron's (1993) comments on appropriate bandwidth grid scale for Monte Carlo simulations are consistent with tuning bandwidth sliders using a logarithmic scale. Fan and Müller (1995) suggested plotting density estimates based on the set of bandwidths $\{h = 1.4^j \hat{h}_{\text{ROT}}, j = -3, -2, \ldots, 2\}$ on the same plot in order to assess the sensitivity to bandwidth choice, where $\hat{h}_{\text{ROT}}$ is a Gaussian rule of thumb bandwidth using (3.7); Scott (1992, p. 161) suggested the set $\{h = \hat{h}_{\text{OS}}/1.05^j, j = 0, 1, 2, \ldots\}$, where $\hat{h}_{\text{OS}}$ is the oversmoothed bandwidth choice. Marron (1995a) proposed a similar idea, centering an equally spaced set of bandwidths (in a log scale) at $\hat{h}_{\text{SJ}}$.

A static display that can show the changes in the fitted density as the bandwidth changes is the mode tree of Minnotte and Scott (1993). In this display, mode locations are plotted against the log of the bandwidth at which the density estimate with those modes is calculated. In this way, the splitting of modes as $h$ decreases is apparent, and the sensitivity of the fitted estimate to the choice of $h$ can be evaluated. The mode tree is based on the use of a Gaussian kernel, since for that kernel the number of zeroes in all derivatives of $\hat{f}$ is monotone decreasing in $h$ (Silverman, 1981). This implies that when using the Gaussian kernel, all modes found at a given value of $h$ remain as $h$ decreases.

The close connection between kernel estimation and estimation based on Fourier series is a major focus of Tarter and Lock (1993). A different, quite promising, orthogonal series method that is locally adaptive is the *wavelet density estimator*; see Kerkyacharian and Picard (1992, 1993), Hall and Patil (1995), and Donoho *et al.* (1996).

## Section 3.2

**3.2.1.** Maguire, Pearson, and Wynn (1952) gave the mine accident data. Gasser and Müller (1979) first examined issues of boundary bias, although in the context of nonparametric regression. Jones (1993a) provided a unified framework for investigation of the problem.

Boundary bias can also affect bandwidth selection methods for kernel density estimators. Plug-in estimators of $R(f'')$ can be inefficient if $f$ has nonzero derivatives at the boundaries, which leads to corresponding inefficiencies and inconsistencies in plug-in bandwidth selectors. See Van Es and Hoogstrate (1994, 1998) (these papers also examined the effects of jumps and kinks in the density $f$ in the interior on bandwidth selection).

**3.2.2.** Dawkins (1989) included the men's marathon data as part of a larger set of values of national racing records. Hurvich and Simonoff (1993) constructed fixed and variable bin width frequency polygons for these data that highlight the possible "round number" threshold effect, while avoiding bumpiness in the upper tail.

Chatterjee, Handcock, and Simonoff (1995) gave the data reporting the proportion of white students in Nassau County school districts (p. 298), along with a detailed description of the data analysis issues involved.

**3.2.3.** Stoker (1993) noted that Jensen's inequality implies that $\hat{f}(x)$ is biased downward where $f$ is concave and biased upward where $f$ is convex, and related this property to derivative estimation.

## Section 3.3

**3.3.1.** Jones (1993a) used the concept of generalized jackknifing to motivate the formulation of boundary kernels using the linear combination of two functions. Gasser and Müller (1979) (and, more recently, Hart and Wehrly, 1992) suggested the boundary kernel (3.12). Rice (1984a) proposed a kernel family based on (3.11), taking $L(x) = cK(cx)$. Jones (1993a) showed that the bias of $\hat{f}_B$ equals

$$\left[\frac{c_1(p)a_2(p) - a_1(p)c_2(p)}{c_1(p)a_0(p) - a_1(p)c_0(p)}\right] \frac{h^2 f''(x)}{2} + o(h^2),$$

or

$$\left[\frac{a_2^2(p) - a_1(p)a_3(p)}{a_0(p)a_2(p) - a_1^2(p)}\right] \frac{h^2 f''(x)}{2} + o(h^2), \tag{3.22}$$

for the particular kernel (3.12). The variance has the form

$$\left\{ \frac{c_1^2(p)b(p) - 2c_1(p)a_1(p)e(p) + a_1^2(p)g(p)}{[c_1(p)a_0(p) - a_1(p)c_0(p)]^2} \right\} \frac{f(x)}{nh} + O(n^{-1}), \tag{3.23}$$

where $e(p) \equiv \int_{-1}^{p} K(u)L(u)du$ and $g(p) \equiv \int_{-1}^{p} L^2(u)du$. Substituting the kernel (3.12) based on a biweight kernel into (3.23) shows that the asymptotic variance is about seven times greater at $p = 0$ than at $p = 1$, which provides theoretical support for the observed pattern in the variability plot presented in Fig. 3.11.

Jones (1993a) established the formal equivalence between using the boundary kernel (3.12) and estimating a density using a weighted local linear estimate. That is, estimate $f$ by $\hat{\beta}_0$, where $\hat{\beta}_0$ and $\hat{\beta}_1$ are the minimizers of

$$h^{-1} \int K\left(\frac{x-u}{h}\right) \left[ n^{-1} \sum_{i=1}^{n} \delta(u - x_i) - \beta_0 - \beta_1(x-u) \right]^2 du, \tag{3.24}$$

and $\delta$ is the Dirac delta function (making $n^{-1} \sum_{i=1}^{n} \delta(u - x_i)$ the empirical density function). See also Sarda (1991), Lejeune and Sarda (1992), Fan, Gijbels, Hu, and Huang (1996), and Cheng, Fan, and Marron (1997). This density estimator will be discussed further in Section 6.4.

Jones and Foster (1996) described a boundary kernel estimator that is nonnegative (but does not integrate to 1).

**3.3.2.** Victor (1976) and Breiman, Meisel, and Purcell (1977) originally proposed the variable kernel estimator. Devroye (1985) and Schilling and Stute (1987) established consistency of the estimator. Victor and Breiman *et al.* proposed choosing $h(x_i)$ to be proportional to the distance to the $k$th nearest neighbor, or $h(x_i) \propto f(x_i)^{-1}$ for univariate data. Schäfer and Trampisch (1982) discussed difficulties with this choice and proposed an *ad hoc* modification of it; see also Tseng and Moret (1990). Abramson (1982) proposed taking $h(x_i) \propto f(x_i)^{-1/2}$ and showed that this is the only function of $f$ where the $O(h^4)$ term of the asymptotic squared bias disappears. Silverman (1986, p. 104), Hall and Marron (1988a), Hall (1990b), and Jones (1990) gave the exact form of the squared bias.

Abramson's (1982) estimate requires bounding the pilot density away from zero. Terrell and Scott (1992) showed that if this is not done (resulting in an estimator they termed the "nonclipped" estimator), the squared bias can be $O([h/\log h]^4)$ rather than the anticipated $O(h^8)$, because of the influence of observations $x_i$ far from the evaluation point $x$. Still, the nonclipped estimator has exhibited good small-sample properties in simulations (Abramson, 1982; Silverman, 1986; Terrell and Scott, 1992; Foster, 1995), presumably because extreme observations are less of a problem in small samples.

McKay (1993) noted that the practical difficulties of tail effects are less important if a kernel estimate is used as the pilot, since it will tend to overestimate $f$ where $f$ is small. McKay also suggested an alternative method of bounding the pilot away from zero that allows a uniformly convergent, higher order bias expansion where $f$ is bounded away from zero and that yields an estimate that is nonnegative and integrates to 1. See Hall, Hu, and Marron (1995) for further discussion of tail effects on the variable kernel estimator and another method of addressing them.

Hall (1992a) showed that the MISE for the variable-bandwidth estimator (3.13) is determined almost completely by the tails of the density $f$, unless these tails are very heavy. He suggested using as an alternative target a weighted version of MISE, where the tails are downweighted. This can be estimated using a weighted cross-validation criterion. One possible weight function is to give zero weight to observations that are too far into the tails, and full weight otherwise.

One way to view the use of a variable bandwidth $h(x_i)$ is that the kernel contributions from each $x_i$ are scaled differently based on local properties of the density. Samiuddin and El-Sayyad (1990) showed that varying the location of each kernel can also lead to improved bias properties. Specifically, a suitably "clipped" version of the estimator

$$\hat{f}(x) = n^{-1}h^{-1}\sum_{i=1}^{n} K\left\{h^{-1}\left[x - x_i - \frac{h^2\sigma_K^2 f'(x_i)}{2f(x_i)}\right]\right\}$$

achieves $O(h^4)$ bias, and hence $O(n^{-8/9})$ mean squared error. Practical application of this estimator requires pilot estimation of $(\log f)' = f'/f$.

Jones, McKay, and Hu (1994) showed that these location- and scale-varying kernel estimators are special cases of a wide family of such estimators

$$\hat{f}(x) = n^{-1}h^{-1}\sum_{i=1}^{n} \alpha(x_i)K\{h^{-1}\alpha(x_i)[x - x_i - h^2 A(x_i)]\},$$

which all solve the differential equation

$$\frac{\sigma_K^2}{2}\left(\frac{f}{\alpha^2}\right)''(u) = (Af)'(u).$$

They also showed how to generalize this family in order to achieve $O(h^6)$ bias, although small-scale Monte Carlo simulations did not indicate any meaningful improvement in performance.

Loftsgaarden and Quesenberry (1965) first proposed the nearest neighbor density estimator, with the kernel function being a uniform density (the estimator based on an arbitrary kernel function $K$ is sometimes called a generalized nearest neighbor estimator and has also been called a balloon estimator, following a suggestion of Tukey and Tukey, 1981). Jones (1990) and Terrell and Scott (1992) examined the properties of estimators of the local-bandwidth form. The best $h(x)$ (in the sense of minimal asymptotic MISE) yields AMISE

$$\text{AMISE}_0 = \frac{5}{4}[\sigma_K R(K)]^{4/5} R\left[(f^2 f'')^{1/5}\right] n^{-4/5},$$

which, by Jensen's inequality, is less than or equal to the corresponding value for the fixed-bandwidth kernel estimator (given in (3.6)). The optimal $h(x)$ is proportional to $\{f(x)/[f''(x)]^2\}^{1/5}$, or inversely proportional to the local roughness of the true density.

While there has been some work on estimating the optimal $h(x)$ at any one point $x_0$ (see, for example, Sheather, 1986; Hall and Schucany, 1989; Mielniczuk, Sarda, and Vieu, 1989; Hall, 1993; and Fan, Hall, Martin, and Patil, 1996), there are no effective data-based methods for determining it at every $x$, as would be necessary for calculation of the local kernel estimate. The nearest neighbor distance, in contrast, is easy to calculate but is far from optimal. Terrell and Scott (1992) showed that for most densities encountered in practice, the integrated squared bias of the estimator is asymptotically infinite, with small-sample roughness as seen in Fig. 3.14 not at all unusual (an example is given in Silverman, 1986, p. 20). Other investigations of consistency and asymptotic properties of the nearest neighbor density estimator include those of Devroye and Wagner (1977), Moore and Yackel (1977), Mack and Rosenblatt (1979), and Rosenblatt (1979).

Marron and Udina (1995) examined the qualitative behavior of both variable and local kernel estimates when the bandwidth functions $h(x_i)$ and $h(x)$, respectively, are allowed to vary widely. They used a cubic spline as the user interface to the smoothing parameter function, with user-controlled knot positions. They found that each of these kernel formulations is consistent with a large family of estimates that contain members with very different appearances, reinforcing the need to choose the function $h$ carefully. Hall, Marron, and Titterington (1995) examined the theoretical properties of such an interface for the local kernel estimator, calling it "partial local smoothing." They examined different ways of estimating $h(x)$ at the knots and found that linear interpolation on as few as $5 - 9$ knots yields most of the potential improved AMISE of the local kernel estimator over the fixed-bandwidth kernel estimator.

**3.3.3.** Parzen (1962) noted that the kernel function can be chosen to lead to improved asymptotic bias of the kernel estimator. Bartlett (1963) noted the necessary conditions for the kernel function. The bounded fourth order kernel that minimizes the asymptotic variance of the estimator is

$$K(x) = \begin{cases} \frac{3}{8}(3 - 5x^2), & |x| < 1, \\ 0, & \text{otherwise} \end{cases}$$

(Deheuvels, 1977a; Müller, 1984). Gasser, Müller, and Mammitzsch (1985) derived higher order (polynomial) kernels that minimize AMISE subject to a fixed number of sign changes in the kernel (it is not possible to minimize AMISE without restrictions for $p > 2$).

Jones and Foster (1993) showed that the principle of generalized jackknifing (Schucany, Gray, and Owen, 1971) can be used to construct many higher (fourth) order kernels from nonnegative (second order) kernels. Let $K$ be a symmetric density function, while $L$ is an alternative symmetric function. Then, if $r_k \equiv \int u^k L(u)du$,

$$\frac{[r_2 K(x) - \sigma_K^2 L(x)]}{r_2 - \sigma_K^2 r_0}$$

is a fourth order kernel (Schucany and Sommers, 1977).

Different choices of $L$ give different fourth order kernels. For example, taking $L(x) = cK(cx)$, $c \in (0,1)$, gives the kernel

$$\frac{K(x) - c^3 K(cx)}{1 - c^2} \tag{3.25}$$

(Schucany and Sommers, 1977; Wand and Schucany, 1990). Taking $L(x) = x^2 K(x)$ gives the locally weighted quadratic kernel

$$\frac{(s_4 - \sigma_K^2 x^2)K(x)}{s_4 - \sigma_K^4},$$

where $s_4 \equiv \int u^4 K(u)du$ (Lejeune and Sarda, 1992). The choice $L(x) = xK'(x)$ gives the kernel

$$\frac{3}{2}K(x) + \frac{1}{2}xK'(x),$$

which also corresponds to (3.25) with $c \to 1$ (Silverman, 1986, p. 69; Wand and Schucany, 1990). Stuetzle and Mittal (1979), Singh (1987), Devroye (1989), Berlinet (1990), Samiuddin and El-Sayyad (1990), Fan and Hu (1992), and Abdous (1995) described other ways to construct higher order kernels.

Marron and Wand (1992) examined the exact MISE for $p$th order kernel estimators of densities that are convex combinations of Gaussian densities. This family of densities is rich enough to produce densities with varying numbers of modes and varying levels of skewness and kurtosis. They found that the asymptotic MISE is sometimes quite different from the exact MISE (with correspondingly different minimizers). Also, they found that higher order kernels only improve on second order kernels for sample sizes of at least $n = 1000$ for simple densities, and as much as $n = 1{,}000{,}000$ for hard-to-estimate densities. Apparently, increased variance balances decreased bias (see also Härdle, 1986, and Foster, 1995).

Marron (1994) showed that higher order kernel estimators are better than nonnegative ones when the true density is well approximated by a parabola over the (effective) window width of the kernel. This will not be true for densities with interesting structure unless the bandwidth is small, which is only appropriate for (very) large sample sizes.

One way to view the bias reduction properties of higher order kernel estimators is that the bias of a second order kernel estimator $\hat{f}(x)$ is estimated and then subtracted from $\hat{f}(x)$ (Jones, 1995b). Bias reduction can also be achieved using a multiplicative correction, as in

$$\tilde{f}(x) = \frac{\hat{f}(x)}{nh}\sum_{i=1}^{n}[\hat{f}(x_i)]^{-1}K\left(\frac{x-x_i}{h}\right)$$

(Jones, Linton, and Nielsen, 1995). This estimator achieves $O(h^4)$ bias, with a bias function different from that of higher order kernels (the asymptotic variance is identical to that of the fourth order kernel estimator based on twicing [Stuetzle and Mittal, 1979]). Monte Carlo simulations reported by Jones *et al.* indicate that this estimator outperforms the fourth order kernel estimator.

Jones (1992) examined the relative error rate of convergence of data-based bandwidth choices for higher order kernels. The choice based on cross-validation is even worse than for second order kernels, as

$$\frac{\hat{h}_{\text{CV}}}{h_0} = 1 + O_p(n^{-1/(4p+2)})$$

(see also Marron, 1986). So, for example, for $p = 4$, the relative error rate is $O_p(n^{-1/18})$. A plug-in estimator analogous to the construction of Sheather and Jones (1991) has

$$\frac{\hat{h}}{h_0} = 1 + O_p(n^{-(2p+1)/(6p+2)}),$$

or relative error rate $O_p(n^{-9/26})$ for $p = 4$ (although the paper did not give any practical plug-in estimate).

The reduced bias of a $p$th order kernel estimator holds only if the density is sufficiently smooth; for example, if the density has only two bounded derivatives, a fourth order kernel estimator should not be used. This suggests choosing the order $p$ in a data-dependent way. Hall and Marron (1988b) proposed doing this using cross-validation and showed that this choice is asymptotically optimal as long as the underlying density is not too smooth.

Hall and Murison (1993) examined making the estimate into a bona fide density using simple "clipping" and normalization, yielding

$$\hat{f}_1(x) = \frac{\hat{f}(x)I(\hat{f}(x) > 0)}{1 + \int |\hat{f}(u)|I(\hat{f}(u) < 0)du},$$

where $I(\cdot)$ is an indicator function. Unless $f$ has very heavy tails (no finite moments), this operation has little effect on the ISE and the choice of $\hat{h}$, so any choice of $\hat{h}$ based on the original estimator can be used for the modified version. Gajek (1986) described more complex corrections for non-bona fide estimates that are guaranteed not to increase the MISE (simply clipping and normalizing cannot increase IAE — see Devroye and Györfi, 1985, Chapter 11).

Konakov (1973) and Davis (1975, 1977) described the Fourier integral estimator. Davis (1977) showed that the estimator achieves the minimal rate of MISE for kernel estimators (as derived in Watson and Leadbetter, 1963) for $f$ sufficiently smooth (if its characteristic function decreases algebraically, exponentially or is compact). Davis (1981) and Hart (1985) discussed data-based choice of the bandwidth. See Ibragimov and Khasminskii (1982), Devroye and Györfi (1985, pp. 133–137), and Abdous (1993) for further discussion of this estimator. Devroye (1992) examined the MIAE properties of estimators of related form, which he called "superkernels."

The geometric combination estimator, with $c = 2$ as in Eq. (3.17), was introduced by Terrell and Scott (1980). Koshkin (1988) described and analyzed the general formulation (3.16). See also Jones and Foster (1993) for generalizations of this estimator.

**3.3.4.** Devroye and Györfi (1985, Chapter 9) and Silverman (1986, Sect. 2.9) described the benefits of transformation-based density estimation. The earthquake data came from the Bulletin of the International Seismological Centre and were discussed in Frohlich and Davis (1990). Yang (1995) suggested the use of the Johnson family for the transformation-based estimate for these data. The observed modes at 0, 33, 100, and 200 km can be explained as follows (Frohlich, 1995): When locations of earthquakes are initially reported, they are likely to have provisionally reported depth in only rough terms, such as "shallow" (zero depth) or to within 100 km (100 or 200 km depths). As further processing continues, some of these quakes are confirmed as definitely shallow, but no more accurate depth can be determined; these are assigned a depth of 33 km by convention (due to seismic modeling that assumes that the Earth's crust is 33 km thick). Others might have the actual depth confirmed, in which case the actual number is assigned to the seismic event. Thus, events at 0, 100, and 200 km never received additional processing and confirmation, while those at 33 km were processed further, but the inherent difficulty in pinpointing the precise depth of shallow events prevented an exact location from being determined.

Wand, Marron, and Ruppert (1991) examined the use of the shifted power transformation for long right-tailed data. They proposed choosing the parameters to minimize $R(\hat{f}''_{Y,a})$, where $\hat{f}''_{Y,a}$ is a pilot kernel estimate with bandwidth $a$. This choice is based on (3.6), since minimizing an estimate of $R(f'')$ is equivalent to minimizing the best possible AMISE in the $Y$ scale, which should translate to accurate estimation in the $X$ scale.

Ruppert and Wand (1992) examined using convex combinations of the identity transformation and the standard normal cumulative distribution function to address thick-tailed (kurtotic) data. Marron and Ruppert

(1994) suggested addressing boundary bias by transforming the data to a density that has its first derivative equal to zero at both boundaries; their parametric family of transformations was based on polynomials and a Beta cumulative distribution function.

Ruppert and Cline (1994) proposed estimating the transformation $g(x)$ nonparametrically as $G[\hat{F}_X(x)]$, where $G$ is the inverse cumulative distribution function of some target distribution, and $\hat{F}_X$ is a kernel-based estimate of the cumulative distribution function of $X$ ($\hat{F}_X(x) = \int_{-\infty}^{x} \hat{f}_X(u)du$, using an initial pilot kernel estimate of $f_X$). Then, the data are smoothed in the transformed scale and then back-transformed to an updated $\hat{f}_X$. They showed that if the target distribution is uniform (that is, $g(x) = \hat{F}_X(x)$), then applying $p$ iterations of this process (using appropriate bandwidths and a boundary kernel) yields an estimate with squared error of order $O_p(n^{-4p/(4p+1)})$. So, for example, two iterations (where the first iteration is just the pilot kernel estimate) yields squared error of order $O_p(n^{-8/9})$, as does a fourth order kernel and the variable kernel estimate (3.13). Such iteration does not improve the MSE rate of the variable kernel estimate past two iterations (Hall and Marron, 1988a), although Ruppert and Cline noted that the MISE of the estimate did seem to improve up to three or four iterations in their Monte Carlo simulations. Their transformation-based estimate and the variable kernel estimate performed similarly in their Monte Carlo simulations and examples. Hössjer and Ruppert (1995) derived exact coefficients for the asymptotic expansion of the estimator.

Jones and Signorini (1997) compared many of the methods that can achieve $O(h^4)$ asymptotic bias, including fourth order kernels, the geometric combination estimator, the variable kernel estimator, and the estimators of Samiuddin and El-Sayyad (1990), Ruppert and Cline (1994), and Jones, Linton, and Nielsen (1995), both theoretically and using Monte Carlo simulations. They showed that all of these methods have asymptotic biases of the same general form, with different constant terms. Their Monte Carlo simulations (based on ISE) indicated roughly similar performance for the variable kernel estimator and the estimators of Samiuddin and El-Sayyad, Ruppert and Cline, and Jones *et al.*, which were better than the fourth-order kernel and geometric combination estimators (with the estimator of Jones *et al.* performing best). They also noted the need for very large samples before qualitative improvement over the ordinary (second order) kernel estimator was evident.

## Section 3.4

Hjort and Jones (1996) laid out the general framework of local likelihood estimation, while Loader (1996) focused on the use of local log-polynomials. Both papers discussed the properties of the estimator, including asymptotic bias and variance representations, and the potential for automatic boundary bias correction. Hjort and Jones showed that the locally closest parametric fit corresponds to the minimizer of local Kullback–Leibler distance. They showed that the bias of the local log-linear estimator (using a Gaussian kernel) provides a bias correction over the local linear (kernel) estimator, giving

$$b(x) = f''(x) - \frac{[f'(x)]^2}{f(x)},$$

which attempts to correct the local slope of the estimator. Higher order local log-polynomials bring in higher order curvature corrections. They also discussed minimizing a squared error measure of distance, rather than Kullback–Leibler distance; this corresponds to (3.24).

Besides theoretical concerns, Loader (1996) discussed computational issues in the construction of local log-polynomial estimation. He showed that the asymptotic relative efficiency of the local log-linear estimator compared with the kernel estimator is $\text{ARE}(x) = |1 - [f'(x)]^2/[f(x)f''(x)]|^{1/2}$ (where values less than 1 mean that the local log-linear estimator is more efficient). For densities that decay exponentially ($f(x) = \exp[-x^\alpha + o(x^\alpha)]$ as $x \to \infty$, with $\alpha > 0$), the relative efficiency is

$$\text{ARE}(x) = |1 - \alpha^{-1}|^{1/2} x^{-\alpha/2} + o(x^{-\alpha/2}).$$

Thus, $\text{ARE}(x) \to 0$ as $x \to \infty$. Very limited Monte Carlo simulation evidence supports the impression of improved bias in the tails.

Copas (1995) described a different (but not dissimilar) local likelihood approach based on ideas from the analysis of censored data.

A different way to incorporate prior feelings for the true density is by using what have been called *semiparametric estimators*. Many variations are possible. Hjort and Glad (1995) proposed using $\hat{f}(x) = f_0(x)\hat{r}(x)$, where $\hat{r}(x)$ is a kernel estimate of $r(x) = f(x)/f_0(x)$, and $f_0$ is a starting guess for $f$. They derived the asymptotic properties of the estimator, which are similar to those of the kernel estimator, except that the squared bias is a function of $[f_0(x)r''(x)]^2$ rather than $f''(x)^2$ (now $f_0$ is the best parametric approximant to $f$). Thus, if $f_0$ is close to $f$, $r(x) \approx 1$, and $[f_0(x)r''(x)]^2$ will be smaller than $f''(x)^2$. The special exponential family estimators of Efron and Tibshirani (1994) reverse the order of estimation, where an ordinary kernel estimate is corrected to match sample moments to a parametric model.

Other semiparametric estimation schemes are also possible. Buckland (1992a) described a method that approximates $r$ as a polynomial, estimating the coefficients of the polynomial using maximum likelihood. The number of terms in the polynomial can be chosen using likelihood ratio tests (see also Fenstad and Hjort, 1995). Elphinstone (1983) described a different polynomial-based method that allows specification of a "target" density.

A semiparametric estimate also can be based on an additive form rather than a multiplicative form; that is,

$$\hat{f}(x) = \pi f_0(x, \hat{\boldsymbol{\theta}}) + (1-\pi)\hat{f}(x), \qquad 0 \le \pi \le 1$$

(Schuster and Yakowitz, 1985; Olkin and Spiegelman, 1987). The parameter $\pi$ is estimated using maximum likelihood. The theoretical properties reported by these authors are valid only if $f$ is defined on a bounded interval and bounded away from zero. Faraway (1990a) showed that this condition is important, as $\hat{\pi}$ is very sensitive to contributions from data in the tails, and he recommended estimating $\pi$ after trimming observations in the tails. See also Jones (1993b) for further discussion of this estimator.

Mixture models are another approach to semiparametric estimation. Consider, for example, a normal mixture density

$$f(x) = \int h^{-1}\phi\left(\frac{x-\theta}{h}\right) dQ(\theta),$$

where $Q$ is a probability measure (termed the mixing distribution). For unrestricted $h$ and $Q$, the likelihood function approaches infinity as $f$ approaches the (unsmoothed) empirical density function (Geman and Hwang, 1982). The class of finite mixtures

$$f(x) = h^{-1}\sum_{j=1}^{q} \pi_j \phi\left(\frac{x-\theta_j}{h}\right)$$

defines a restricted parametric family, but it is not amenable to likelihood-based analysis (see, for example, McLachlan and Basford, 1988). Roeder (1990, 1992) proposed estimating $f$ based on the observed sample spacings and proved almost sure convergence of the estimator. Priebe (1994) proposed an adaptive mixture family, where the estimate is recursively updated as new observations are added to the sample, and established strong consistency. Marchette *et al.* (1996) used a mixture model indirectly by first fitting it and then estimating the density using a linear combination of kernel estimates, where the parameters of the linear combination are estimated based on the original estimated mixture model, calling this *filtered kernel density estimation.*

A natural alternative to a semiparametric estimation scheme if some parametric family is considered a good guess for the true density is to incorporate that feeling into a prior distribution and use a Bayesian approach. Hjort (1996) described various Bayesian approaches to density estimation. One approach is to bin the data by dividing the interval of interest into cells. Then, a Dirichlet distribution constitutes a conjugate prior family. Ferguson (1973) introduced the Dirichlet process, which avoids the need to bin the data. Unfortunately, the Dirichlet process does not incorporate smoothness in a useful sense, but this can be overcome by smoothing the prior. The resultant posterior is a mixture of Dirichlet processes (Ghorai and Rubin, 1982; Ferguson, 1983; Lo, 1984).

A different Bayesian estimator is based on treating the log-density rather than the density. Lenk (1988, 1991, 1993) assumed a logistic transform of a Gaussian process as the prior process for a Bayesian analysis and described specific functional forms for the covariance function of the process (and the resultant estimation process). See also Leonard (1978) and Thorburn (1986). Another Bayesian approach uses a normal mixture model; see Escobar and West (1995) and the references therein.

### Section 3.5

Good and Gaskins (1971, 1972) introduced maximum penalized likelihood estimation. They suggested enforcing nonnegativity by operating on the square root of the density $\gamma = f^{1/2}$ and then squaring the result. They proposed the penalties

$$\Phi_1(f) = \alpha \int_{-\infty}^{\infty} \frac{f'(u)^2}{f(u)} du$$

and

$$\Phi_2(f) = 4\alpha \int_{-\infty}^{\infty} \gamma'(u)^2 \, du + \beta \int_{-\infty}^{\infty} \gamma''(u)^2 \, du.$$

de Montricher, Tapia, and Thompson (1975) examined in detail the properties of the estimators, including issues of existence and uniqueness, and showed that the MPLE based on a penalty involving derivatives is a spline with knots at the order statistics. Klonias (1982, 1984) examined a general class of penalized likelihood estimators and suggested choosing the smoothing parameter(s) based on cross-validation. Cox and O'Sullivan (1990) provided asymptotic analysis of penalized likelihood estimators.

Penalized likelihood estimation can be viewed as a Bayesian method, with the prior for the density having the form $\exp[-\Phi(f)]$ and the posterior mode being the final estimate. Good and Gaskins (1980) used this Bayesian framework to suggest a way to evaluate the importance of individual modes ("bump-hunting") through the logarithm of the Bayes factor on the odds that the bumps would be present in a sample of infinite size.

In order to avoid computational difficulties, Scott, Tapia, and Thompson (1980) converted the penalized likelihood to one on discrete data by binning the observations (see also Tapia and Thompson, 1978, reprinted in Thompson and Tapia, 1990). They called this the discrete maximum penalized likelihood estimator (DMPLE) and gave conditions where the DMPLE converges to the MPLE as the bins narrow. Granville and Rasson (1995) also proposed binning the observations, and they examined an approximation to the MPLE based on a Taylor Series expansion around a uniform set of binned counts. Ghorai and Rubin (1979), Good and Gaskins (1980), Ishiguro and Sakamoto (1984), and Klonias and Nash (1987) discussed other MPLE computational methods.

Silverman (1982b, 1984) examined the use of penalties based on the log of the density, as in (3.20), and derived the equivalent kernel for the spline estimator.

Since a cubic spline is a piecewise cubic polynomial that has continuous second derivatives, this suggests approximating the MPLE by a linear combination of cubic polynomial pieces, anchored at each data value, that are restricted to have continuous second derivatives. Such functions are called *B-splines.* So, an approximate MPLE is defined by the set of coefficients $\mathbf{a}$ that minimize (3.20), where $g(x_i)$ has the form

$$g(x) = \sum_{j=1}^{n} a_j B_j(x),$$

where $B_j$ is the B-spline anchored at the $j$th knot (O'Sullivan, 1988). This penalized spline estimator is called a *P-spline.* Since the cubic B-spline $B_j$ is only positive over the five consecutive knots beginning at the $j$th knot, this minimization can be accomplished relatively easily.

A related estimator defines the penalty directly in terms of the coefficients of adjacent B-splines, encouraging smoothness by forcing the coefficients to be close. For example, the P-spline corresponding to second order differences is determined by the minimizer $\hat{\mathbf{a}}$ of

$$L(\mathbf{a}) = -\sum_{i=1}^{n}\sum_{j=1}^{n} a_j B_j(x_i) + \int \exp\left[\sum_{j=1}^{n} a_j B_j(u)\right] du + \alpha \sum_{j=3}^{n} (a_j - 2a_{j-1} + a_{j-2})^2$$

(Eilers and Marx, 1994). A density estimate can be calculated by binning the data and choosing $\alpha$ by a model selection criterion such as $AIC$.

A different way to enforce the constraint $\int f = 1$ in (3.19) is to use the logistic transform $f = \exp(g)/\int \exp(g)$. This is equivalent to minimizing

$$L(g) = -n^{-1}\sum_{i=1}^{n} g(x_i) + \alpha \int [g''(u)]^2\, du + \log \int e^{g(u)} du$$

and then taking $\hat{f} = \exp(\hat{g})/\int \exp(\hat{g})$. This *logistic spline* estimate is then defined through a side condition on $g$, such as $\int g = 0$ or $g(0) = 0$; in either case, it is a cubic spline estimator. While the minimizer is not computable, the minimizer over an adaptive restricted space is, and it shares the same convergence rates. See Leonard (1978), Gu (1993), and Gu and Qiu (1993).

Stone and Koo (1986) first described the fitting of logspline models. Stone (1990) described the theory underlying such models. Kooperberg and Stone (1991) described the practical aspects of constructing logspline density estimates.

Boneva, Kendall, and Stefanov (1971), Lii and Rosenblatt (1975), Wahba (1975a, 1976), and Barron and Sheu (1991) discussed other spline-based density estimators.

Mächler (1995a) proposed a penalized likelihood estimator designed to control the maximum number of modes of the final estimate. The roughness penalty is related to the relative change of curvature of the underlying density. The MPLE solves an ordinary differential equation with boundary conditions; the solution requires a crude approximate solution to be provided, and Mächler used logspline estimates for this purpose.

## Computational issues

Many computer packages provide kernel density estimation as a standard option; examples include NCSS, SAS/INSIGHT, S–PLUS, SOLO, Stata, STATGRAPHICS PLUS, and Systat. IMSL provides a Fortran subroutine to calculate a kernel density estimate. APL2STAT, a set of APL2 programs for statistical analysis (Friendly and Fox, 1994), includes code for kernel density estimation and is available via the World Wide Web at the address `http://www.math.yorku.ca/SCS/friendly.html`. The collections `ash` and `haerdle` in the `S` directory of `statlib` contain S–PLUS functions that calculate kernel estimates based on WARPing. The package XploRe (Härdle, Klinke, and Turlach, 1995) also uses WARPing to calculate density estimates. The collection `fan-marron` in the `jcgs` directory of `statlib` includes code to calculate kernel estimates based on binning and updating algorithms described in Fan and Marron (1994). JMP and the R–code of Cook and Weisberg (1994) include sliders, which allow interactive control of the bandwidth and presentation of its effect on the form of the estimate (see also the XLISP–STAT code of Tierney, 1990, p. 305).

Park and Turlach (1992) gave pseudo-code algorithms for the bandwidth selectors of Hall *et al.* (1991), Jones, Marron, and Park (1991), Sheather and Jones (1991), and Hall, Marron, and Park (1992), as well as unbiased and biased cross-validation. They have made GAUSS code available to calculate these selectors. Venables and Ripley (1994) gave S–PLUS code to calculate the Sheather–Jones and cross-validation (unbiased and biased) selectors. XploRe includes code to calculate the selectors of Silverman (1986, Section 3.4.2), Jones, Marron, and Park (1991), Sheather and Jones (1991), Hall, Marron, and Park (1992), and Park and Marron (1992). Fortran and C code to calculate a bandwidth in the spirit of that of Engel, Herrmann, and Gasser (1994) can be obtained using a WWW browser at the URL `http://www.unizh.ch/biostat/software.html`.

S–PLUS code to calculate and display the mode tree is available via anonymous `ftp` at the address `ftp.stat.rice.edu` and can be found in the directory `pub/scottdw/Mode.Tree`.

Wavelet density estimation is provided in XploRe. Such estimates can also be calculated using the C code (with S–PLUS drivers) included in the `wavethresh` collection in the `S` directory of `statlib`.

Scott (1991) gave pseudo-code for a WARPing (average shifted histogram) version of the variable kernel estimate and has made S–PLUS code available for implementation of the algorithm.

Udina (1994) described a general software package to calculate and display many variations of kernel density estimation using XLISP–STAT. The code calculates kernel estimates based on direct, binning, and updating algorithms. It allows the setting of the bandwidth manually using sliders, as well as automatic choice using the normal reference rule, cross-validation, and the automatic bandwidth selectors of Park and Marron (1990), Hall *et al.* (1991), and Sheather and Jones (1991). The code can also calculate variable- and local-bandwidth estimates, allowing the user to control the level of local smoothing interactively. The code (and a Postscript version of the paper describing the code) is available via anonymous `ftp` at the address `libiya.upf.es` and can be found in the directory `pub/stat/kde`.

C code (and S–PLUS interfaces) to calculate local log polynomial (constant, linear, and quadratic) estimates is available at `http://cm.bell-labs.com/stat/project/locfit` using a World Wide Web browser (this location also includes a description of the code and extensive description of the method).

Buckland (1992b) gave Fortran code to calculate the semiparametric, polynomial-based estimate of Buckland (1992a). C code to calculate adaptive mixture density estimates can be obtained using a World Wide Web browser at `http://irisd.nswc.navy.mil/Code/am.tar.Z`, while the address `http://irisd.nswc.navy.mil/Code/fke.tar.Z` is the location of code for filtered kernel density estimation.

IMSL provides a Fortran subroutine to calculate a discrete maximum penalized likelihood density estimate.

The collection `logspline` in the `S` directory of `statlib` contains S–PLUS functions that calculate logspline density estimates based on methods described in Kooperberg and Stone (1991).

Eilers and Marx (1996) discussed computational details of calculating P-splines using finite differences of coefficients of adjacent B-splines. Paul Eilers contributed S–PLUS code to calculate the estimates to the S–news electronic mailing list, which can be found in the collection `digest153` in the `S-news` directory of `statlib` (June 21, 1994, with correction June 23, 1994).

RKPACK–II, a collection of Ratfor routines for penalized likelihood density estimation, is available using a World Wide Web browser at the URL `http://www.stat.purdue.edu/~chong/software.html`.

## Exercises

**Exercise 3.1.** Investigate the relationship between the success of different bandwidth selectors for the kernel density estimator and sample size by calculating the associated density estimate for different methods for

(a) a small sample — the percentage of silica in chondrite meteors ($n = 22$),

(b) a moderately sized sample — velocities of galaxies ($n = 82$),

(c) a large sample — earthquake depths ($n = 2178$).

Do you think that any methods are more effective (in a comparative sense) for small samples than for large ones?

**Exercise 3.2.** Use the Sheather–Jones-based estimate of $R(f'')$ to determine bin widths for fixed bin-width frequency polygons for the data sets examined in Chapter 2. Are any of the resultant bin widths very different from the Gaussian-based rule? Do any of the resultant frequency polygons seem to reflect the underlying structure more effectively?

**Exercise 3.3.** Use a goodness-of-fit test of your choice to test whether the mine accident data are consistent with being a sample from an exponential distribution. Would you expect such a test to be statistically significant?

**Exercise 3.4.** Construct variability plots corresponding to kernel estimation for the CD rate data (Fig. 3.2). For what values of CD rate does the variability of the estimator increase?

**Exercise 3.5.** A natural estimate of the derivative of a density $f'(x)$ is the derivative of a kernel estimate of the density; that is,

$$\hat{f}'(x) = n^{-1}h^{-2}\sum_{i=1}^{n} K'\left(\frac{x - x_i}{h}\right)$$

(assuming differentiability of $K$). Calculations similar to those leading to (3.6) imply that the optimal bandwidth is $O(n^{-1/7})$, with optimal AMISE of order $O(n^{-4/7})$. Compare "reasonable" choices of $\hat{h}$ for estimation of $f'$ (by eye) for the CD rate and Swiss bank note data with those for estimation of $f$, and construct variability plots for the resultant estimates. Are the density derivative estimates less precisely determined than the density estimates, as the asymptotics would suggest?

**Exercise 3.6.** Verify that, in (3.12), if $K$ is the biweight kernel,

$$a_0(p) = (3p^5 - 10p^3 + 15p + 8)/16,$$
$$a_1(p) = (5p^6 - 15p^4 + 15p^2 - 5)/32,$$
$$a_2(p) = (15p^7 - 42p^5 + 35p^3 + 8)/112.$$

**Exercise 3.7.** Transformation-based kernel estimators can be iterated, in the sense that once-transformed data can be transformed again, smoothed, and then back-transformed twice to the original scale (and so on). Investigate this possibility for the earthquake depth data. Does fitting a second Johnson family transformation to the once-transformed data improve the final density estimate?

**Exercise 3.8.** Construct a variability plot for the variable kernel estimate for the marathon record data (Fig. 3.12). Which parts of the estimated density are less precisely determined?

**Exercise 3.9.** In Chapter 2, the baseball salary data were analyzed after a logarithmic transformation, while the hockey shooting percentage data were analyzed after a logistic transformation. Do these transformations improve kernel estimation for these data sets?

**Exercise 3.10.** Compare the properties of local likelihood density estimation using constant bandwidth to using nearest neighbor variable bandwidth (as in Fig. 3.17) for the racial distribution data. Which estimate highlights the underlying structure better?

**Exercise 3.11.** The local log-polynomial density estimate for the mine accident data in Fig. 3.18 is based on a quadratic model, but the somewhat exponential nature of the data suggests that a local log-linear model might be more sensible. Is that true?

**Exercise 3.12.** Construct variability plots corresponding to local likelihood density estimation (Fig. 3.17) and logspline density estimation (Fig. 3.20) for the racial distribution data. Do the variability envelopes have similar shapes?

Chapter 4

# Multivariate Density Estimation

## 4.1 Simple Density Estimation Methods

Exploring and identifying structure is even more important for multivariate data than univariate data, given the difficulties in graphically presenting multivariate data and the comparative lack of parametric models to represent it. Unfortunately, such exploration is also inherently more difficult.

An obvious way to present bivariate data is using a scatter plot. Figure 4.1 gives a scatter plot of the number of points scored per minute (`PPM`) and number of assists credited per minute (`APM`) for the 96 National Basketball Association (NBA) players who played the guard position during the 1992–1993 season and played an average of at least 10 minutes per game. While a scatter plot can be effective as a representation of the relationship between two variables, it is not at all effective as a representation of the distribution of the observations, in the sense of locating high density regions. The reason is that the overprinting of observations on the plot is not a good indicator of relative density.

Scatter plots can be improved for this purpose by representing the number of (near) replicates by numbers or symbols, such as in a sunflower plot (where the number of petals of the sunflower equals the number of replicates). Still, a direct representation designed to reflect density estimation is likely to be more evocative of structure.

The histogram can be easily generalized to multiple dimensions. Let $\{\mathbf{x}_1, \ldots, \mathbf{x}_n\}$ be a random sample from $f(\mathbf{x})$, where $x \in \mathbb{R}^d$. If the region of interest is divided into hyperrectangles of size $h_1 \times \cdots \times h_d$, with $n_k$ observations falling in the hyperrectangular bin $B_k$, then the histogram estimator has the form

$$\hat{f}(\mathbf{x}) = \frac{n_k}{nh_1 \cdots h_d}$$

for $\mathbf{x} \in B_k$.

Let $\dot{f}_j = \partial f(\mathbf{x})/\partial x_j$ and $h = \min_j h_j$, and assume that $f$ is sufficiently smooth (roughly speaking, $f$ should be twice continuously differentiable with bounded integrable partial derivatives). Then multivariate Taylor Series expansions and integration approximations imply that

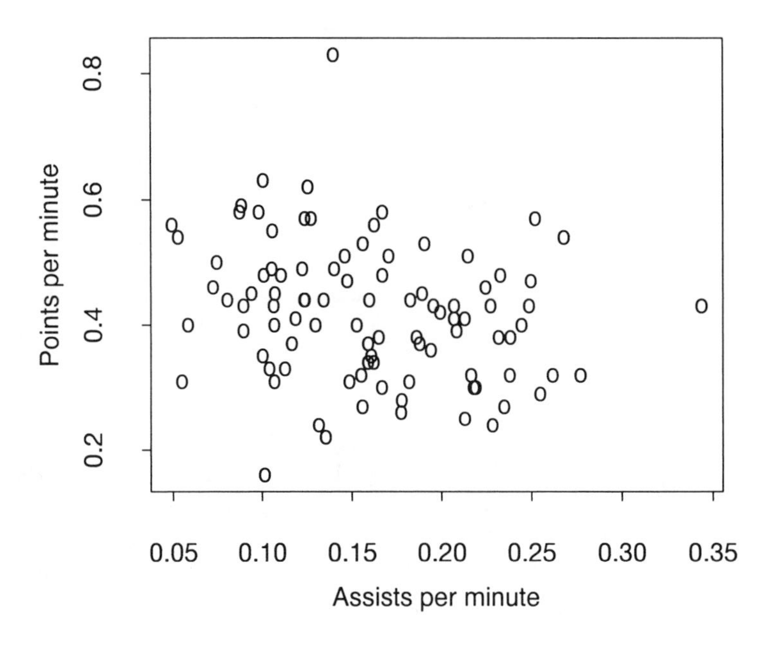


**Fig. 4.1.** Scatter plot of points scored per minute versus assists credited per minute for 96 NBA guards.

$$\text{AMISE} = \frac{1}{nh_1 \cdots h_d} + \frac{1}{12} \sum_{i=1}^{d} h_i^2 R(\dot{f}_i). \tag{4.1}$$

The first term in (4.1) is the integrated asymptotic variance, while the second term refers to the integrated squared bias. The bin widths that minimize AMISE are thus

$$h_{j0} = R(\dot{f}_j)^{-1/2} \left[ 6 \prod_{i=1}^{d} R(\dot{f}_i)^{1/2} \right]^{1/(d+2)} n^{-1/(d+2)}, \tag{4.2}$$

with minimized AMISE equaling

$$\text{AMISE}_0 = \frac{1}{4} \left[ 36 \prod_{i=1}^{d} R(\dot{f}_i) \right]^{1/(d+2)} n^{-2/(d+2)}. \tag{4.3}$$

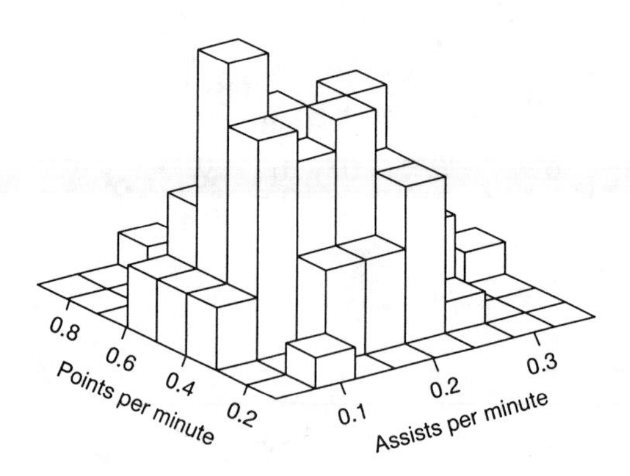


**Fig. 4.2.** Perspective view of bivariate histogram for basketball data.

A simple data-based rule for choosing $h_j$ is to substitute a reference distribution $f$ into (4.2). So, for example, if the reference distribution is multivariate normal, with the different variables being independent with possibly different standard deviations $\sigma_j$, (4.2) becomes

$$h_{j0} = (2)(3^{1/(d+2)})\pi^{d/(2d+4)}\sigma_j n^{-1/(d+2)},$$

or roughly $\hat{h}_{j0} \approx 3.5\hat{\sigma}_j n^{-1/(d+2)}$.

Figure 4.2 gives a perspective view of a bivariate histogram for the basketball data, taking $h_{\text{APM}} = .04$ and $h_{\text{PPM}} = .1$. It is almost impossible to identify any structure past a general high density region because of the blockiness of the estimate. That is, the estimate is too rough to be useful as a representation of the true density.

A natural improvement on the histogram is linear interpolation, forming a multivariate frequency polygon. There is no unique way to do this, but a reasonable approach is to define the density estimate surface as determined by the centers of the $2^d$ adjacent histogram bins. Figure 4.3 gives this *linear blend frequency polygon* (LBFP) from the histogram of Fig. 4.2. The estimate is smoother than the histogram, although it is still difficult to identify the underlying structure.

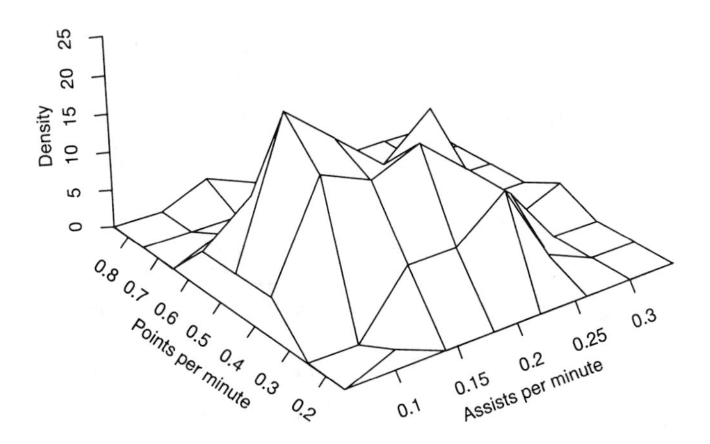


**Fig. 4.3.** Perspective view of bivariate linear blend frequency polygon for basketball data.

Not surprisingly, the multivariate frequency polygon provides significant improvement over the multivariate histogram in terms of accuracy. Define an LBFP bin

$$B_{k_1,\ldots,k_d} = \prod_{j=1}^{d} [t_{k_j}, t_{k_j} + h_j),$$

where $h_j$ is the bin width in the $j$th dimension. Then, for $\mathbf{x} \in B_{k_1,\cdots,k_d}$, the LBFP is

$$\hat{f}(\mathbf{x}) = \frac{1}{nh_1 \ldots h_d}$$
$$\times \sum_{j_1,\ldots,j_d \in \{0,1\}^d} \left[ \prod_{i=1}^{d} \left( \frac{x_i - t_{k_i}}{h_i} \right)^{j_i} \left( 1 - \frac{x_i - t_{k_i}}{h_i} \right)^{1-j_i} \right] n_{k_1+j_1,\ldots,k_d+j_d}.$$

The AMISE for this estimator is

$$\text{AMISE} = \frac{2^d}{3^d n h_1 \cdots h_d} + \frac{49}{2880} \sum_{j=1}^{d} h_j^4 R(\ddot{f}_{jj}) + \frac{1}{32} \sum_{i<j} h_i^2 h_j^2 R(\sqrt{\ddot{f}_{ii} \ddot{f}_{jj}}), \tag{4.4}$$

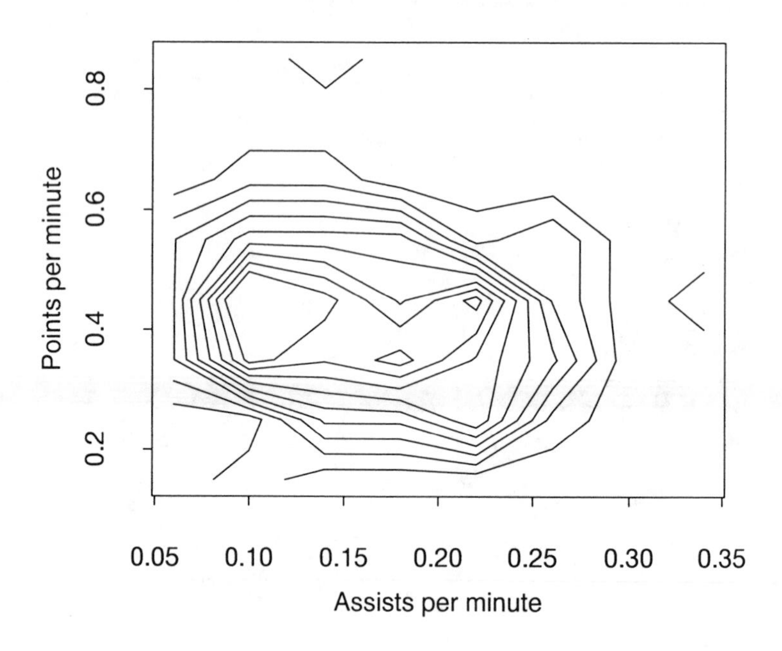


**Fig. 4.4.** Contour plot of bivariate linear blend frequency polygon for basketball data.

where

$$\ddot{f}_{ij} = \frac{\partial^2 f(\mathbf{x})}{\partial x_i \partial x_j}.$$

The optimal choice of $h_j$ is then $h_{j0} = O(n^{-1/(d+4)})$, with optimal $\text{AMISE}_0 = O(n^{-4/(d+4)})$.

The difference between (4.1) and (4.4) represents the improvement achievable through better smoothing. The optimal AMISE rate for the histogram is $O(n^{-2/(d+2)})$, while that of the frequency polygon is $O(n^{-4/(d+4)})$, which is necessarily smaller. For $d = 1$, $\text{AMISE}_0^{\text{hist}} = O(n^{-2/3})$ while $\text{AMISE}_0^{\text{fp}} = O(n^{-4/5})$, as noted in Chapter 2. For $d = 2$, $\text{AMISE}_0^{\text{hist}} = O(n^{-1/2})$ and $\text{AMISE}_0^{\text{fp}} = O(n^{-2/3})$; for $d = 3$, $\text{AMISE}_0^{\text{hist}} = O(n^{-2/5})$ and $\text{AMISE}_0^{\text{fp}} = O(n^{-4/7})$.

Figure 4.4 is a contour plot of the frequency polygon of Fig. 4.3. The density contours correspond to density values $(1.35, 3.35, \ldots, 17.35)$. The contour plot avoids the problems in detecting structure in the perspective plot because of peaks blocking others behind them. The piecewise linear nature of the density contours comes from the construction of the frequency polygon. Despite that, a bimodal character of the density is apparent, corresponding to guards averaging roughly 0.1 assists per minute played, and those averaging roughly twice that level. Teams usually play two guards at a time, with one having primary responsibility as a scorer (a "shooting guard") and the other having primary responsibility as a passer (a "point guard"), so this bimodality (which was not at all apparent in the original scatter plot) is quite natural.

In succeeding sections of this chapter, smoother (and more effective) density estimators will be described, but Figure 4.4 already shows the kinds of exploratory analysis possible using density estimation. Unfortunately, the frequency polygon also illustrates three problems that go with multivariate density estimation:

(1) Multivariate density estimates are necessarily more complicated than univariate ones. There are more possibilities among which to choose in implementation and more smoothing parameters that have to be set.

(2) Multivariate density estimates are difficult to visualize graphically. While a contour plot for a bivariate density is easy to understand, for trivariate data this is no longer true, and compromises must be made (such as "slicing" the density by fixing the value of one variable, or plotting three-dimensional contour volumes inside each other, perhaps in different colors). Past three dimensions, any ability to look at the entire density is lost, and only slicing is possible.

(3) As the dimension of the data increases, density estimation gets progressively more difficult. This is apparent from the frequency polygon's AMISE rate $O(n^{-4/(d+4)})$, which approaches zero more slowly as $d$ gets larger. The need for progressively larger sample sizes in higher dimensions to achieve comparable accuracy is called the *curse of dimensionality.* An important consequence of this pattern is the somewhat paradoxical fact that in high dimensions, "local" neighborhoods are almost surely empty, and neighborhoods that are not empty are almost surely not "local."

Examples of this paradox are not difficult to find. For example, consider a uniform sample over the hypercube $[-1, 1]^d$. When $d = 2$, roughly 79% of the observations will fall in the unit circle centered at the origin, but for $d = 5$ this proportion falls to 16%, and for $d = 10$ it is 0.25%. That is, large neighborhoods have virtually no data in them, which means that any local character is lost. For the multivariate normal density with $d = 10$, over half of the observations will fall (on average) in regions where the density is less than one-hundredth of its maximum value, and over 99% will fall outside the hypersphere centered at the origin with radius 1.16. That is, most of the data accumulation occurs in the tails of the density, in contrast to the pattern in low dimensions. Unfortunately, the tails are precisely the part of the density that is least likely to be of any great interest to the data analyst.

## 4.2 Kernel Density Estimation

### 4.2.1 Properties of the kernel estimator

Kernel density estimation can be easily generalized from univariate to multivariate data, in theory if not always in practice. The general form of the estimator is

$$\hat{f}(\mathbf{x}) = \frac{1}{n|H|} \sum_{i=1}^{n} K_d[H^{-1}(\mathbf{x} - \mathbf{x}_i)], \tag{4.5}$$

where $|H|$ is the absolute value of the determinant of the matrix $H$. Here $K_d : \mathbb{R}^d \to \mathbb{R}$ is the kernel function, often taken to be a $d$-variate probability density function, and $H$ is a nonsingular $d \times d$ bandwidth matrix. A popular technique for generating $K_d$ from a univariate kernel $K$ is by using a product kernel,

$$K_d(\mathbf{u}) = \prod_{j=1}^{d} K(u_j).$$

The AMISE of the estimator is derived using multivariate Taylor Series expansions. Assume that all second partial derivatives of $f$ are piecewise continuous and square integrable, and that the kernel $K_d$ satisfies the conditions

$$\int K_d(\mathbf{u})d\mathbf{u} = 1, \int \mathbf{u}K_d(\mathbf{u})d\mathbf{u} = \mathbf{0}, \int \mathbf{u}\mathbf{u}'K_d(\mathbf{u})d\mathbf{u} = I_d,$$

where these multivariate integrals are over $\mathbb{R}^d$, and $I_d$ is the $d \times d$ identity matrix (these conditions are generalizations of those required for univariate kernels, with the third corresponding to scaling the kernel to have unit variance). Define $h > 0$ and the $d \times d$ matrix $A$ to satisfy $H = hA$, where $A$ has unit determinant. Then, if $h \to 0$ and $nh^d \to \infty$ as $n \to \infty$, the AMISE has the form

$$\text{AMISE} = \frac{R(K)}{nh^d} + \frac{h^4}{4} \int \{\text{trace}[AA' \nabla^2 f(\mathbf{u})]\}^2 \, d\mathbf{u}, \tag{4.6}$$

where $\nabla^2 f(\mathbf{u})$ is the $d \times d$ Hessian matrix,

$$\nabla^2 f(\mathbf{u})_{ij} = \frac{\partial^2 f(\mathbf{u})}{\partial u_i \partial u_j}.$$

The optimal $H$ is not generally available in closed form, but (4.6) shows that $h$ should be taken to be $O(n^{-1/(d+4)})$, yielding $\text{AMISE}_0 = O(n^{-4/(d+4)})$.

Figure 4.5 gives a contour plot of a kernel estimate for the NBA data corresponding to that for the frequency polygon in Fig. 4.4. The kernel used is the multivariate normal (Gaussian) density with $H = \text{diag}(.025, .05)$. The plot highlights the bimodal structure clearly but also suggests even more detail. It appears that the left mode displays negative correlation, while

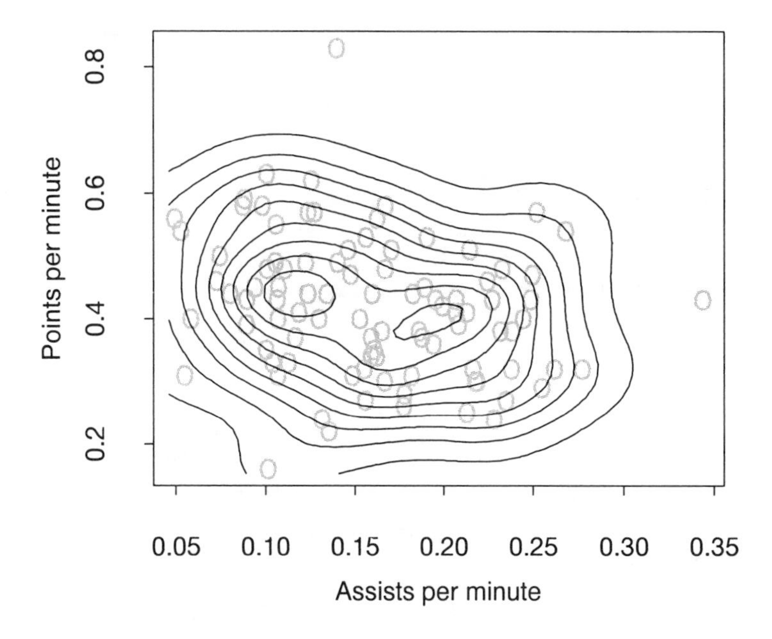


**Fig. 4.5.** Contour plot of kernel estimate for basketball data.

the right mode exhibits positive correlation. Other statistics support this pattern, as the observed correlation between the two variables for players averaging less than 0.2 assists per minute is −0.19, while that for players averaging at least 0.2 assists per minute is 0.15. Thus, while there is apparently a tradeoff between points and assists for shooting guards (the left mode), for point guards the better players are better at both.

The density contours also strongly identify two observations as very unusual. These observations correspond to a remarkably prolific scorer (Michael Jordan) and a remarkably prolific passer (John Stockton). Both of these players are among the best ever at their positions, so the observed pattern is not surprising.

The stability of the observed patterns can be assessed using a variability plot (Fig. 4.6). For these bivariate data, this takes the form of two contour plots (a lower and upper limit plot), but otherwise the construction is the same as for the univariate estimates of Chapter 3. The two modes are apparent in both halves of the plot, and their heights are only 35% lower (lower limit) and 45% greater (upper limit) than the estimated value in Fig. 4.5, respectively, which supports the existence of the observed modes.

As was true for univariate data, the choice of kernel function $K_d$ has

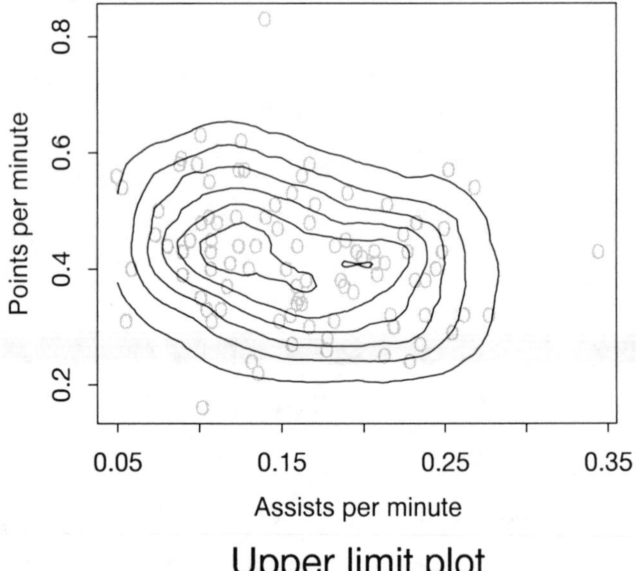


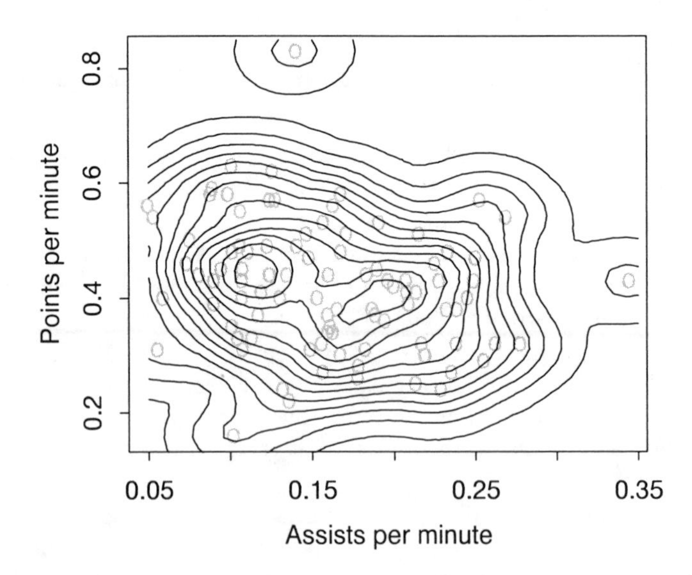


**Fig. 4.6.** Variability plot of kernel estimate of NBA data.

little effect on AMISE (although the effect does increase with $d$) and can be chosen based on theoretical or computational concerns.

### 4.2.2 Choosing the bandwidth matrix

The general kernel (4.5) requires specification of the bandwidth matrix $H$, which has $d(d+1)/2$ distinct entries. This number becomes unmanageable very quickly, which suggests restricting $H$ to have some simpler form. Three possibilities have typically been considered:

(1) $H = hI$. Setting the smoothing parameter to be constant for every variable implies that the amount of smoothing in each direction is the same. This is sensible only if the scales of all variables are roughly constant, so this would be done only after each variable was standardized to be on a common scale.

(2) $H = \text{diag}(h_1, \ldots, h_d)$. This parameterization allows different amounts of smoothing in each coordinate direction. This approach is also the "practical" version of approach (1); if $s_j$ is the scaling constant for the $j$th variable, (1) is equivalent to using $H = h\,\text{diag}(s_1, \ldots, s_d)$.

(3) $H = hS^{1/2}$, where $S$ is an estimate of the covariance matrix of $\mathbf{x}$. This is the multivariate generalization of coordinatewise-scaling, since it is equivalent to linearly transforming the data to have unit estimated covariance (often called *sphering* the data), using a constant bandwidth $H = hI$, and then transforming back to the original scale. The idea is to use a kernel that is the same general shape as the density.

Unfortunately, none of these parameterizations is rich enough to be able to handle all possible density shapes. Both scaling and sphering based on the sample standard deviations and covariance matrix, respectively, can be arbitrarily poorly behaved, since (as was noted in Chapters 2 and 3) these values do not measure scale in a meaningful way in terms of density estimation accuracy. Using a diagonal $H$ is often good enough, although sometimes a full matrix is necessary.

Data-based choices of $H$ are based on the same principles as are used for univariate data. The obvious reference distribution to use is the multivariate normal; using a Gaussian kernel gives optimal bandwidth matrix

$$H = \left(\frac{4}{d+2}\right)^{1/(d+4)} \Sigma^{\frac{1}{2}} n^{-1/(d+4)}$$

(this is a justification for sphering the data). The constant ranges between 0.924 and 1.059, equaling 1 for $d = 2$, so a rough Gaussian-based rule takes $\hat{H} = \hat{\Sigma}^{1/2} n^{-1/(d+4)}$. For a diagonal bandwidth matrix (which is optimal for independent $\mathbf{x}$), this corresponds to $\hat{H} = \text{diag}(\hat{\sigma}_1 n^{-1/(d+4)}, \ldots, \hat{\sigma}_d n^{-1/(d+4)})$.

More sophisticated bandwidth selectors are also possible. The least squares cross-validation score generalizes easily to

$$\mathrm{CV} = \int \hat{f}(\mathbf{u})^2\, d\mathbf{u} - \frac{2}{n}\sum_{i=1}^{n} \hat{f}_{-i}(\mathbf{x}_i).$$

Interestingly, the relative accuracy of the cross-validated bandwidth improves as the dimension increases. If $H = hI$, then

$$\frac{\hat{h}_{\mathrm{CV}}}{\hat{h}_0} - 1 = O_p(n^{-d/(2d+8)}),$$

where $\hat{h}_0$ is the minimizer of ISE. That is, as $d \to \infty$, the convergence rate of $\hat{h}_{\mathrm{CV}}$ approaches the optimal rate $O_p(n^{-1/2})$. Despite this, the cross-validation bandwidth matrix often undersmooths in practice.

Plug-in selectors also can be generalized to multivariate density estimation, based on the representation of AMISE (4.6). One such generalization, related to the univariate selector $\hat{h}_{\mathrm{SJ}}$, can achieve relative error rates

$$\frac{\hat{h}}{\hat{h}_0} - 1 = O_p(n^{-\alpha}),$$

where $\alpha = 5/14$ if $d = 1$ and $\alpha = 2/(d+4)$ if $d \geq 2$. This rate is faster than that of cross-validation for $d \leq 3$, equal for $d = 4$, and slower for $d \geq 5$.

Figure 4.7 illustrates the use of the plug-in selector for the earthquake data described in Chapter 3. The latitude and longitude of each earthquake form the bivariate data set, so the geographic distribution of earthquake events is the focus of study. A kernel estimate using a Gaussian kernel based on $H = \mathrm{diag}(h_{\mathrm{lat}} = 5.40, h_{\mathrm{long}} = 13.33)$ is superimposed on a world map in the figure, representing the distribution of earthquakes worldwide. The dominant feature is the high density of quakes along the Pacific rim, with modes centering in Chile, Japan, and New Guinea. Three other notable "hot spots" are the South Sandwich Islands (off the southeastern coast of South America), Kazakhstan (in Central Asia), and Novaya Zemlya (off Russia's Arctic coast). The latter isolated mode actually represents 22 underground nuclear explosions by the former U.S.S.R.

The estimator (4.5) can be applied to higher dimensional data, of course, but the difficulties of presenting the estimate become more serious. Figure 4.8 illustrates the problem. It is a contour shell of a kernel estimate for the three-dimensional version of the earthquake data. The base of the cube represents latitude and longitude, while the vertical axis is the logarithm of the depth of the quake, where the view is from the surface looking down (145 quakes at the surface, with zero depth, are not included). The shell corresponds to $f = .08\max(\hat{f})$. It is apparent that it is virtually impossible to sort out the patterns in the density.

For these data, a much better representation of a density estimate is a contour plot, fixed at different depths, of the conditional bivariate density of latitude and longitude. Figure 4.9 gives four such views of a Gaussian-based kernel using (close to) the Gaussian reference rule $H = \mathrm{diag}(h_{\mathrm{lat}} = 10, h_{\mathrm{long}} = 15, h_{\log(\mathrm{depth})} = .1)$ at selected depths.

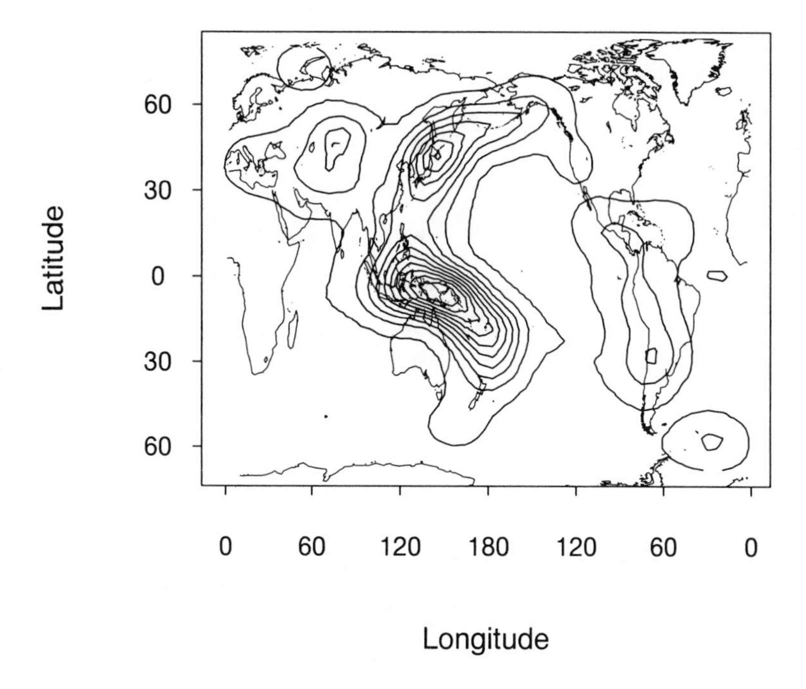


**Fig. 4.7.** Contour plot of kernel estimate of earthquake distribution based on plug-in selector.

The four contour plots illustrate the change in location of earthquakes as the depth increases. Closer to the surface (50 km depth), the dominant locations are the Pacific rim, including Alaska, with prominent modes at Japan and New Guinea, and smaller modes at the South Sandwich Islands and Kazakhstan. The relative likelihood of a quake at a depth of 200 km occurring in Kazakhstan is much higher, as is one in Chile, while the density around New Guinea splits into two modes. There is no indication of earthquake activity at this depth in the South Sandwich Islands and Alaska, and this is the case, as the maximum observed earthquake depths in those locations are around 140 km and 160 km, respectively.

The bifurcation around New Guinea becomes more pronounced as the depth increases, as do the modes in Japan and Chile. On the other hand, the relative likelihood of quakes in Kazakhstan lessens (the maximum observed depth of quakes there is around 230 km). In this instance, where depth of the quake can be intuitively viewed as a conditional factor related to geographic distribution of earthquakes, these conditional contour views are very effective at describing the three-dimensional pattern in the data.

Two-dimensional perspective views of the density at the given depth

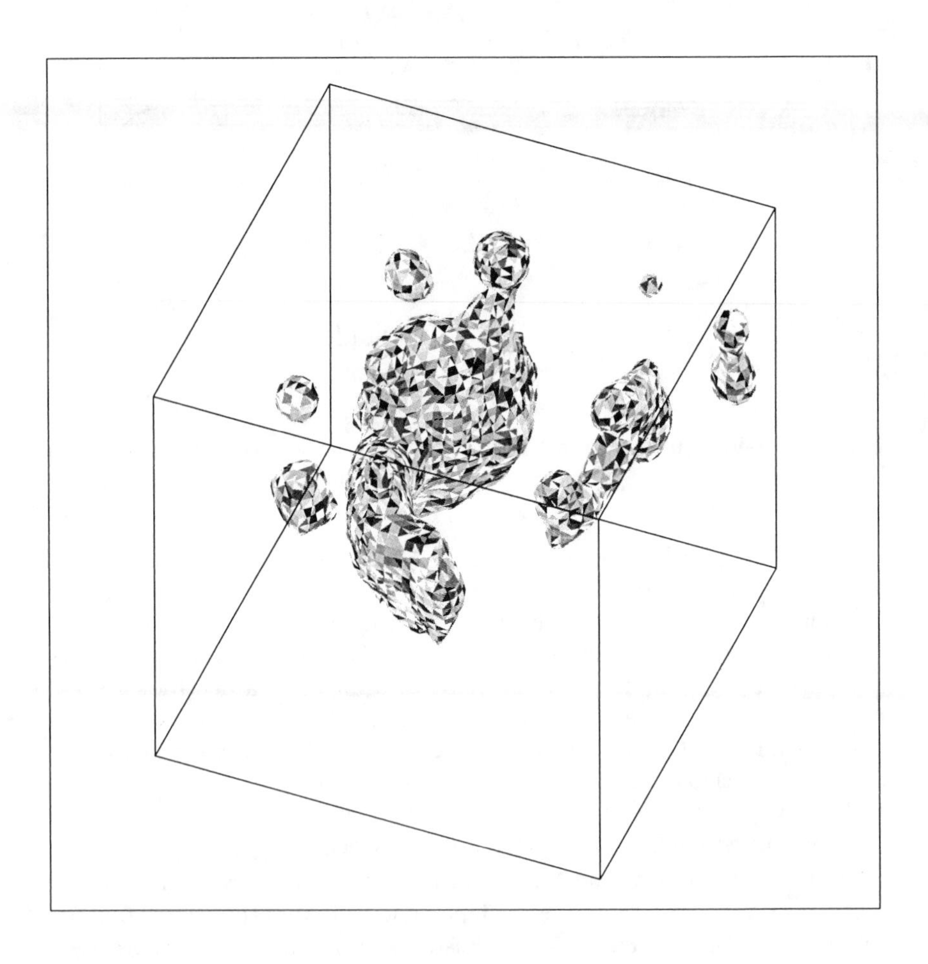

**Fig. 4.8.** Three-dimensional perspective plot of one contour shell of kernel estimate of earthquake data.

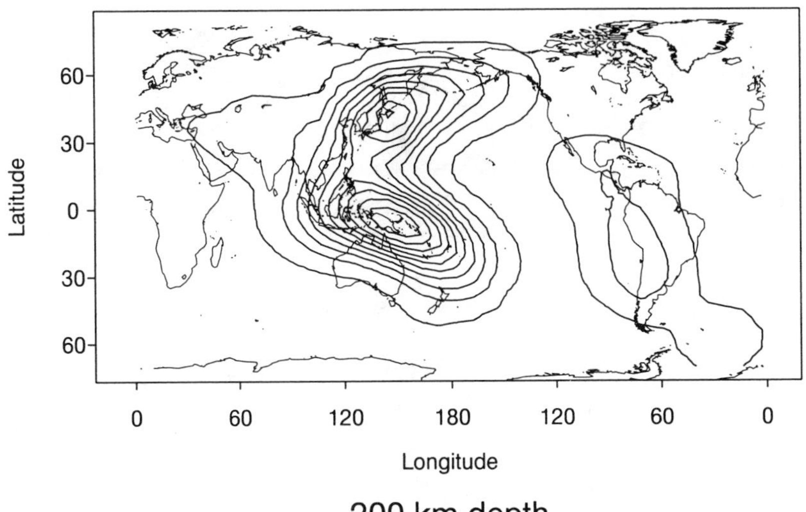


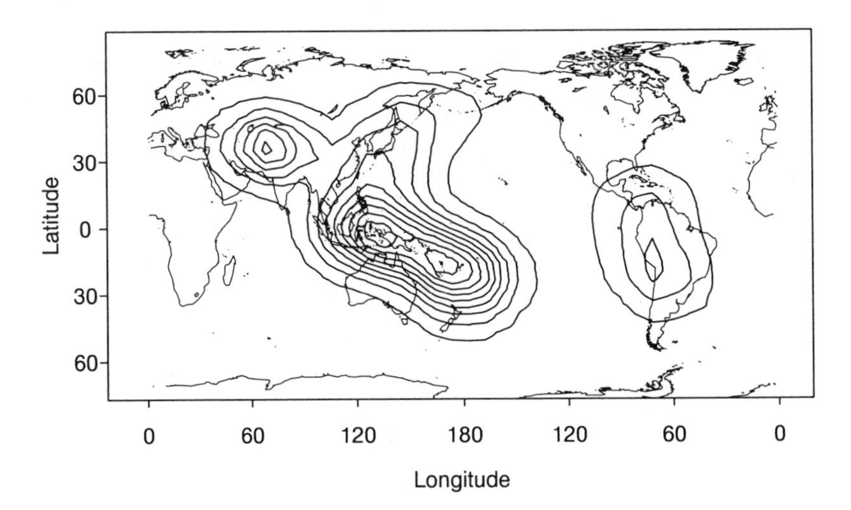


**Fig. 4.9.** Views of conditional density of latitude and longitude earthquake distribution given (log) depth: (a) 50 km depth; (b) 200 km depth.

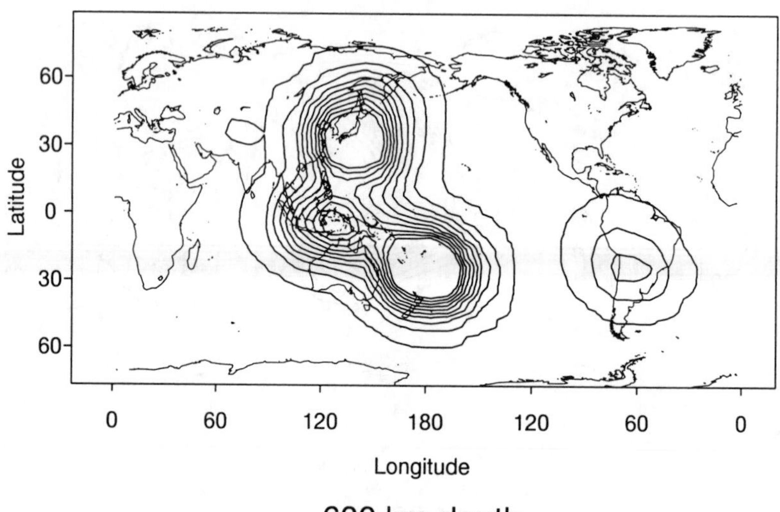


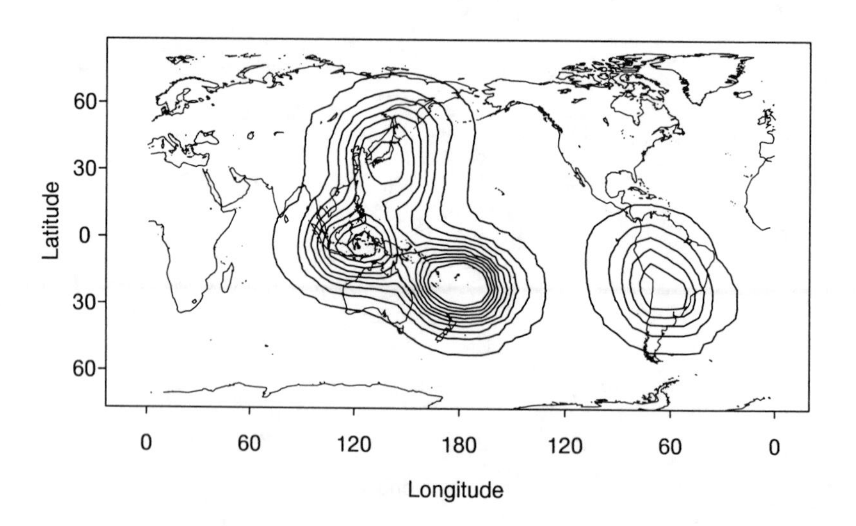


**Fig. 4.9 (cont.).** Views of conditional density of latitude and longitude earthquake distribution given (log) depth: (c) 400 km depth; (d) 600 km depth.

values could be used to determine the overall probability of earthquakes occurring (rather than the conditional probability given that depth). The previously constructed marginal density estimate for earthquake depth (Fig. 3.16) implies that most earthquakes occur at depths less than about 100 km, so the modes in the conditional contour plots at greater depths are representing rare events in an absolute sense.

## 4.3 Other Estimators

The multivariate kernel estimator carries with it the problems noted earlier for the univariate estimator — that is, boundary bias and a lack of sensitivity to local variations in smoothing. Many of the proposed univariate solutions also carry over, along with some new approaches.

### 4.3.1 Local variation in smoothing

Consider Fig. 4.10. These contour plots represent kernel density estimates for the joint distribution of the bottom margin and diagonal length of the bills in the Swiss bank note data discussed in chapters 2 and 3. The bills with shorter diagonal correspond to forged bills, while those with longer diagonal are real bills. The top plot is based on the plug-in choice of $H = \text{diag}(h_{\text{bot}} = .397, h_{\text{diag}} = .233)$ and is oversmoothed. The low-density mode at the lower left, which corresponds to the mode at 8 mm (bottom margin) in the univariate forged bills plot in Fig. 3.4, has been smoothed over. Further, there is no indication of any bimodal structure at the top of the plot, despite the noticeable bulge at 9 mm in the univariate real bills plot in Fig. 3.4. Reducing the bandwidths to $H = \text{diag}(.1, .2)$ supports more structure, but at the expense of severe undersmoothing.

The problem is that while the degree of smoothing implied by the plug-in choice is correct for the right side of the plot, it is too large for the left side. What is needed is an estimator that allows local variation in smoothing. One such estimator for bivariate data that takes advantage of the connection between the joint density and conditional densities is as follows. Let $f_1(x_1)$ and $f_2(x_2)$ be the marginal densities of $x_1$ and $x_2$, respectively, and let $f_{2|1}(x_2|x_1)$ and $f_{1|2}(x_1|x_2)$ be the conditional densities of $x_1$ given $x_2$ and $x_2$ given $x_1$, respectively. Then, by the definition of a conditional density function,

$$f(x_1, x_2) = [f_{2|1}(x_2|x_1)f_1(x_1)f_{1|2}(x_1|x_2)f_2(x_2)]^{1/2}.$$

An estimate of $f(x_1, x_2)$ can be constructed by estimating the two marginal and two conditional densities. Any of the univariate estimates described in Chapter 3 could be used to estimate the marginal densities,

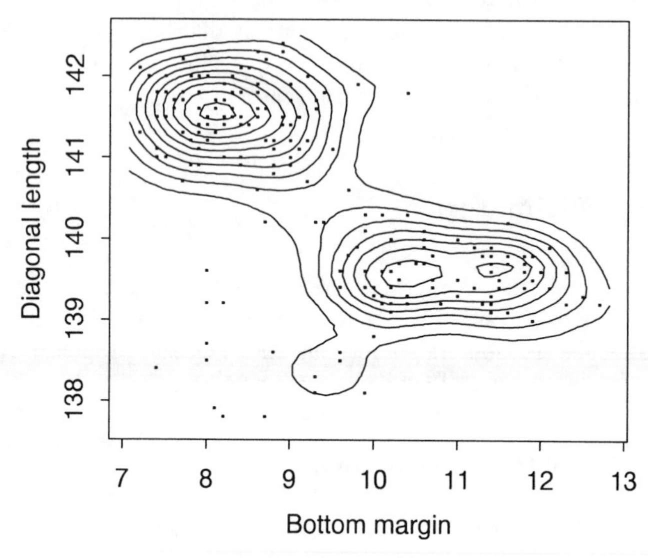


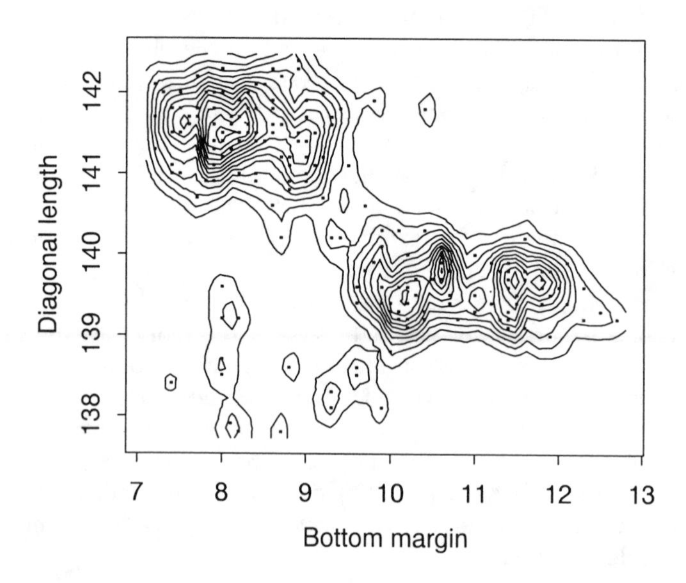


**Fig. 4.10.** Kernel estimates of Swiss bank note data based on plug-in bandwidth choice and undersmoothed bandwidth choice.

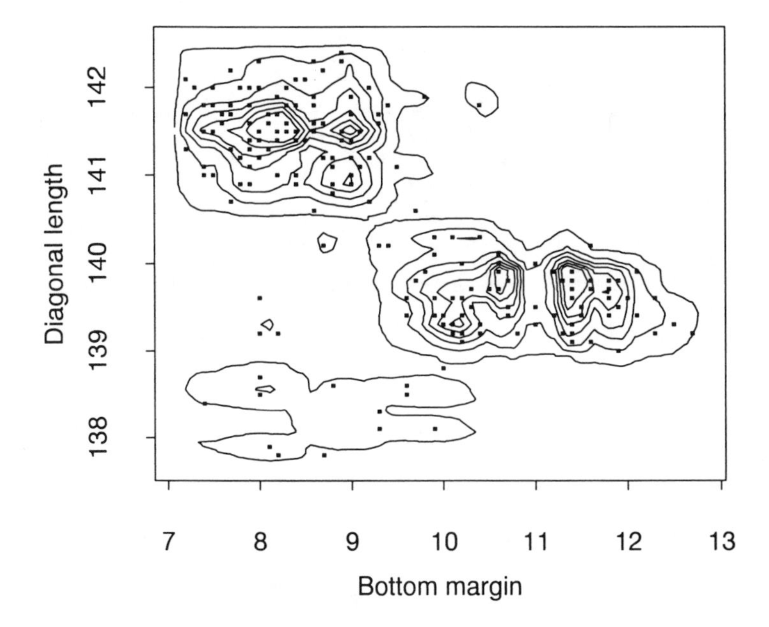


**Fig. 4.11.** Contour plot of marginal/conditional estimate of Swiss bank note data.

but estimating the conditional densities is more challenging. Consider dividing the bivariate region of interest into a rectangular grid, based on $B_1$ bins $\{I_{1i}\}$ (for $x_1$) and $B_2$ bins $\{I_{2j}\}$ (for $x_2$). Then $f_{2|1}(x_2|x_1 \in I_{1i})$ can be estimated using a univariate density estimate based on the set of $x_2$ values corresponding to $x_1 \in I_{1i}$ (with corresponding construction for $f_{1|2}(x_1|x_2 \in I_{2j})$). Then, estimating the marginal and conditional densities with the usual accuracy obtained from kernel estimates (for example), and choosing $B_1$ and $B_2$ appropriately, implies that the mean squared error of the estimate

$$\hat{f}(x_1, x_2) = [\hat{f}_{2|1}(x_2|x_1 \in I_{1i})\hat{f}_1(x_1)\hat{f}_{1|2}(x_1|x_2 \in I_{2j})\hat{f}_2(x_2)]^{1/2},$$
$$x_1 \in I_{1i},\ x_2 \in I_{2j},$$

converges to zero at the rate $O(n^{-4/7})$, which is slightly slower than that of the kernel estimator (which has convergence rate $O(n^{-2/3})$).

Figure 4.11 illustrates the use of this "marginal/conditional" estimator. The contour plot highlights all the interesting structure in the data, including the bimodality for longer diagonal bills, the bimodality for medium diagonal length bills, and the small mode for shortest diagonal bills. Since

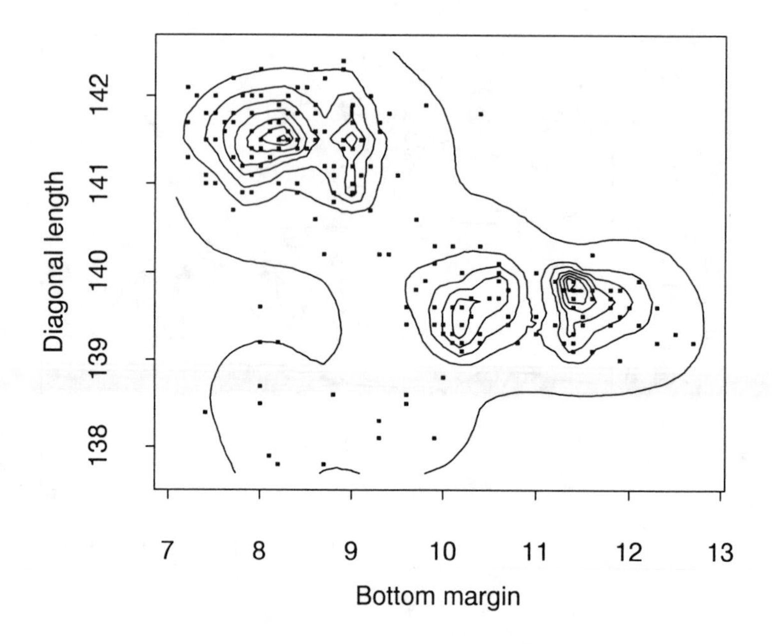


**Fig. 4.12.** Contour plot of variable kernel estimate of Swiss bank note data.

the estimate at any point is based on two conditional estimates, which vary based on location, a good deal of local adaptivity is possible, allowing all the structure to come through.

The estimate in Fig. 4.11 appears undersmoothed, and a smoother adaptive estimate would be preferable. This can be easily accomplished by using the marginal/conditional estimate as a pilot estimate for a variable kernel estimate, as in Chapter 3. The estimator has the form

$$\hat{f}(\mathbf{x}) = \frac{1}{n}\sum_{i=1}^{n}\frac{1}{|H_i|}K_d[H_i^{-1}(\mathbf{x}-\mathbf{x}_i)]. \tag{4.7}$$

Taking $H_i = Hf(\mathbf{x}_i)^{-1/2}$ (as in the univariate case) removes the $O(h^2)$ bias term, resulting in an estimate with MSE $= O(n^{-8/(d+8)})$. Thus, using the marginal/conditional estimate as a bivariate pilot estimate (suitably clipped away from zero, if necessary), and then determining a variable kernel estimate (4.7), yields an estimator with MSE $= O(n^{-4/5})$.

Figure 4.12 is a contour plot of the variable kernel estimate using the marginal/conditional estimate for the pilot and $H = \text{diag}(.05, .05)$. The previously noted structure is still evident, and the estimate is considerably

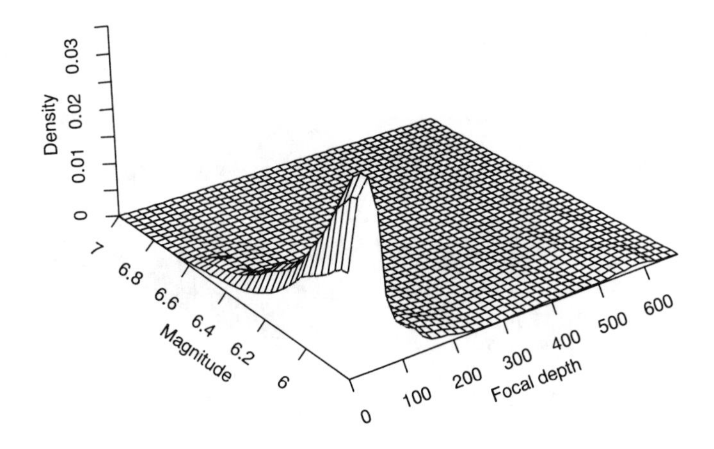


**Fig. 4.13.** Perspective plot of bivariate kernel estimate of earthquake data.

smoother than that of Fig. 4.11, providing a compelling argument for five modes in the data.

### 4.3.2 Boundary bias

Multivariate kernel estimators suffer from the same boundary bias problems that univariate estimators do. Indeed, the problems can be more severe in higher dimensions, as the boundary region is a larger proportion of the region of interest. Figure 4.13 is a perspective plot of a bivariate kernel estimate of the density of the focal depth and magnitude of the earthquake data examined earlier. Here $H = \text{diag}(9.75, .05)$, which corresponds to the Gaussian reference rule. Even though 31% of all the quakes had magnitude less than or equal to 5.8, and 43% occurred at depths less than or equal to 33 km, the estimate drops towards zero at the lower boundary in both dimensions due to boundary bias.

In principle, boundary kernel estimates can be defined for multivariate kernels, but determining the boundary region gets progressively harder in higher dimensions. This makes methods that accomplish automatic boundary bias correction, such as local likelihood estimation, particularly attractive. The estimator has the form $\hat{f}_\ell(\mathbf{x}) = f(\mathbf{x}, \hat{\boldsymbol{\theta}}(\mathbf{x}))$, where $\hat{\boldsymbol{\theta}}(\mathbf{x})$ maximizes

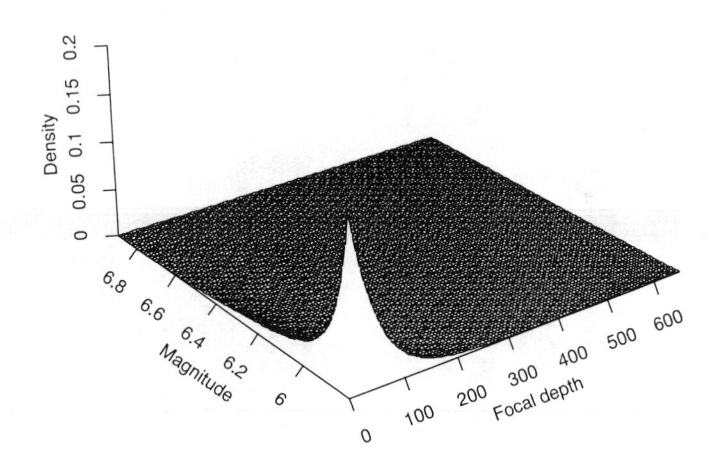


**Fig. 4.14.** Perspective plot of local log-quadratic estimate of earthquake data.

the multivariate version of the local likelihood (3.18). Once again, the dimension of $\boldsymbol{\theta}$ determines the asymptotic properties of the estimator, although now it is the minimum number of parameters in each direction that matters. So, for example, fitting a local log-linear density

$$f = a \exp[b_1(t_1 - x_1) + b_2(t_2 - x_2)]$$

to bivariate data (modeling local level and local slopes) yields $O_p(n^{-1/3})$ convergence both in the interior and at the boundary.

Figure 4.14 is a perspective plot of a local log-quadratic density estimate corresponding to the estimate in Fig. 4.13. The bandwidth is $3s_i$ in each direction, where $s_i$ is the sample standard deviation for the $i$th variable. The boundary bias is gone, but the estimate cannot be viewed as completely satisfactory, as the multimodality of the focal depth variable (Fig. 3.16) has been smoothed over.

Figure 4.15 shows a more successful application of the local log-quadratic estimate. This estimate of the density for the Swiss bank note data (using a nearest neighbor bandwidth covering 25% of the observations) captures the multimodal structure well and is even a bit smoother than the variable kernel estimate in Fig. 4.12.

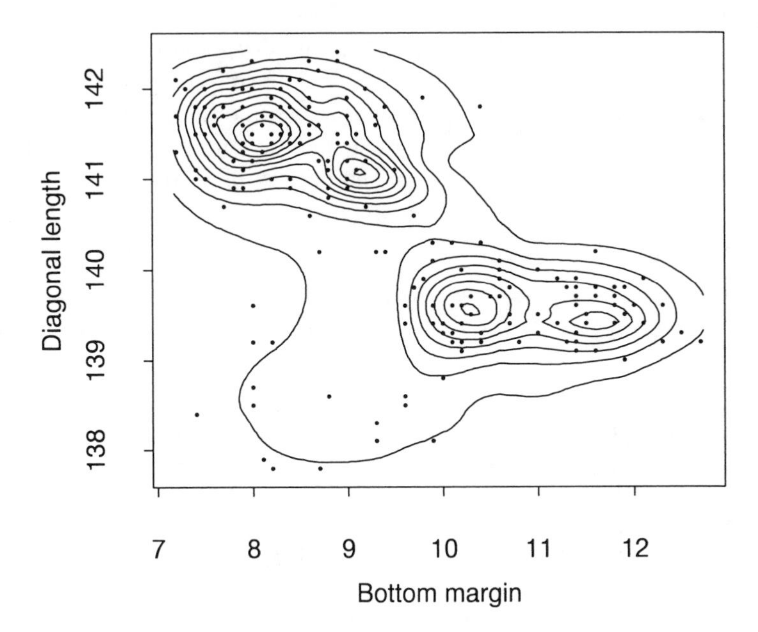


**Fig. 4.15.** Contour plot of local log-quadratic estimate of Swiss bank note data.

## 4.4 Dimension Reduction and Projection Pursuit

The existence of large empty regions in multidimensional space suggests that collapsing the data down to a smaller number of dimensions can lead to better density estimation. Such dimension reduction should be driven by the goal of preserving any interesting structure in the data in the lower-dimensional data while removing uninteresting attributes. This is the idea behind *projection pursuit.*

Since the most efficient way to discover structure is graphically, it makes sense to focus on linear projections (which are easiest to understand) to one or two dimensions. The projection is chosen to maximize some numerical index that gauges "interestingness." Given the general impression that the multivariate normal distribution is uninteresting, and that most linear combinations of variables will be distributed roughly normally (due to the Central Limit Theorem), constructing projections to be as nonnormal as possible seems sensible.

Thus, any statistic for testing normality is a candidate to be the basis of a projection index. Those that are computationally easy to construct, and focus on identifying multimodality, are particularly attractive. One such index can be constructed in the following way. First, sphere the data in order to remove location, scale, and correlation structure (call a sphered data value $\mathbf{z}$). Let $X = \boldsymbol{\alpha}'\mathbf{z}$ be a one-dimensional candidate linear projection, and define $R = 2\Phi(X) - 1$, where $\Phi(\cdot)$ is the Gaussian cumulative distribution function. If $X \sim N(0,1)$, then $R \sim \text{Uniform}[-1,1]$, so a measure of nonuniformity of $R$ corresponds to a measure of nonnormality of $X$.

A useful projection index can be defined as an approximation to the integrated squared distance between the density of $R$ and .5, the uniform density over $[-1,1]$. Approximating this using Legendre polynomials gives the index

$$I(\boldsymbol{\alpha}) = \frac{1}{2}\sum_{j=1}^{J}(2j+1)\left\{\frac{1}{n}\sum_{i=1}^{n}P_j[2\Phi(\boldsymbol{\alpha}'\mathbf{z}_i) - 1]\right\}^2, \tag{4.8}$$

where $J$ is the order of the Legendre approximation, and the Legendre polynomials $P_j(R)$ satisfy the recursive relationship

$$P_j(R) = \begin{cases} 1, & \text{if } j = 0, \\ R, & \text{if } j = 1, \\ [(2j-1)RP_{j-1}(R) - (j-1)P_{j-2}(R)]/j, & \text{if } j \geq 2. \end{cases}$$

The linear projection $\boldsymbol{\alpha}$ maximizes $I(\boldsymbol{\alpha})$ under the condition $\boldsymbol{\alpha}'\boldsymbol{\alpha} = 1$. Bivariate projections are defined in the same way, with a bivariate projection being transformed to be close to uniform on the square $(-1,1) \times (-1,1)$.

Since it is likely that more than one projection would be needed to represent the structure in the data, a way to remove the structure found at any iteration must be devised so that a new projection can be constructed that is (in some sense) orthogonal to what has already been found. Since a projection has no interest if it is normal, structure can be removed by applying a transformation that results in a normal distribution in the projected subspace but does not affect orthogonal directions.

Figure 4.16 illustrates the potential for projection pursuit to find structure in high-dimensional data. The data consist of information about 93 new 1993 model automobiles, concerning price, city miles per gallon, highway miles per gallon, engine size, horsepower, fuel tank capacity, and weight. The figure gives the bivariate projections for the first two projections (based on fourth order Legendre polynomials), along with a bivariate kernel estimate superimposed on the projected points.

The first projection has the form

$$\text{Horizontal}: 5 \times \text{Engine size} - \text{Fuel capacity},$$
$$\text{Vertical}: \text{Price} + 5 \times \text{Engine size}.$$

The density estimate (based on a Gaussian product kernel with $H = \text{diag}(.6, .4)$) supports three high density regions: the bulk of cars (in the center right of the plot), expensive autos with large engines (at the upper

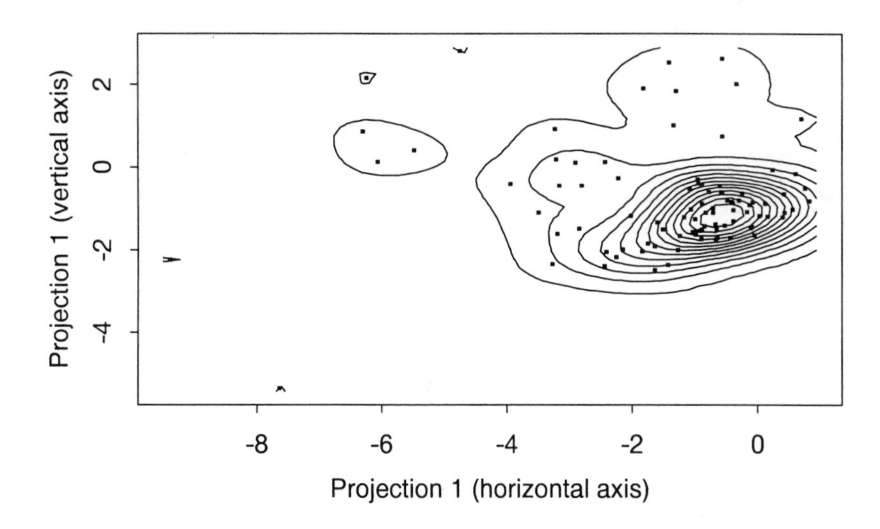


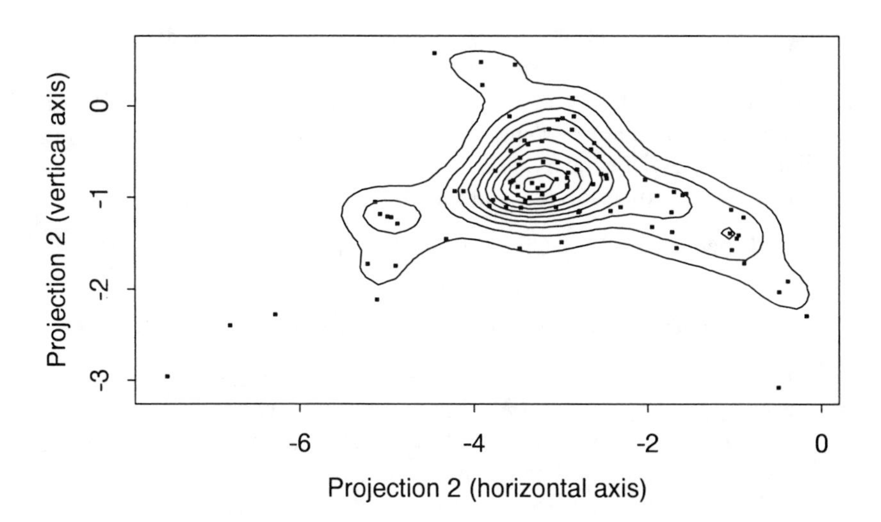


**Fig. 4.16.** Bivariate projections of first two projections from application of projection pursuit to 1993 auto data, with density estimates superimposed on projected points.

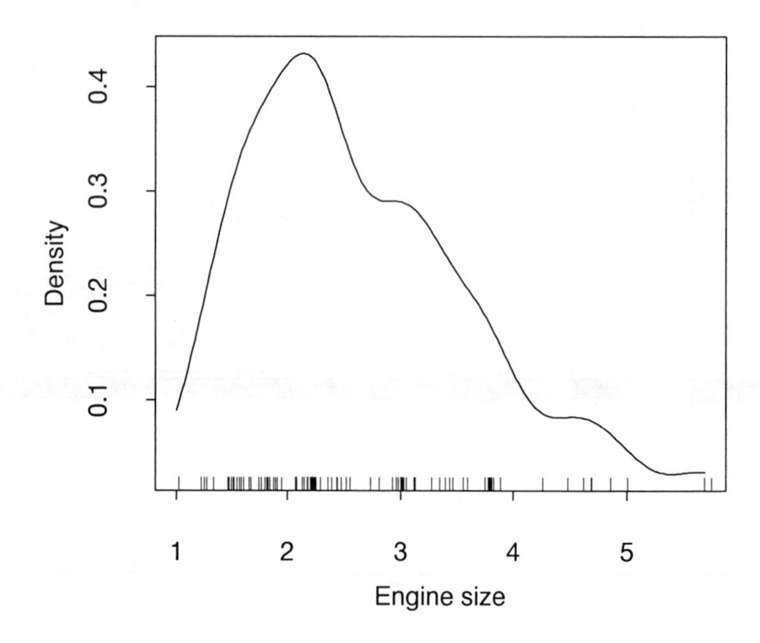


**Fig. 4.17.** Kernel estimate of engine size for 1993 auto data.

right of the plot) such as the Lincoln Town Car and Cadillac Deville, and expensive autos with smaller engines (at the center of the plot) such as the Cadillac Seville and Lexus SC300.

The second projection has the form

$$\text{Horizontal}: -10 \times \text{City MPG} + 5 \times \text{Highway MPG} \\ + 16 \times \text{Engine size} - 4 \times \text{Fuel capacity},$$
$$\text{Vertical}: -\text{City MPG} - 6 \times \text{Engine size}.$$

The density estimate (using $H = \text{diag}(.35, .2)$) identifies three separate directions away from the center (which corresponds to the bulk of the autos): larger cars (to the right) such as the Lincoln Continental and Pontiac Bonneville, small cars (to the left) such as the Dodge Colt and Ford Festiva, and small low mileage cars (to the top) such as the Mazda RX-7 and Toyota Previa.

These patterns suggest that the interesting structure in these data comes from engine size in particular, along with price, miles per gallon, and fuel tank capacity. Figures 4.17 and 4.18 confirm this. Figure 4.17, a kernel estimate of engine size, shows three modes, at around 2.2 liters (small cars), 3 liters (midsize cars), and 4.5 liters (large cars).

The bivariate density estimates in Fig. 4.18 show that while price increases with engine size, some midsize cars (around 3 liter engine size) can be expensive (note the bulge in the center of the density estimate), while very large engines are associated with both expensive and relatively inexpensive cars.

Not surprisingly, there is a negative association between engine size and mileage, but the density estimate suggests three high density regions: very small, high mileage cars; moderate size, good mileage cars; and large, lower mileage cars. Note also that the mileage flattens out as engine size increases, not falling below 15 miles per gallon. Finally, there is a positive association between engine size and fuel tank capacity, with the tank size increasing more slowly as engine size increases.

The power of projection pursuit is its ability to "home in" on interesting patterns in the data and point the data analyst toward them. The projections themselves can define indices that highlight such patterns, and (as happens for these data) the variables that define the projections can be examined directly as ones that are likely to show interesting relationships.

## 4.5 The State of Multivariate Density Estimation

The gap between theory and practice is probably widest for multivariate density estimation, compared with other smoothing problems. Although many estimation schemes, including kernel and local likelihood estimation, directly generalize to higher dimensions, practical implementation lags behind this theoretical fact. The "empty space phenomenon" in higher dimensions described in Section 4.1, whereby most local neighborhoods are almost surely empty, argues against very effective direct density estimation in more than four or five dimensions. Even with the use of color and dynamic graphics, more than four dimensions are almost impossible to represent, and contour plots, which can be highly evocative of structure, can only represent two-dimensional slices of data. Asymptotic arguments require massive amounts of data in high dimensions, making it very difficult to understand completely what various estimators are actually doing.

Further progress on effective implementation of direct density estimators on two- and three-dimensional data (such as how to regulate the degree of smoothing, and the usefulness of local variations in smoothing) would be welcome, but it seems likely that the most useful approach for higher-dimensional data is dimension reduction of some sort, such as that accomplished using projection pursuit. It is likely that higher-dimensional data will fall into lower-dimensional subspaces, which can be identified and (it is hoped) understood.

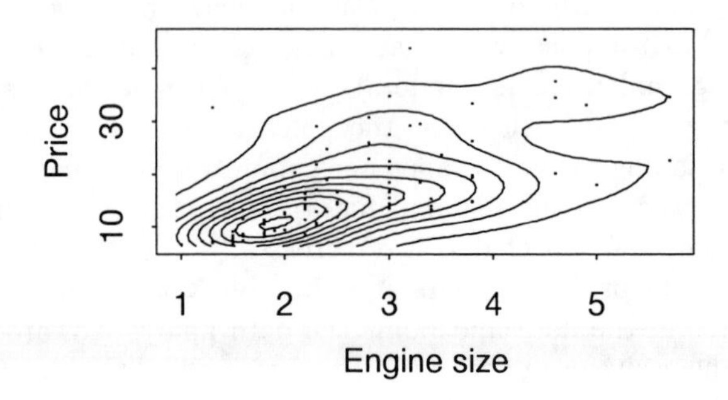


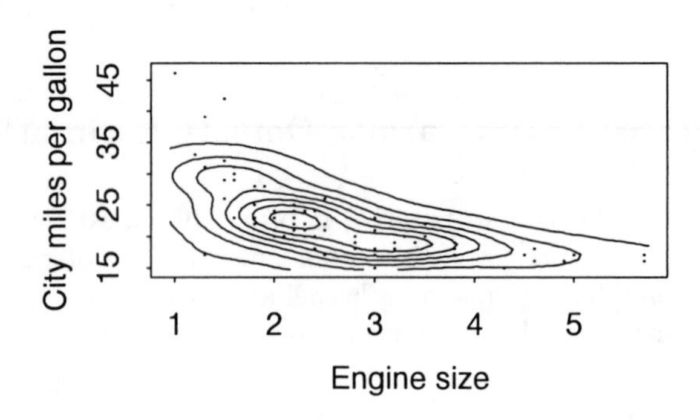


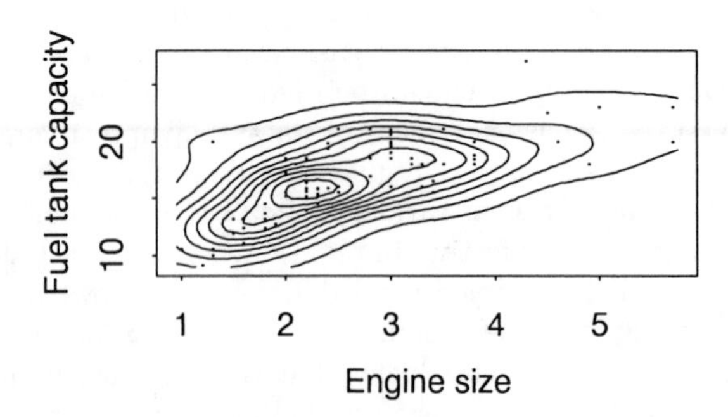


**Fig. 4.18.** Bivariate kernel estimates of engine size and price, city miles per gallon, and fuel tank capacity, respectively, for 1993 auto data.

## Background material

### Section 4.1

Chatterjee, Handcock, and Simonoff (1995, p. 299) gave a superset of the NBA data and analyzed it using regression methods. Cleveland and McGill (1984) and Schilling and Watkins (1994) discussed the construction and properties of sunflower plots.

Geffroy (1974) and Abou-Jaoude (1976a) examined the consistency properties of the multivariate histogram. Lecoutre (1985) derived the AMISE of the histogram estimator for $d$-dimensional data, assuming constant bin width in all dimensions. Scott (1992, Chapter 3) derived corresponding results for general choices of $h_j$. Scott (1988), Hüsemann and Stevens (1991), and Hüsemann and Terrell (1991) examined the use of nonrectangular, rotated and nonregular bins. Hexagonal bins are the optimal bin shape for bivariate histograms using a regular grid pattern and also can be better on the grounds of graphical perception (Carr *et al.*, 1987), but their AMISE is only about 2% less than that using square bins, as in (4.3).

Scott (1992, Section 4.2) discussed multivariate frequency polygons. Terrell (1983) and Hjort (1986) derived the properties of the linear blend frequency polygon. Hjort showed that the LBFP integrates to 1, and derived the AMISE of the estimator.

Scott (1985a,b) described a different way to construct a frequency polygon from a multivariate histogram. Triangular meshes are formed by interpolating values at the centers of $d + 1$ adjacent histogram bins (the construction is not unique). Scott (1985a) derived similar (but more complex) AMISE results to (4.4) and proposed a rough guideline for bin width choice as $\hat{h}_j = 2\hat{\sigma}_j n^{-1/(d+4)}$.

Bellman (1961) first used the term "curse of dimensionality" when describing the computational effort required in combinatorial optimization over many dimensions. Silverman (1986, Sect. 4.5.1) and Scott (1992, Chapter 7) discussed its implications for density estimation in some detail. In particular, Silverman described the properties of the multivariate normal density in this context, and Scott showed the effect on the uniform density over a hypercube.

### Section 4.2

**4.2.1.** Cacoullos (1966) introduced multivariate kernel density estimation, taking $H = hI$, and studied its consistency properties. Devroye and Györfi (1985, Sect. 3.1) examined almost sure convergence and convergence in $L^p$ norm. Epanechnikov (1969) examined the estimator using a vector of bandwidths $H = \text{diag}(h_1, \ldots, h_d)$, while Deheuvels (1977b) first treated the case of an arbitrary bandwidth matrix $H$. Scott (1992, Sect. 6.3.2) suggested the use of the parameterization $H = hA$ in the derivation of the

AMISE of the estimator, while Wand (1992b) derived these results using a different representation.

An alternative to the use of product kernels to construct a multivariate kernel function is to use a spherically symmetric kernel,

$$K_d(\mathbf{x}) = \frac{K[(\mathbf{x}'\mathbf{x})^{1/2}]}{\int K[(\mathbf{u}'\mathbf{u})^{1/2}]d\mathbf{u}}.$$

In general, this kernel is different from the corresponding product kernel (except for the multivariate normal kernel). For example, the product Epanechnikov kernel is

$$K_d(\mathbf{x}) = \begin{cases} \left(\frac{3}{4}\right)^d \prod_{j=1}^d (1 - x_j^2), & \text{if } |x_j| \le 1 \text{ for all } j, \\ 0, & \text{otherwise,} \end{cases}$$

while the spherically symmetric version is

$$K_d(\mathbf{x}) = \begin{cases} \left[\frac{d(d+2)}{4}\right] \Gamma\left(\frac{d}{2}\right) \pi^{-d/2}(1 - \mathbf{x}'\mathbf{x}), & \text{if } \mathbf{x}'\mathbf{x} \le 1 \\ 0, & \text{otherwise.} \end{cases}$$

Each of these kernels minimizes AMISE within their respective classes (Epanechnikov, 1969; Fukunaga and Hostetler, 1975). Wand and Jones (1995, Table 4.1) showed that the spherically symmetric versions of kernels based on the Beta density function (such as the uniform, quadratic (Epanechnikov) and quartic (biweight)) are more efficient than the product versions, although the drop in efficiency is less than 10% for the quadratic and quartic kernels for $d \le 4$. Cline (1988) showed that asymmetric kernels are inadmissible for multivariate density estimation.

Calculation of multivariate density estimates should not be based on (4.5) for large samples because of the computational burden involved. Scott (1985b) and Härdle and Scott (1992) showed that WARPing can be generalized to multivariate kernel estimation. Wand (1994b) compared different binning rules for multivariate data and showed the usefulness of the Fast Fourier Transform in this context.

One way to quantify the curse of dimensionality is by determining the increase in the number of observations needed to achieve a given level of accuracy. Epanechnikov (1969), Silverman (1986, p. 94), and Scott and Wand (1991) investigated this for multivariate normal data and a Gaussian kernel. The following table gives the sample sizes needed to achieve AMISE = .393 and AMIAE = .5987:

| *Dimension* | *AMISE sample size* | *AMIAE sample size* |
|---|---|---|
| 1 | 4 | 4 |
| 2 | 17 | 11 |
| 3 | 52 | 32 |
| 4 | 155 | 98 |
| 5 | 480 | 312 |
| 6 | 1563 | 1020 |
| 7 | 5382 | 3415 |
| 8 | 19558 | 11719 |
| 9 | 74746 | 41203 |
| 10 | 299149 | 148366 |

The sample sizes necessary to achieve the given $AMIAE$ are 35%–50% smaller than those necessary to achieve the given AMISE, but in either case they increase rapidly with $d$.

**4.2.2.** Fukunaga (1972, p. 175) proposed the sphering technique when smoothing multivariate data.

Wand and Jones (1993) examined various parameterizations for $H$ for bivariate data using normal mixture densities. They showed that scaling and sphering using sample standard deviations and covariances can lead to poor performance. They found that using diagonal $H$ is generally reasonable, although it can be made arbitrarily inefficient for certain densities.

Scott (1992, p. 152), Wand (1992b), and Wand and Jones (1995, p. 111) derived Gaussian-based rules for bandwidth selection. Hall (1985), Marron (1986), Jones (1992), and Sain, Baggerly, and Scott (1994) investigated the properties of least squares cross-validation-based bandwidth selection. Sain *et al.* also proposed a biased cross-validation selector that achieves the same convergence rate as (unbiased) cross-validation. Results of Monte Carlo simulations up to three dimensions show that the cross-validated choice often undersmooths, while biased cross-validation is much better behaved.

Wand and Jones (1994) investigated plug-in methods for multivariate data. If $f$ is sufficiently smooth, the terms in AMISE that are a function of $f$ can be written using terms of the form

$$\psi_r = \int f^{(r)}(\mathbf{u})f(\mathbf{u})d\mathbf{u},$$

which can be estimated using a kernel estimator (Wand, 1994, and González-Manteiga, Sánchez-Sellero, and Wand, 1996, described binned approximations to these functionals). Their estimated bandwidths performed well in small-scale simulations for bivariate densities, exhibiting low variability with a tendency to oversmooth slightly. Difficulties in estimating (4.6) for large $d$ are an obvious impediment to plug-in methods for high-dimensional data.

Hall and Wand (1988a) described an algorithm to choose $h$ to minimize $MIAE$. Scott and Wand (1991) showed that the minimizer of mean absolute error at any value $\mathbf{x}$ is slightly smaller (within 4%) than the minimizer of mean squared error, so a rule based on minimizing absolute error is not appreciably different from one based on minimizing squared error.

Minnotte and Scott (1993) generalized the mode tree diagram to bivariate data, allowing the effects of bandwidth choice on the estimate to be represented graphically.

Frohlich and Davis (1990) analyzed the three-dimensional earthquake data using single linkage clustering. Their results were similar to those reported here.

Scott (1992, Sect. 1.4 and Plates 1–16) discussed and illustrated various ways of presenting higher-dimensional density estimates, including the use of color and different "banding" styles of contour shells.

## Section 4.3

**4.3.1.** Flury and Riedwyl (1988, Fig. 2.4) noted the multimodal character of the distribution of bottom margin values for both the real and forged Swiss bank notes.

Simonoff (1995b) proposed the marginal/conditional estimator and studied its properties theoretically and via Monte Carlo simulations. He found that the number of bins $B_1$ and $B_2$ should be chosen to increase with $n$ at the rate $n^{2/7}$, with $B_1 = B_2 = \lceil .75n^{2/7} \rceil$ working well in his examples ($\lceil \cdot \rceil$ means round up to the nearest greater integer). Each marginal density and the $B_1 + B_2$ conditional densities were estimated using oversmoothed kernel estimates with bandwidths $2.5\hat{h}_{\mathrm{SJ}}$, which were found to provide a better graphical display. Since $B_1$ and $B_2$ increase very slowly with $n$, the estimate can be calculated very quickly, as it is based on only $B_1 + B_2 + 2$ univariate density estimates.

Abramson (1982) formulated the variable kernel estimate for arbitrary dimensions and showed that choosing the bandwidth at $\mathbf{x}_i$ to be proportional to $f(\mathbf{x}_i)^{-1/2}$ removes the $O(h^2)$ bias term. Devroye and Penrod (1986) examined using the $k$th nearest neighbor distance as the bandwidth, which corresponds to using $H_i \propto f(\mathbf{x}_i)^{-1/d}$, and established the uniform consistency of the resultant estimate under weak conditions on $f$, $n$, and $k$. The nearest neighbor variable kernel corresponds to Abramson's choice for bivariate ($d = 2$) data but otherwise will not achieve reduced bias. Devroye (1985) established conditions for weak consistency of estimates of the form (4.7) for all $f$.

Worton (1989) compared the fixed and variable bandwidth estimators for bivariate data, using an idealized version of the variable kernel estimator with the pilot density being the true density (without clipping away from zero). He compared the estimators with respect to both finite sample and asymptotic properties and found that the variable kernel estimator can give appreciably smaller MISE than the fixed-bandwidth estimator.

Bowman and Foster (1993) also studied an idealized version of the estimator (4.7), using the true density in $H_i = Hf(\mathbf{x}_i)^{-1/p}$, but without clipping away from zero. They found that while taking $p = 2$ does reduce bias (as expected), the increase in variability more than offsets this compared with taking $p = d$ (as in nearest neighbor variable kernel estimation).

Local kernel (balloon) estimation also can be generalized to multiple dimensions, with

$$\hat{f}(\mathbf{x}) = \frac{1}{n|H(\mathbf{x})|}\sum_{i=1}^{n} K[H(\mathbf{x})^{-1}(\mathbf{x}-\mathbf{x}_i)].$$

Devroye and Wagner (1977), Moore and Yackel (1977), Mack and Rosenblatt (1979), and Hall (1983b) investigated convergence properties of this estimator when $H(\mathbf{x})$ is (a function of) the $k$th nearest neighbor distance. Terrell and Scott (1992) showed that this estimator can achieve a faster rate of convergence of asymptotic MSE compared with the fixed kernel estimator for certain $\mathbf{x}$, corresponding to certain properties of the matrix of second partial derivatives of $f$ at $\mathbf{x}$. Unfortunately, such improved convergence rates are not possible at modes, and the estimate will not generally be a density function.

**4.3.2.** Dong and Simonoff (1995) described the construction of $d$-dimensional boundary kernels based on the product Epanechnikov kernel. The kernels take the form of cubic polynomials, with coefficients that solve $d$ systems of four equations, and the solutions are given in the paper.

Hjort and Jones (1996) and Loader (1996) discussed application of local likelihood estimation to multidimensional data. The estimator, as described by Hjort and Jones, is $\hat{f}_\ell(\mathbf{x}) = f(\mathbf{x}, \hat{\boldsymbol{\theta}}(\mathbf{x}))$, where the local estimate $\hat{\boldsymbol{\theta}}$ solves

$$n^{-1}\sum_{i=1}^{n} K_H(\mathbf{x}_i - \mathbf{x})v(\mathbf{x}_i, \boldsymbol{\theta}) - \int K_H(\mathbf{u}-\mathbf{x})v(\mathbf{u}, \boldsymbol{\theta})f(\mathbf{u}, \boldsymbol{\theta})d\mathbf{u} = \mathbf{0},$$

where $v(\cdot)$ is a weight function and $K_H(\mathbf{u}) = |H|^{-1}K(H^{-1}\mathbf{u})$. The estimate thus involves solving $p$ nonlinear integral equations, where $p$ is the number of parameters in $\boldsymbol{\theta}$. Choosing $v$ appropriately then yields the expected convergence rate in both the interior and boundary regions. For example, if all derivatives of $g = \log f$ exist up to order $p+1$, and a local log-polynomial up to order $p$ is fit,

$$\hat{g}(\mathbf{x}) - g(\mathbf{x}) = O_p(n^{(p+1)/(2p+2+d)})$$

(Loader, 1996).

As local likelihood estimation involves solving integral equations, computational issues become important. In particular, multidimensional numerical integration should be avoided if possible. Loader (1994, 1996) discussed methods to speed up computation of the estimate.

The semiparametric estimator proposed by Hjort and Glad (1995) can be generalized to multivariate data in a straightforward way as

$$\hat{f}_s(x) = n^{-1} f(\mathbf{x}, \hat{\boldsymbol{\theta}}) \sum_{i=1}^{n} K_H(\mathbf{x}_i - \mathbf{x}) / f(\mathbf{x}_i, \hat{\boldsymbol{\theta}}).$$

The properties of $\hat{f}_s$ carry through as before. For a product kernel, for example, the variance is identical to that of the multivariate kernel (4.5), while the bias is approximately

$$\frac{1}{2} \sum_{j=1}^{d} \left[ \int u^2 K_j(u) du \right] h_j^2 f_0(\mathbf{x}) \ddot{r}_{jj}(\mathbf{x}),$$

where $f_0(\mathbf{x})$ is the best parametric approximant and $r(\mathbf{x}) = f(\mathbf{x})/f_0(\mathbf{x})$. Thus, if $f_0 \ddot{r}_{jj}$ is small compared with $\ddot{f}_{jj}$, the semiparametric estimator has smaller bias than the kernel estimator. The estimator still suffers from boundary bias, however, so the difficulties in constructing multidimensional boundary kernels persist.

Penalized likelihood and spline-based methods also can be generalized to multivariate data, although computational issues become much more crucial. Natural roughness penalties take the form of sums of squared derivative terms (one for each dimension), such as (for bivariate data)

$$\Phi = \int \int \left[ \left( \frac{\partial f}{\partial u} \right)^2 + \left( \frac{\partial f}{\partial v} \right)^2 \right] du\, dv$$

(Scott, Tapia, and Thompson, 1978, who use a discrete version of this penalty) or

$$\Phi = \int \int \left[ \left( \frac{\partial^2 \gamma}{\partial u^2} \right)^2 + \left( \frac{\partial^2 \gamma}{\partial v^2} \right)^2 \right] du\, dv,$$

where $\gamma = \sqrt{f}$ (Good *et al.*, 1989). Granville and Rasson (1995) gave a general formulation of the penalized likelihood approach using a Bayesian framework, and proposed a penalty based on $g = \log f$.

Another possible approach is to mimic the sphering technique of Fukunaga (1972), transforming the data to have zero mean and diagonal covariance matrix, then smoothing in each dimension using a univariate MPLE, and transforming back to the original scale (Bennett, 1974; Scott, Tapia, and Thompson, 1978; Thompson and Tapia, 1990, pp. 141–145). This estimator is called the *pseudo-independent* estimator.

Gu (1993, 1995b) and Gu and Qiu (1993) defined the logistic spline estimator for multivariate, as well as univariate, data. In this context, the spline estimate can take the form of thin plate splines and tensor product splines (Gu and Wahba, 1993a).

## Section 4.4

Survey articles on projection pursuit include those of Huber (1985), Jones and Sibson (1987), and Li and Cheng (1993). The idea behind projection pursuit, defining an index that measures the "interestingness" of lower-dimensional projections of multivariate data, is due to Kruskal (1969, 1972). Friedman and Tukey (1974) coined the term and defined their index so as to identify clustering in the data. Huber (1985) and Jones and Sibson (1987) showed that the index can be written as an estimate of $R(f) = \int f^2$ based on a kernel density estimate of the projected points; thus, maximizing the index looks for projections away from a parabolic (quadratic) density (this is the same argument as the one that yields the Epanechnikov kernel as the optimal nonnegative kernel function).

Since, generally speaking, most projections look normal (Diaconis and Freedman, 1984), a more sensible target for noninterestingness is the normal distribution. This also coincides with the idea of minimizing randomness, as measured by entropy, since the normal distribution maximizes entropy (Huber, 1985). Using the negative Shannon entropy $\int f \log f$ as a projection index will therefore identify nonnormal projections. This index is computationally intensive, and Jones and Sibson (1987) derived an approximation based on Hermite orthogonal polynomials that is a simple function of third and fourth moments of the projected data (that is, skewness and kurtosis).

A direct approach to choosing a projection index is to measure the distance between the observed projected data and normality. Cook, Buja, and Cabrera (1993) formulated this general approach as follows. Let $T : \mathbb{R} \to \mathbb{R}$ be an arbitrary, strictly monotone, and smooth transformation on the random variable $X$ so that $Y = T(X)$. If $X$ has density $f(x)$, let $Y$ have density $g(y)$. Define a null version of the $f$ density to be $\phi(x)$ (such as the normal distribution), with the corresponding null version of $g$ being $\psi(y)$. A general family of indices is then

$$I = \int_{\mathbb{R}} [g(y) - \psi(y)]^2 \psi(y) dy.$$

Taking $T = 2\Phi(X) - 1$, as in Friedman (1987), yields the *exploratory projection pursuit* method described in Section 4.4, as in (4.8). The density $g(y)$ is then expanded in terms of the Legendre polynomial family. In the original scale, this index corresponds to

$$\int_{\mathbb{R}} \frac{[f(x) - \phi(x)]^2}{2\phi(x)} dx$$

and thus puts more weight on the tails of $f$. Morton (1989) proposed a version of this index that is affine invariant, based on terms of a Fourier series and Laguerre polynomials.

Hall (1989a) noted this higher weight in the tails and proposed measuring distance from $\phi$ in the original scale; that is,

$$\int_{\mathbb{R}} [f(x) - \phi(x)]^2 \, dx.$$

The density $f(x)$ is then expanded in terms of the Hermite polynomial family. This corresponds to using the transformation $T(X) \propto \Phi_{\sigma=\sqrt{2}}(X)$. Cook, Buja, and Cabrera (1993) proposed using the identity transformation $T(X) = X$, yielding the index

$$\int_{\mathbb{R}} [f(x) - \phi(x)]^2 \phi(x) dx,$$

which they termed the natural Hermite index (the density $f(x)$ is then also expanded in terms of the Hermite polynomial family). They showed that this index puts less weight on the tails than either the Legendre or Hermite indices. In particular, the lowest order terms of the Hermite and natural Hermite indices correspond to identifying holes in the center of the data, while the second order term of all three indices focuses on skewness (the first order term of the Legendre index is identically zero). Posse (1995a) proposed an index based on using a $\chi^2$ statistic to test normality of the projected data.

Posse (1995b) compared the ability of various implementations of bivariate exploratory projection pursuit to discover underlying structure. He found that the optimization algorithm used can have a large effect on the ability of projection pursuit to identify structure, regardless of the index used, and recommended a random search method for locating the global maximum of the projection index (see Posse, 1990). He found that the Legendre, Hermite, and $\chi^2$ indices are sensitive to departures from normality in the center of the distribution, while the Laguerre–Fourier and natural Hermite indices are useful for identifying clusters. These properties do not necessarily coincide with the theoretical distances that the indices estimate, reinforcing the importance of examining the indices themselves, rather than the theoretical distances. See also Sun (1993) for comparisons of indices.

Nason (1995) discussed projection pursuit using three-dimensional indices. His method was based on generalizing the index of Jones and Sibson (1987), and he described computational issues and the treatment of outliers.

Projection pursuit also can be used to estimate directly the multivariate density $f$; while this is of limited importance graphically (since multivariate densities past two dimensions cannot be directly graphed very easily), it can be useful if the density is used indirectly in some statistical functional (such as in discriminant analysis). See Friedman, Stuetzle, and Schroeder (1984), Huber (1985), Buja and Stuetzle (1985), Jee (1987), and

Jones and Sibson (1987). Stahel (1981) and Donoho (1982) used a projection pursuit construction to derive multivariate location and dispersion estimators that are affinely equivariant and highly robust, having a breakdown point approaching .5. See also Li and Chen (1985), Donoho and Gasko (1992), and Ammann (1993).

In practice, the data are usually centered and sphered before the application of projection pursuit, so that any location and scale effects are removed. For high-dimensional data, it is often useful to remove irrelevant and redundant information. Extracting the largest $q$ principal component axes (with $q$ chosen to account for most of the variation in the data), and then operating on this $q$-dimensional data can improve the ability to discover interesting structure (see, for example, Jee, 1987; Friedman, 1987; and Scott, 1992, Sect. 7.3). Note that principal components analysis is itself a form of projection pursuit, with the projection index being the proportion of total variance accounted for by the projected data.

The 1993 automobile data are part of a larger set given in Lock (1993).

Huber (1985) and O'Sullivan and Pawitan (1993) discussed computerized tomography, another approach to multivariate density estimation that is based on using low- (one-) dimensional density smoothers. O'Sullivan and Pawitan described implementation of the estimator (including on a parallel computer) using penalized B-spline estimates and showed that the estimator achieves the same MISE convergence rate as kernel and spline estimators.

## Computational issues

S–PLUS code to construct a bivariate histogram is available via anonymous `ftp` at the address `ftp.stat.rice.edu` and can be found in the directory `pub/scottdw/Multi.Den.Est/hist.2d`. Exponent Graphics, S–PLUS, STATGRAPHICS PLUS, and STATISTICA also provide bivariate histograms.

Two-dimensional kernel estimation is available in Systat, XploRe, and JMP (the latter package includes a slider to control the amount of smoothing). The collection `kde2d` in the `S` directory of `statlib` contains S–PLUS functions that implement bivariate kernel estimation, also allowing sphering of the data. The collection `ash` in the `S` directory of `statlib` contains S–PLUS functions that calculate average shifted histogram (WARPing) estimates up to 10 dimensions, and includes functions to represent three-dimensional contour volumes graphically. The author of Wand (1994) has made available S–PLUS and Fortran code to calculate two- and three-dimensional estimates using the binning methods discussed in the paper.

Fortran code to calculate the marginal/conditional density estimate of Simonoff (1995b) can be obtained using a World Wide Web browser at `http://www.stern.nyu.edu/~jsimonof/bivar.f`.

C code (and S–PLUS interfaces) to calculate multivariate local log polynomial (up to quadratic) estimates is available using a WWW browser at the URL `http://cm.bell-labs.com/stat/project/locfit`.

RKPACK–II, a collection of Ratfor routines for multivariate penalized likelihood density estimation, is available using a WWW browser at the URL `http://www.stat.purdue.edu/~chong/software.html`.

The collection `projpurs` in the `general` directory of `statlib` contains Fortran code to implement the exploratory projection pursuit algorithm described in Friedman (1987). The collection `cook-b-c` in the `jcgs` directory of `statlib` contains S–PLUS code to calculate one- and two-dimensional Hermite, Legendre, and natural Hermite projection indices for a given projection, one-dimensional indices using kernel methods based on entropy and the index of Friedman and Tukey (1974), and several auxiliary plotting and utility routines. XploRe provides a similar range of projection pursuit approaches. XGobi, which is available as the collection `xgobi` in the `general` directory of `statlib`, also includes projection pursuit functionality, with the ability to use projection pursuit to guide "tours" through the data (Cook *et al.*, 1995). Fortran and S–PLUS code to implement the three-dimensional projection pursuit method of Nason (1995) is available by anonymous `ftp` at `ftp.stats.bris.ac.uk` in the directory `/pub/software/pp3`, in the file `pp3.shar.gz`.

O'Sullivan and Pawitan (1993) have made available Fortran and S–PLUS code to calculate the computerized tomography multivariate density estimate.

## Exercises

**Exercise 4.1.** Construct bivariate histograms, using whatever bin widths seem appropriate, for the earthquake distribution data (Fig. 4.7) and Swiss bank note data (Fig. 4.12). Do the histograms bring out the structure in the data? Now try linear blend frequency polygons, representing the density estimate using a contour plot. Are the frequency polygons better at illuminating the structure? Are they as useful as smoother estimates?

**Exercise 4.2.** Construct a variability plot for the kernel estimate of the earthquake distribution data (Fig. 4.7), as was done for the basketball data in Fig. 4.6. Does the plot support the qualitative impressions in Fig. 4.7?

**Exercise 4.3.** Construct a bivariate kernel density estimate for the earthquake distribution data, for events occurring at zero depth. Is the distribution different from those given in Fig. 4.9? Does the plug-in bandwidth matrix choice give an appealing estimate in this case? What about the Gaussian reference rule?

**Exercise 4.4.** Construct bivariate kernel estimates using the sphering choice of bandwidth matrix $H = hS^{1/2}$ for the basketball data, earthquake distribution data, and Swiss bank note data. Are the resultant estimates similar to those on unsphered data using the plug-in bandwidth choice?

**Exercise 4.5.** Construct a three-dimensional kernel estimate of the earthquake data (as in Fig. 4.9), transforming the focal depth variable using a Johnson family transformation rather than a logarithmic one. Are slices at 50, 200, 400, and 600 km similar to those in the figure? What does a slice at zero depth look like? How does it compare to the estimate constructed in Exercise 4.3?

**Exercise 4.6.** Construct bivariate kernel estimates for the Swiss bank note data, separating the real from the forged bills, using the plug-in choice of bandwidth for each estimate. Do the two estimates reinforce the impression of five modes in the (joint) data set?

**Exercise 4.7.** Construct marginal/conditional estimates for the two parts of the Swiss bank note data set. Do they identify the underlying structure? Is it helpful to use these estimates as pilots in constructing variable kernel estimates? Does local likelihood estimation give better estimates?

**Exercise 4.8.** Construct pseudo-independent estimates of the densities for the basketball data, earthquake distribution data, and Swiss bank note data using any of the nonkernel univariate density estimates described in Chapter 3 (such as local likelihood, semiparametric, P-spline, logistic spline or logspline estimates). How do the estimates compare to those of Exercise 4.4? How do they compare to the bivariate kernel estimates in Figs. 4.5, 4.7, and 4.12, respectively?

**Exercise 4.9.** Construct the first two principal components for the 1993 auto data, and plot the corresponding component scores, along with a superimposed bivariate density estimate of your choice. Do the principal components highlight the same structure as exploratory projection pursuit does? Which method seems to do a better job?

**Exercise 4.10.** Perform a projection pursuit analysis on the four-dimensional (longitude, latitude, focal depth, and magnitude) earthquake data. Do you discover anything new? Do the previously noted patterns come through? Is the analysis more effective if any variables are transformed first?

**Exercise 4.11.** Perform a projection pursuit analysis on the 1993 auto data using an index different from the Legendre index. Are the results the same, or are different patterns highlighted?

Chapter 5

# Nonparametric Regression

## 5.1 Scatter Plot Smoothing and Kernel Regression

The most widely used general statistical procedure is (linear) regression. Regression models are powerful tools for modeling a target variable $y$ as a function of a set of predictors $\mathbf{x}$, allowing prediction for future values of $y$ and the construction of tests and interval estimates for predictions and parameters.

Regression models are also susceptible to the same problems as any other parametric model. Consider the simple linear regression model,

$$y_i = \beta_0 + \beta_1 x_i + \epsilon_i, \qquad i = 1, \ldots, n, \tag{5.1}$$

with the errors $\epsilon$ usually taken to be independent and identically (roughly Gaussian) distributed with zero mean and variance $\sigma^2$. If this model is a good representation of reality, least squares estimates of $\boldsymbol{\beta}$ can be calculated, and inference and prediction follow.

Figure 5.1 is a scatter plot of the eruption duration (horizontal axis) and following eruption time interval (vertical axis) of 222 eruptions of the "Old Faithful Geyser" in Yellowstone National Park. It is apparent that a positive association exists between these variables, with a longer interval until the next eruption following longer eruptions. The National Park Service would like to predict the time until the next eruption, so that tourists can be sure to see it; the superimposed least squares regression line can be used for that purpose.

What if the linear model (5.1) is not appropriate? Fitting a linear model to a nonlinear relationship can give results that are worse than useless, implying a degree of certainty that is not realistic. A more general alternative to (5.1) is the nonparametric regression model

$$y_i = m(x_i) + \epsilon_i. \tag{5.2}$$

The regression curve $m(x)$ is the conditional expectation $m(x) = E(Y|X = x)$, with $E(\epsilon|X = x) = 0$, and $V(\epsilon|X = x) = \sigma^2(x)$ not necessarily constant.

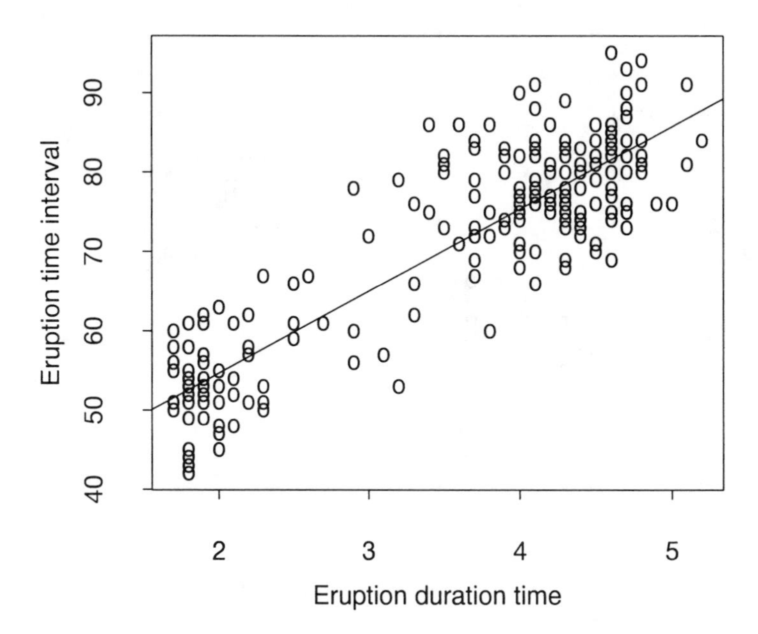


**Fig. 5.1.** Scatter plot of "Old Faithful Geyser" eruption time interval versus previous eruption duration, with least squares linear regression line superimposed.

The model (5.2) removes the parametric restrictions on $m(x)$ and allows (perhaps unexpected) alternative structure to come through. An estimate of $m(x)$ superimposed on the scatter plot (a so-called scatter plot smoother) thus can be a highly effective way to check the appropriateness of the model. The multivariate density estimation results of Chapter 4 provide guidance on how to estimate $m$. By definition,

$$\begin{aligned} m(x) &= E(Y|X=x) \\ &= \int y f(y|x) dy \\ &= \int y \frac{f(x,y)}{f_X(x)} dy, \end{aligned} \tag{5.3}$$

where $f_X(x)$, $f(x,y)$, and $f(y|x)$ are the marginal density of $X$, the joint density of $X$ and $Y$, and the conditional density of $Y$ given $X$, respectively. A product kernel estimate of $f(x,y)$ is

$$\hat{f}(x,y) = \frac{1}{nh_x h_y} \sum_{i=1}^{n} K_x\left(\frac{x-x_i}{h_x}\right) K_y\left(\frac{y-y_i}{h_y}\right),$$

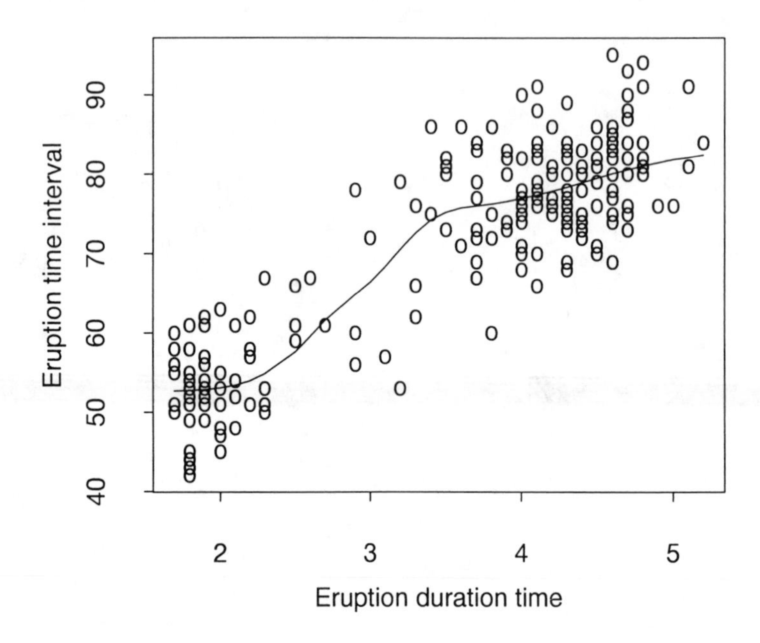


**Fig. 5.2.** Scatter plot of "Old Faithful" data, with Nadaraya–Watson kernel estimate superimposed.

while a kernel estimate of $f_X(x)$ is

$$\hat{f}_X(x) = \frac{1}{nh_x}\sum_{i=1}^{n} K_x\left(\frac{x - x_i}{h_x}\right).$$

Substituting into (5.3), and noting that $\int K_y(u)du = 1$ and $\int uK_y(u)du = 0$, yields the *Nadaraya–Watson kernel estimator*,

$$\hat{m}_{\mathrm{NW}}(x) = \frac{\sum_{i=1}^{n} K\left(\frac{x-x_i}{h}\right) y_i}{\sum_{i=1}^{n} K\left(\frac{x-x_i}{h}\right)} \equiv \sum_{i=1}^{n} w_i y_i,$$

a linear function of $\mathbf{y}$ with weights

$$w_i = (nh)^{-1}\frac{K\left(\frac{x-x_i}{h}\right)}{\hat{f}_X(x)}$$

(this estimator is sometimes called an *evaluation kernel estimator*).

Figure 5.2 again gives the "Old Faithful Geyser" eruption data, with a

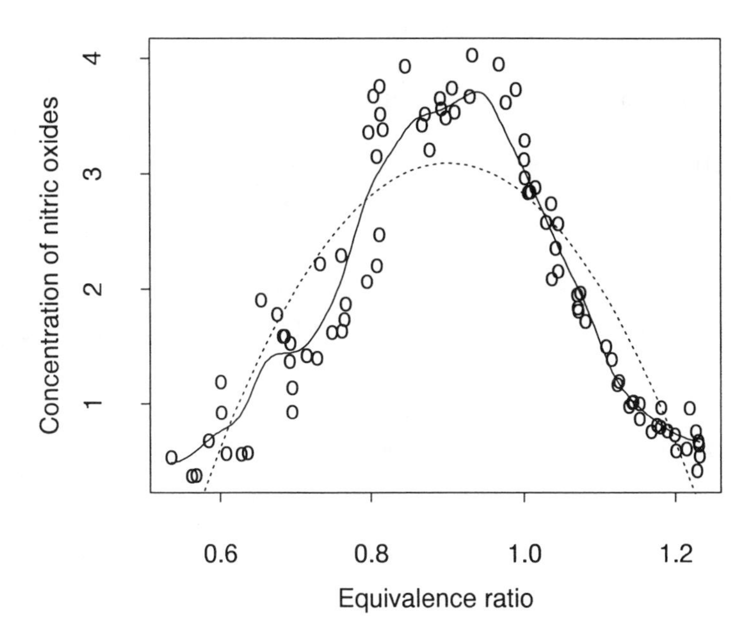


**Fig. 5.3.** Scatter plot of nitric oxide levels versus equivalence ratio, with Nadaraya–Watson kernel estimate (solid line) and quadratic least squares regression line (dashed line) superimposed.

Nadaraya–Watson kernel estimate superimposed (using a Gaussian kernel, with $h = .25$). The curve shows that a reasonable alternative model to the simple regression line of Fig. 5.1 is to treat the two high density regions separately (shorter eruption duration followed by shorter eruption time interval, and longer eruption duration followed by longer eruption interval, respectively), with the average eruption interval for the former group being around 50 minutes and that for the latter group being 75 – 80 minutes.

Figure 5.3 illustrates another data set where the smooth curve argues against a simple parametric model. The data relate the concentration of nitric oxides in engine exhaust to the equivalence ratio, a measure of the richness of the air/ethanol mix, for a burning of ethanol in a single-cylinder automobile test engine. The curve shows an increase in nitric oxides up to an equivalence ratio of about 1, followed by a steady decrease. This pattern might suggest a parabolic (quadratic) relationship, but the superimposed quadratic least squares fit shows that this is not the case.

The Nadaraya–Watson kernel estimator is most natural for data using a random design, as in (5.3) (that is, when the design is a random sample from some distribution having density $f_X$). If $f_X$ is known *a priori*, an obvious alternative weighting is

$$w_i = (nh)^{-1} \frac{K\left(\frac{x-x_i}{h}\right)}{f_X(x)}. \tag{5.4}$$

If the design is not random, but is rather a fixed set of ordered nonrandom numbers $x_1, \ldots, x_n$, the intuition of (5.3) is lost, and a different form of kernel estimator could be considered. One approach is to estimate a "density function" at the design points using the spacings between the design points,

$$\hat{f}_X(x_i) = \frac{1}{n(x_i - x_{i-1})}.$$

Substituting this into the weights $w_i$ (changing from $f_X(x)$ to $f_X(x_i)$) yields the *Priestley–Chao kernel estimator*,

$$\hat{m}_{\text{PC}}(x) = h^{-1} \sum_{i=1}^{n} (x_i - x_{i-1}) K\left(\frac{x - x_i}{h}\right) y_i$$

(a closely related version of the estimator, $\hat{m}_{\text{PC1}}$, substitutes $(x_{i+1} - x_{i-1})/2$ for $x_i - x_{i-1}$). Another estimator intended for the fixed design case is the *Gasser–Müller kernel estimator*,

$$\hat{m}_{\text{GM}}(x) = h^{-1} \sum_{i=1}^{n} \left[ \int_{s_{i-1}}^{s_i} K\left(\frac{x - u}{h}\right) du \right] y_i,$$

where $x_{i-1} \le s_{i-1} \le x_i$ (a common choice being $s_{i-1} = (x_{i-1} + x_i)/2$, with $s_0$ and $s_n$ being the upper and lower limits of the range of $x$, respectively). Estimators of this type are often called *convolution estimators.*

## 5.2 Local Polynomial Regression

### 5.2.1 Local polynomial estimation

Basic calculus shows that $\hat{m}_{\text{NW}}$ is the solution to a natural weighted least squares problem, being the minimizer $\hat{\beta}_0$ of

$$\sum_{i=1}^{n} (y_i - \beta_0)^2 K\left(\frac{x - x_i}{h}\right).$$

That is, $\hat{m}_{\text{NW}}$ corresponds to locally approximating $m(x)$ with a constant, weighting values of $y$ corresponding to $x_i$s closer to $x$ more heavily.

This suggests fitting higher order local polynomials, since a local constant usually makes sense only over a very small neighborhood. Such a ($p$th order) *local polynomial regression estimator* is the minimizer of

$$\sum_{i=1}^{n}[y_i - \beta_0 - \cdots - \beta_p(x - x_i)^p]^2 K\left(\frac{x - x_i}{h}\right).$$

Let $X_x$ be the design matrix

$$\begin{pmatrix} 1 & x - x_1 & & (x - x_1)^p \\ \vdots & \vdots & \cdots & \vdots \\ 1 & x - x_n & & (x - x_n)^p \end{pmatrix}$$

(the estimator can be written using terms of the form $x_i - x$ rather than $x - x_i$, but this does not affect any of its properties, and using $x - x_i$ allows connections with kernel estimation to be more apparent), and let

$$W_x = h^{-1} \operatorname{diag}\left[K\left(\frac{x - x_1}{h}\right), \ldots, K\left(\frac{x - x_n}{h}\right)\right]$$

be the weight matrix; then, if $X_x' W_x X_x$ is invertible,

$$\hat{\boldsymbol{\beta}} = (X_x' W_x X_x)^{-1} X_x' W_x \mathbf{y}.$$

The estimator $\hat{m}_p(x)$ is then the intercept term $\hat{\beta}_0$, or (in matrix notation)

$$\hat{m}_p(x) = \mathbf{e}_1'(X_x' W_x X_x)^{-1} X_x' W_x \mathbf{y} \equiv S_x \mathbf{y}, \tag{5.5}$$

where $\mathbf{e}_r$ is the $(p+1) \times 1$ vector having the value 1 in the $r$th entry and zero elsewhere and $S_x = \mathbf{e}_1'(X_x' W_x X_x)^{-1} X_x' W_x$. More generally, $q! \times \hat{\beta}_q = q! \mathbf{e}_{q+1}'(X_x' W_x X_x)^{-1} X_x' W_x \mathbf{y}$ is an estimate of the $q$th derivative of $m(x)$, $m^{(q)}(x)$. The local linear ($p = 1$) estimator of $m(x)$ can be written as

$$\hat{m}_1(x) = \frac{1}{nh} \sum_{i=1}^{n} \frac{[\hat{s}_2(x,h) - \hat{s}_1(x,h)(x - x_i)]K[(x - x_i)/h]y_i}{\hat{s}_2(x,h)\hat{s}_0(x,h) - \hat{s}_1(x,h)^2}, \tag{5.6}$$

where

$$\hat{s}_r(x,h) = \frac{1}{nh} \sum_{i=1}^{n} (x - x_i)^r K\left(\frac{x - x_i}{h}\right).$$

### 5.2.2 Properties of local polynomial estimators in the interior

The distinction between the fixed and random design cases complicates the asymptotic analysis of $\hat{m}_p$. Problems occur because the matrix $X_x' W_x X_x$ in (5.5) might not be invertible (at least $p+1$ different points with positive weight are required). Taking $h$ large enough can guarantee invertibility for fixed designs, but for a random design, bounded kernel function $K$, and any fixed $h$, there is positive probability that invertibility does not hold. Conditioning on the observed predictor values makes the analysis similar to the fixed design case and hence easier (results for fixed designs take

$x_i = F^{-1}(i/n)$, with $F$ the cumulative function of the "density" $f$ of the design).

Consider a point $x$ away from the boundary region (boundary effects are discussed in the next subsection). Then, assuming appropriate smoothness of $m$ and $f$, the asymptotic conditional bias of $\hat{m}_p(x)$ is

$$E(\hat{m}_p(x) - m(x)|x_1, \ldots, x_n) = \frac{h^{p+1} m^{(p+1)}(x) \mu_{p+1}(K_{(p)})}{(p+1)!} + o_p(h^{p+1}) \quad (5.7a)$$

if $p$ is odd, and

$$E(\hat{m}_p(x) - m(x)|x_1, \ldots, x_n) = h^{p+2} \left[ \frac{m^{(p+1)}(x) f'_X(x)}{f_X(x)(p+1)!} + \frac{m^{(p+2)}(x)}{(p+2)!} \right] \mu_{p+2}(K_{(p)}) + o_p(h^{p+2}) \quad (5.7b)$$

if $p$ is even, where

$$\mu_q(K_{(p)}) = \int u^q K_{(p)}(u) du$$

and $K_{(p)}$ is a $(p+1)$th order kernel function when $p$ is odd and a $(p+2)$th order kernel function when $p$ is even. These kernels are related to the generalized jackknifing higher order kernels described in Chapter 3 (as in (3.25)). The conditional variance equals

$$V(\hat{m}_p(x)|x_1, \ldots, x_n) = \frac{R(K_{(p)}) \sigma^2(x)}{nh f_X(x)} + o_p[(nh)^{-1}]. \quad (5.8)$$

Thus, the conditional MSE equals

$$\text{MSE}(\hat{m}_p(x)|x_1, \ldots, x_n) = \left[ \frac{h^{p+1} m^{(p+1)}(x) \mu_{p+1}(K_{(p)})}{(p+1)!} \right]^2 + \frac{R(K_{(p)}) \sigma^2(x)}{nh f_X(x)} + o_p[h^{2p+2} + (nh)^{-1}] \quad (5.9)$$

if $p$ is odd, and

$$\text{MSE}(\hat{m}_p(x)|x_1, \ldots, x_n) = h^{2p+4} \left[ \frac{m^{(p+1)}(x) f'_X(x)}{f_X(x)(p+1)!} + \frac{m^{(p+2)}(x)}{(p+2)!} \right]^2 \mu_{p+2}(K_{(p)})^2 + \frac{R(K_{(p)}) \sigma^2(x)}{nh f_X(x)} + o_p[h^{2p+4} + (nh)^{-1}]$$

if $p$ is even.

Several implications can be drawn from (5.7) and (5.8). First, the degree of the polynomial being fit determines the order of the bias of $\hat{m}_p$, with polynomials of adjacent pairs of degree being conceptually similar. For example, local constant ($p = 0$) and local linear ($p = 1$) estimation both yield $O_p(h^2)$ bias, local quadratic ($p = 2$) and local cubic ($p = 3$) estimation both yield $O_p(h^4)$ bias, and so on.

Given this, odd degree polynomials have a simpler asymptotic bias expression, similar to that of kernel density estimators (substituting $m$ for $f$). The bias of even degree local polynomial estimators depends on the design density, while that of odd degree estimators does not (odd degree estimators can be viewed as "design adaptive" in that sense).

Equations (5.7) and (5.8) determine the asymptotic rate for $h$ that will minimize the conditional MSE. For $p$ odd, $h = O(n^{-1/(2p+3)})$, yielding MSE $= O_p(n^{-(2p+2)/(2p+3)})$, while for $p$ even, $h = O(n^{-1/(2p+5)})$, yielding MSE $= O_p(n^{-(2p+4)/(2p+5)})$. So, as expected, for $p = 0$ or 1, the optimal MSE $= O_p(n^{-4/5})$; for $p = 2$ or 3, the optimal MSE $= O_p(n^{-8/9})$, and so on.

In an asymptotic sense, local linear estimation combines the good properties of other kernel estimators while avoiding their faults. While $\hat{m}_{\text{PC}}$, $\hat{m}_{\text{PC1}}$, and $\hat{m}_{\text{GM}}$ have the simple bias expansion as in (5.7$a$),

$$\text{Bias} = \frac{h^2 m''(x)\mu_2(K)}{2} + o_p(h^2),$$

$\hat{m}_{\text{NW}}$ does not (as it corresponds to $p = 0$ in (5.7$b$)). On the other hand, the variance of $\hat{m}_{\text{PC1}}$ and $\hat{m}_{\text{GM}}$ for random designs equals

$$\text{Variance} = \frac{3}{2} \times \frac{R(K)\sigma^2(x)}{nhf_X(x)} + o_p[(nh)^{-1}];$$

that is, 1.5 times the local linear value from (5.8) (also attained by $\hat{m}_{\text{NW}}$). For fixed designs, local polynomial and convolution estimators are asymptotically equivalent. Thus, $\hat{m}_1$ combines the simple bias of the convolution kernel estimators with the smaller random design variance of the evaluation kernel estimator.

Practical application of local polynomial estimators, however, reinforces the potential dangers of trusting the implications of (5.7) and (5.8) uncritically. For example, while the quadratic (Epanechnikov) kernel is asymptotically optimal (with respect to conditional MSE), its use leads to estimators with infinite unconditional (finite sample) variance. This is because of the kernel's boundedness, and the possibility of sparse regions in the design, and can translate into roughness of the estimate.

Using an unbounded kernel, such as the Gaussian, corrects this problem, yielding local linear estimators with finite conditional and unconditional variance. Another way to achieve this is to require at least four points in the interval covered by the kernel, such as by using a bandwidth based on nearest neighbors. Local linear and quadratic estimators with bandwidth based on nearest neighbors are often called *loess* (or the older term *lowess*) estimators.

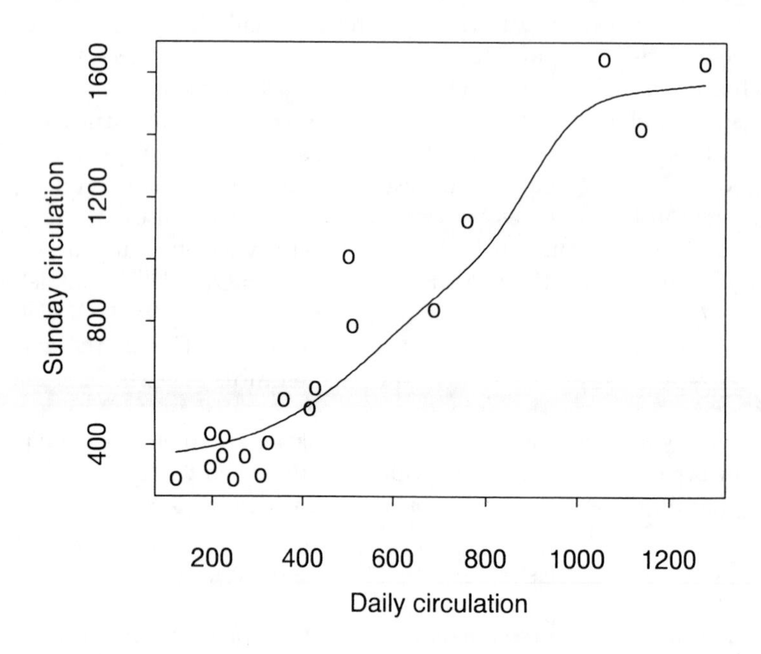


**Fig. 5.4.** Scatter plot of Sunday circulation versus daily circulation for 19 newspapers, with Nadaraya–Watson kernel estimate superimposed.

### 5.2.3 Properties of local polynomial estimators near the boundary

As was true in the density estimation context (Section 3.2.1), one of the most serious problems when using kernel regression estimators is boundary bias, because of the asymmetric contribution of observations to the kernel summation near the boundary. Figure 5.4 gives a scatter plot of the Sunday circulation versus the daily circulation (in thousands) for 19 newspapers, with $\hat{m}_{\text{NW}}$ superimposed (using a Gaussian kernel and $h = 150$). Boundary bias is evident here, as the curve "flattens out" at the left boundary, with most of the observations falling below the curve for daily circulation less than 400,000 and then lying above it for daily circulation between 400,000 and 800,000.

Local polynomial estimation can automatically provide boundary bias correction. By fitting local polynomials at values in the boundary region, the estimator does not "flatten out" because of the lack of available data past the boundary the way the kernel estimator (which fits a local constant) does.

Consider local linear estimation, and assume (without loss of generality) that the support of $f_X$ is $[0,1]$. Then, for left boundary points $x = ch$, $c \geq 0$ (for kernels supported on $[-1,1]$, $c \leq 1$), the conditional bias of $\hat{m}_1$ is

$$E(\hat{m}_1(x) - m(x)|x_1, \ldots, x_n) = \frac{\alpha_K(c)m''(x)h^2}{2} + o_p(h^2),$$

where

$$\alpha_K(c) = \frac{s_{2,c}^2 - s_{3,c}s_{1,c}}{s_{2,c}s_{0,c} - s_{1,c}^2}$$

and $s_{\ell,c} = \int_{-\infty}^{c} u^\ell K(u)du$, $\ell = 0,1,2,3$. This is exactly analogous to the bias achieved by the generalized jackknifing kernel (3.12) discussed in Chapter 3, as given in (3.22). That is, the local linear estimator is, in this sense, asymptotically an automatic boundary kernel estimator. In contrast, the conditional bias of $\hat{m}_{\text{NW}}$ satisfies

$$E(\hat{m}_{\text{NW}}(x) - m(x)|x_1, \ldots, x_n) = -\frac{m'(x)s_{1,c}h}{s_{0,c}} + o_p(h),$$

which is only $O_p(h)$ unless $m'(x) = 0$, rather than $O_p(h^2)$.

The practical benefits of this boundary bias correction can be seen in Fig. 5.5, where a local linear estimate (with Gaussian kernel and $h = 150$) is superimposed on the scatter plot of Fig. 5.4. The local linear estimate avoids the bias at the left boundary and highlights much more effectively the close-to-linear relationship between daily and Sunday circulation.

Figure 5.6 illustrates bias correction at both boundaries. The plot relates the electricity usage at an all-electric residential home (in kilowatt-hours) to the average daily temperature for 55 months. The local constant ($\hat{m}_{\text{NW}}$) and local linear estimates (using a Gaussian kernel and $h = 9$) are very similar in the interior, but $\hat{m}_{\text{NW}}$ is severely negatively biased at the left boundary and moderately positively biased at the right boundary because of a flattening-out effect. The local linear estimate automatically corrects this and highlights the nonlinear (inverse, as would be expected in this northern North American home) relationship between electricity usage and temperature.

Naturally, bias correction comes with a price — that is, increased variance. The asymptotic conditional variance of $\hat{m}_1$ near the boundary is

$$V(\hat{m}_1(x)|x_1, \ldots, x_n) = \frac{\beta_K(c)\sigma^2(x)}{nhf_X(x)} + o_p[(nh)^{-1}], \tag{5.10}$$

where

$$\beta_K(c) = \frac{\int_{-\infty}^{c}(s_{2,c} - us_{1,c})^2 K^2(u)du}{(s_{2,c}s_{0,c} - s_{1,c}^2)^2}.$$

For the Gaussian kernel, for example, the asymptotic conditional variance of $\hat{m}_1$ is about 3.17 times that of $\hat{m}_{\text{NW}}$ at the boundary if the same bandwidth is used; for the biweight kernel, it is about 3.58 times that of $\hat{m}_{\text{NW}}$.

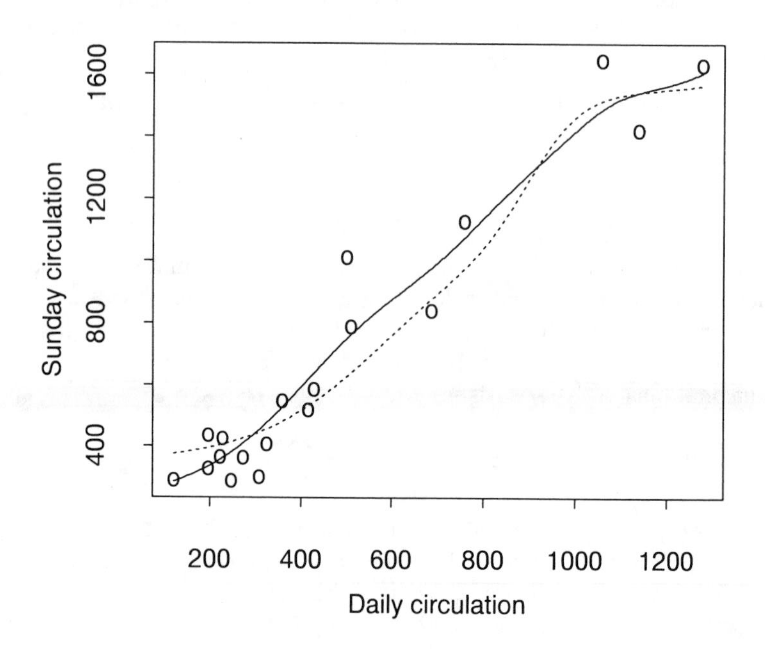


**Fig. 5.5.** Scatter plot of newspaper circulation data, with local linear estimate (solid line) and Nadaraya–Watson kernel estimate (dashed line) superimposed.

Figure 5.7 is a variability plot of the local linear estimator for the electricity usage data. The plot is constructed by repeatedly sampling with replacement from the data and fitting the local linear estimate to the resamples. The dashed lines in the figure are the upper and lower pointwise 2.5% points for 200 resamples. Note that the envelope does not represent a 95% confidence region for $m(x)$, since its construction ignores the bias of $\hat{m}_1$. The envelope is narrow, suggesting a good deal of stability in the estimate, but it noticeably widens at the boundaries. Equation (5.10) provides support for this pattern, since the asymptotic conditional variance is more than six times greater at the boundary than it is in the interior for the Gaussian kernel, assuming the same values of $h$ and $f_X$.

The automatic boundary bias correction of $\hat{m}_p$ for $p = 1$ versus $p = 0$ persists for higher order polynomials, with odd values of $p$ asymptotically improving on the boundary bias of even values. So, for example, while local quadratic and local cubic estimators both achieve $O_p(h^4)$ conditional bias in the interior, local quadratic estimators have $O_p(h^3)$ conditional bias in the boundary region.

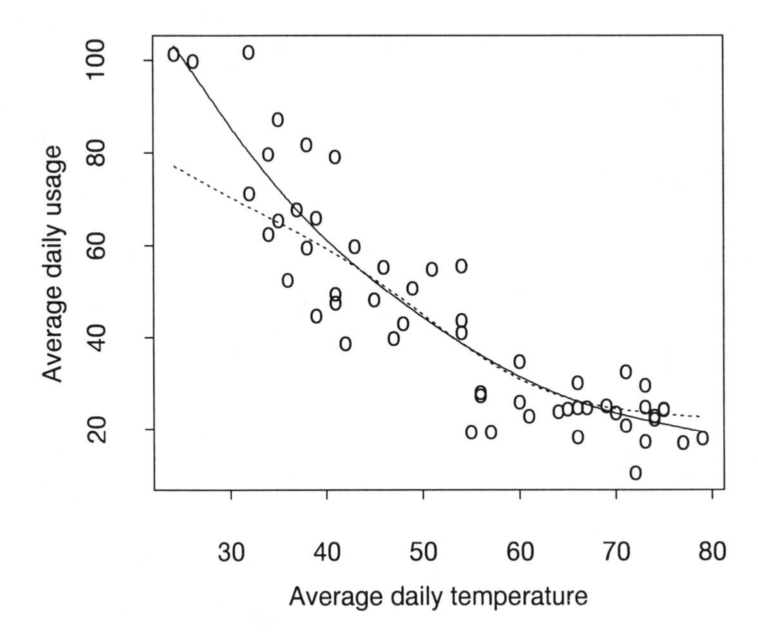


**Fig. 5.6.** Scatter plot of electricity usage versus temperature in an all-electric residence, with local linear estimate (solid line) and Nadaraya–Watson kernel estimate (dashed line) superimposed.

### 5.2.4 Choosing the degree of the polynomial fit

The results of Sections 5.2.1 and 5.2.2 provide some guidance on how to choose the degree of polynomial $p$ being fit when constructing $\hat{m}_p$. If the regression curve $m(x)$ changes rapidly, $|m''(x)|$ will be large, and the $O_p(h^2)$ conditional bias term in (5.7$a$) for $p = 1$ can be large (that is, the local linear estimate can flatten out sharp peaks and troughs). In that circumstance, removing the $O_p(h^2)$ bias term by going to $p = 2$ or $p = 3$ can be beneficial, at least asymptotically. This is the same argument as was used in Chapter 3 regarding higher order kernels and density estimation, of course (since higher degree local polynomial fitting is asymptotically equivalent to higher order kernel fitting), where it was argued that the benefits of using such kernels were minor.

Experience with regression models appears to be more positive, however. Consider, for example, Fig. 5.8, which refers to the nitric oxide production data examined in Fig. 5.3. The dashed line is a local linear estimate using a Gaussian kernel and $h = .0253$, while the solid line is a local cubic

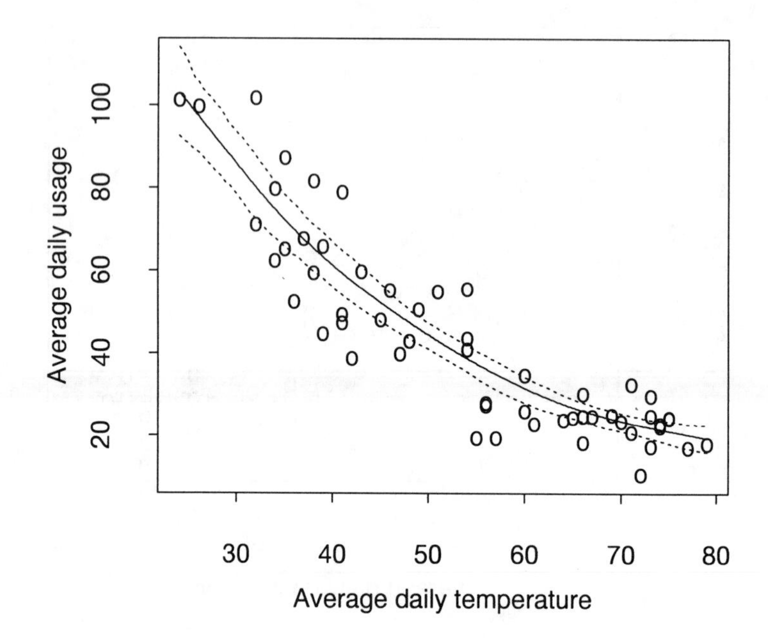


**Fig. 5.7.** Variability plot of local linear estimate for electricity usage data.

estimate with $h = .04$. Despite the wider bandwidth, the local cubic estimate seems to follow the contours of the regression curve better, particularly around the center of the plot, where it is changing most rapidly.

Figures 5.9 and 5.10 also show the possible advantages of using higher order polynomials. The plots refer to data from a vineyard on a small island in Lake Erie. The vineyard is divided into 52 rows, and the 52 observations in the data set correspond to the sum of the yields of the harvests in 1989, 1990, and 1991 separated by row, as measured by the total number of lugs (a *lug* is a basket that is used to carry the harvested grapes, which holds roughly 30 pounds of grapes). The row numbers are naturally ordered, with increasing row number reflecting movement from northwest to southeast.

Figure 5.9 gives the data, with two local linear estimates (each using a Gaussian kernel) based on $h = 3$ and $h = 1.5$, respectively. Both curves show the general pattern of the yield being higher in the middle of the vineyard than at the edges. This is because of weather (wind) and animal (birds and raccoons) damage at the outer, exposed, parts of the vineyard. Rows 31–52 are shorter than rows 1–30 (100 yards long versus 120 yards), which accounts for the steeper drop in yield at the right side of the plot compared with the left side.

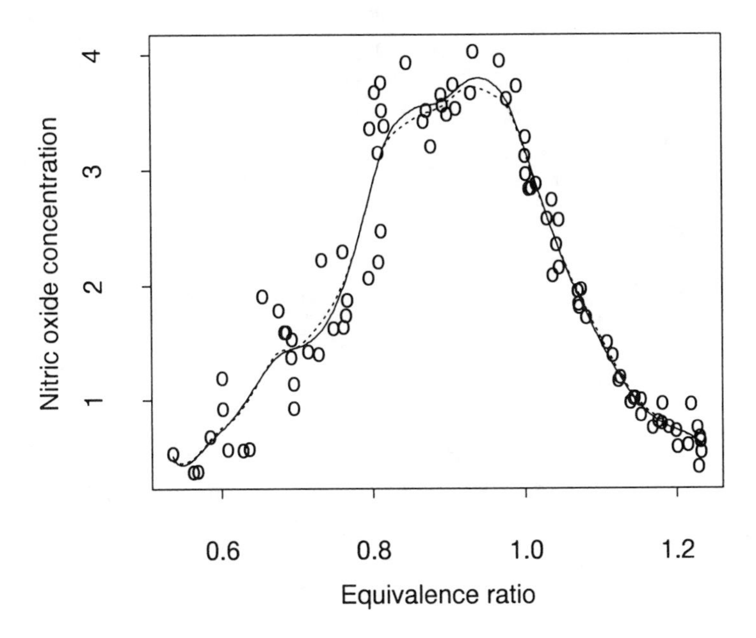


**Fig. 5.8.** Scatter plot of ethanol data, with local linear (dashed line) and local cubic (solid line) estimates superimposed.

The local linear estimate using $h = 3$ is very reasonable for rows 10–30 and 40–52 but seems to be oversmoothed for the other rows, exhibiting noticeable negative bias at the left boundary and pronounced positive bias at the dip in yield in rows 30–40 (there is a farmhouse directly opposite those rows, which could possibly account for this dip). The local linear estimate with $h = 1.5$ picks up the left boundary and the dip quite nicely but is undersmoothed for the other rows, with many spurious bumps in the estimated curve.

Figure 5.10 gives a local cubic estimate ($h = 3$) that apparently addresses the difficulties of the local linear estimate. The curve is pleasingly smooth, without spurious bumpiness, yet picks up the structure well, including the dip in rows 30–40.

Asymptotically, the choice between even and odd $p$ is easy — odd values (1 versus 0, 3 versus 2, and so on) have a clear advantage. The conditional bias for odd values of $p$ is simpler than that for even values and does not depend on the design $f_X$. Even more importantly, local $p$th degree polynomial estimators have $O_p(h^{p+1})$ conditional bias in the boundary region, which is higher order for the odd member of the pair, apparently

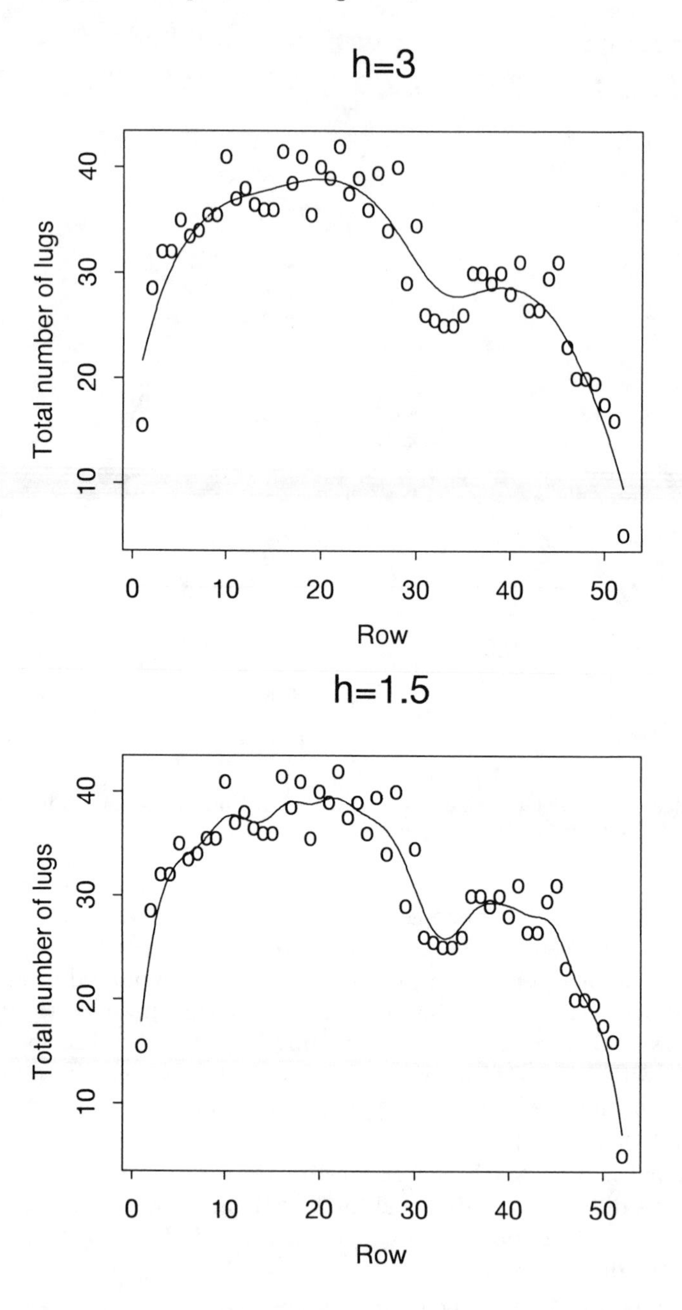


**Fig. 5.9.** Scatter plots of total lug counts versus row for three vineyard harvests, with local linear estimates superimposed, using $h = 3$ and $h = 1.5$, respectively.

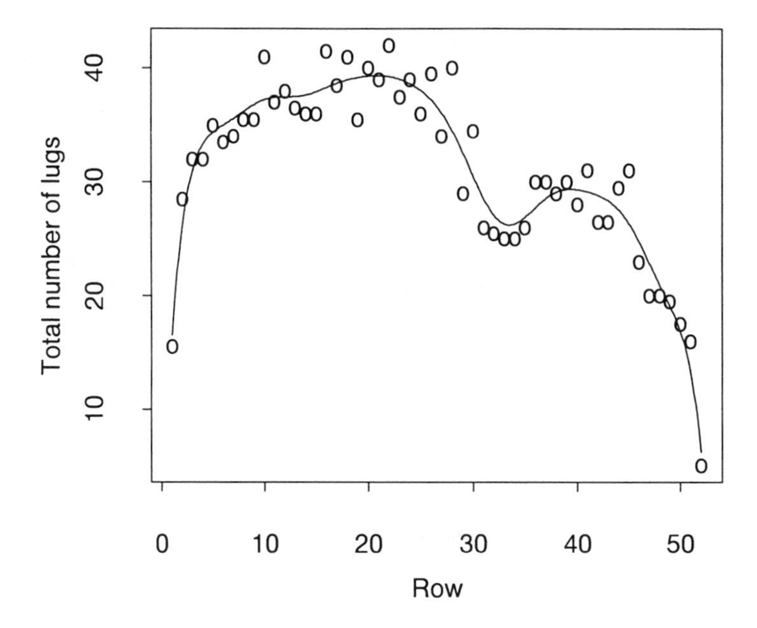


**Fig. 5.10.** Scatter plot of vineyard data, with local cubic estimate superimposed.

deciding the matter convincingly.

As with all asymptotic arguments for smoothing methods, however, things aren't quite that straightforward for finite samples. Figure 5.11 gives variability plots for the electricity usage data for local quadratic and local cubic estimates (each using a Gaussian kernel and $h = 3$) that are directly comparable to Fig. 5.7. The fitted regression estimates themselves are virtually identical (despite potential gains in bias from going to higher $p$), but the increased variability in the boundaries associated with higher $p$ is obvious from the variability envelopes (particularly at the left boundary). There is little reason to go past $p = 1$ for these data, as the worse variance properties for $p = 3$ versus $p = 2$ can actually overwhelm the better boundary bias properties.

Figure 5.12 is a variability plot for the local cubic estimate given in Fig. 5.10 for the vineyard data. The narrow envelope for most of the range of the data argues in favor of using $p = 3$, but even here, the marked increase in variability at the boundary that comes with higher order bias correction should be recognized and taken into account in any discussion of the implications of the fitted curve.

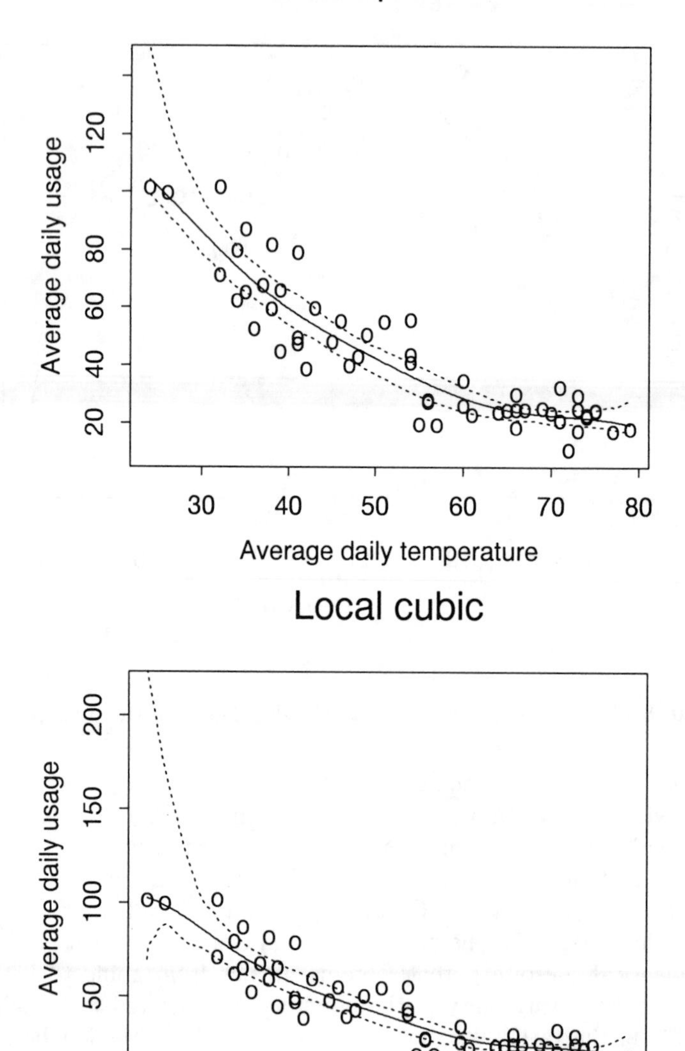


**Fig. 5.11.** Variability plots of local quadratic and local cubic estimates for electricity usage data.

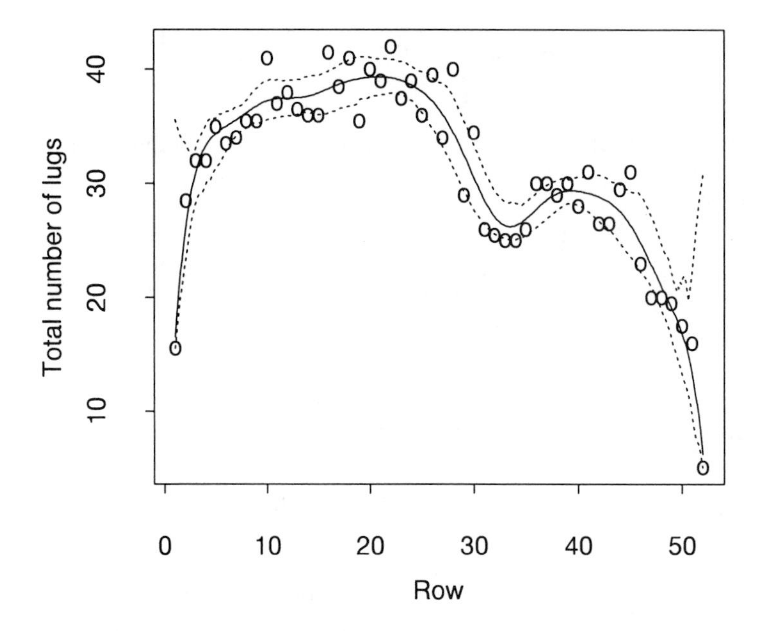


**Fig. 5.12.** Variability plot of local cubic estimate for vineyard data.

## 5.3 Bandwidth Selection

The appearance of a local polynomial regression estimate depends strongly on the bandwidth $h$, and an automatic choice is desirable (further tuning of the bandwidth by the data analyst is often beneficial as well). Each of the bandwidth selectors described in Chapter 3 has an analogous formulation in the regression context and often exhibits similar behavior.

Assume that the support of $f_X$ is $[0, 1]$ and that the errors are homoscedastic with variance $\sigma^2$ (nonconstant variance suggests varying the bandwidth with $x$, which will be discussed in the next section). The weighted conditional MISE of $\hat{m}_p$ is a global measure of accuracy of $\hat{m}_p$ and equals

$$\begin{aligned}\text{MISE}(\hat{m}_p|x_1,\ldots,x_n) &\equiv E[\text{ISE}(\hat{m}_p|x_1,\ldots,x_n)]\\ &= E\left\{\int [\hat{m}_p(u) - m(u)]^2 f_X(u)du|x_1,\ldots,x_n\right\}. \qquad (5.11)\end{aligned}$$

The weighting by $f_X$ puts more emphasis on accuracy in regions with more data, since the mean averaged squared error (mean ASE, or MASE)

$$\text{MASE} = E\left\{n^{-1}\sum_{i=1}^{n}[\hat{m}_p(x_i) - m(x_i)]^2\right\}$$

is a discrete approximation to (5.11). For $p$ odd, (5.9) implies that

$$\begin{aligned}\text{MISE}(\hat{m}_p|x_1,\ldots,x_n) \\ &= \left[\frac{h^{p+1}\mu_{p+1}(K_{(p)})}{(p+1)!}\right]^2 \int m^{(p+1)}(u)^2 f_X(u)du + \frac{R(K_{(p)})\sigma^2}{nh} \\ &+ o_p[h^{2p+2} + (nh)^{-1}]. \qquad (5.12)\end{aligned}$$

Cross-validation proceeds directly, using the idea of "leave-one-out" prediction. The criterion has the form

$$\text{CV}(h) = \frac{1}{n}\sum_{i=1}^{n}[y_i - \hat{m}_p^{(i)}(x_i)]^2,$$

where $\hat{m}_p^{(i)}(x_i)$ is the estimate $\hat{m}_p$ based on the data with $x_i$ removed, evaluated at $x_i$. The cross-validatory choice of $h$ ($\hat{h}_{\text{CV}}$) is the minimizer of CV($h$).

Unfortunately, $\hat{h}_{\text{CV}}$ suffers from the same unappealing characteristics in the regression context as it does in the density estimation context; that is, it is highly variable and tends to undersmooth in practice, yielding curves that are too "wiggly." For local linear estimation, for example, $\hat{h}_{\text{CV}}$ converges very slowly to the minimizers of ASE and MASE, at the relative rate $O_p(n^{-1/10})$. This suggests also trying a plug-in selector for $h$.

The minimizer of the MISE (5.12) is asymptotically

$$h_0 = \left[\frac{(p+1)(p!)^2 R(K_{(p)})\sigma^2}{2n\mu_{p+1}(K_{(p)})^2 \int m^{(p+1)}(u)^2 f_X(u)du}\right]^{1/(2p+3)}, \qquad (5.13)$$

as long as $\int m^{(p+1)}(u)^2 f_X(u)du$ is nonzero. A plug-in choice of $h$ relies on estimating $\sigma^2$ and replacing the integral in the denominator with an estimate. Note that a plug-in selector will fail if the true regression relationship is linear, since then the denominator in (5.13) equals zero (the optimal choice of $h$ for linear $m$ is $h = \infty$, which will give the least squares regression line). Practically speaking, a plug-in method cannot be expected to do very well for near-linear regression relationships, as it will likely undersmooth (the same is true for plug-in selectors for kernel density estimation based on (3.9), but for [more unusual] straight-line densities).

Consider the local linear estimate $\hat{m}_1$. Equation (5.13) then simplifies to

$$h_0 = \left[ \frac{R(K)\sigma^2}{n\mu_2(K)^2 \int m''(u)^2 f_X(u)du} \right]^{1/5}. \tag{5.14}$$

A natural estimate of $\int m''(u)^2 f_X(u)du$ is $n^{-1}\sum_{i=1}^n \widehat{m''(x_i)}^2$, where

$$\widehat{m''(x_i)} = 2\mathbf{e}_3'(\tilde{X}_{x_i}'\tilde{W}_{x_i}\tilde{X}_{x_i})^{-1}\tilde{X}_{x_i}'\tilde{W}_{x_i}\mathbf{y}$$

and $\tilde{X}$ and $\tilde{W}$ are based on a local cubic estimate using an appropriately chosen bandwidth $g$.

A plug-in estimator for $\sigma^2$ is based on a residual sum of squares,

$$\hat{\sigma}^2 = \nu^{-1}\sum_{i=1}^n [y_i - \tilde{m}_1(x_i)]^2, \tag{5.15}$$

where $\tilde{m}_1$ uses an appropriately chosen bandwidth $\lambda$. By analogy with ordinary linear regression, $\nu$ can be thought of as the error degrees of freedom. If $S_{\mathbf{x}}$ is defined to be the matrix with $i$th row $S_{x_i}$, $\nu$ equals

$$\nu = \text{trace}[(I - S_{\mathbf{x}})(I - S_{\mathbf{x}})'] = n - 2\,\text{trace}(S_{\mathbf{x}}) + \text{trace}(S_{\mathbf{x}}S_{\mathbf{x}}'). \tag{5.16}$$

This choice of $\nu$ makes $\hat{\sigma}^2$ conditionally unbiased for $\sigma^2$ if $m$ is linear (further analogy with the usual parametric analysis of variance implies that $2\,\text{trace}(S_{\mathbf{x}}) - \text{trace}(S_{\mathbf{x}}S_{\mathbf{x}}')$ plays the role of the number of "parameters" fit by the smoother). This construction then yields a plug-in choice $\hat{h}$ that satisfies $\hat{h}/h_0 - 1 = O_p(n^{-2/7})$, a considerable improvement over the $O_p(n^{-1/10})$ rate of $\hat{h}_{\text{CV}}$.

Figure 5.8 gave an example of the use of this plug-in selector. The local linear estimate given there (using $\hat{h} = .0253$) uses the plug-in selector and gives a reasonable view of the regression relationship. Figure 5.13 is comparable, presenting a local linear estimate for the vineyard data of Figs. 5.9 and 5.10. The estimate uses the plug-in bandwidth $\hat{h} = 1.8$, a compromise between the two values used in Fig. 5.9, which captures the structure almost as well as the local cubic estimate of Fig. 5.10 (except for some slight bumpiness between rows 10 and 20).

Figure 5.14 presents another data set related to the vineyard. There are three response values for each row, corresponding to the 1989, 1990, and 1991 harvests, which are the differences between the number of lugs for that row and harvest year and the average number of lugs per row for that harvest year. Correction by the average number of lugs per row for each harvest removes the harvest effect and allows the row-to-row variation to come through, while also reflecting year-to-year variation around the annual mean. The plug-in choice for bandwidth is $\hat{h} = 1.8$, and the plot gives the resultant local linear estimate. The pattern is very similar to the one in Figs. 5.10 and 5.13, and the local linear estimate captures it well, with no need to go to higher degree polynomials.

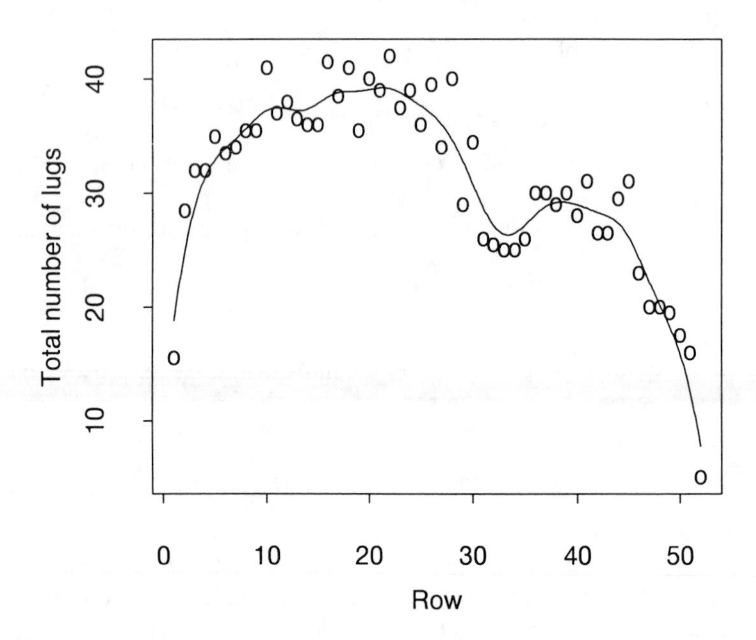


**Fig. 5.13.** Local linear estimate for vineyard data using plug-in bandwidth.

## 5.4 Locally Varying the Bandwidth

The form of the conditional MSE of $\hat{m}_p$ at a given value $x$, as given (for $p$ odd) in (5.9), shows that it is dependent on three factors: the design density $f_X(x)$, the curvature $m^{(p+1)}(x)$, and the local variance $\sigma^2(x)$. Consider, for example, the local linear estimator $\hat{m}_1$. The conditional MSE for this estimator equals

$$\text{MSE}(\hat{m}_1(x)|x_1, \ldots, x_n) = \left[\frac{h^2 m''(x)\mu_2(K)}{2}\right]^2 + \frac{R(K)\sigma^2(x)}{nhf_X(x)} + o_p[h^4 + (nh)^{-1}]. \tag{5.17}$$

The minimizer of the asymptotic conditional MSE is thus

$$h_{0,x} = \left[\frac{R(K)\sigma^2(x)}{n\mu_2(K)^2 m''(x)^2 f_X(x)}\right]^{1/5}. \tag{5.18}$$

Equations (5.17) and (5.18) imply that in order to minimize the conditional MSE at any given point, a locally varying bandwidth $h(x)$ should be

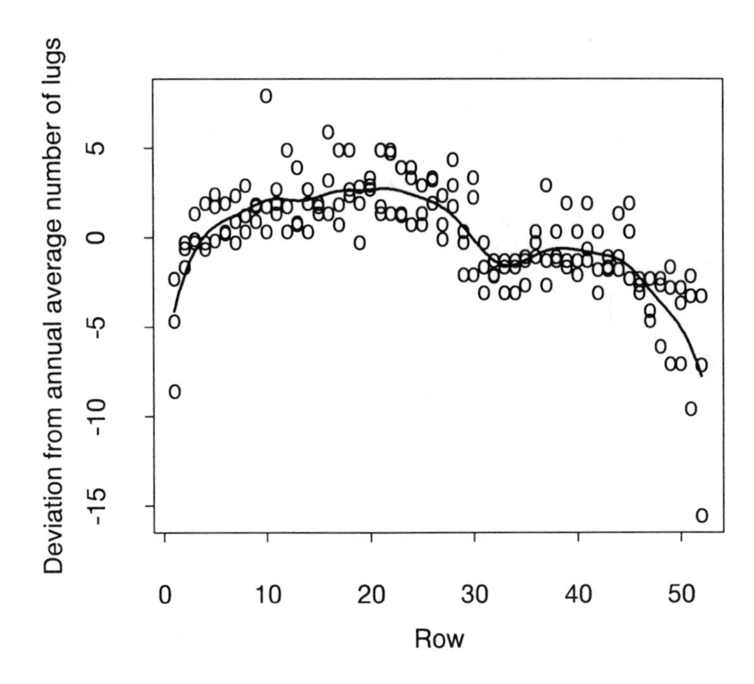


**Fig. 5.14.** Scatter plot of lug counts by row for three years, removing year effect, with local linear estimate based on plug-in selector superimposed.

used that varies directly with the local variance and inversely with the local curvature and local design density. Each of these results makes intuitive sense: a larger local variance requires a larger bandwidth to smooth over the local increase in responses farther from $m(x)$; greater local curvature requires a smaller bandwidth to pick up the sharp peak or trough; and a sparser local design (lower design density) requires a larger bandwidth to avoid the roughness that comes from smoothing over too few observations. Unfortunately, it is exceedingly difficult to estimate $h_{0,x}$ at each $x$, since the available information regarding the unknown values $\sigma^2(x)$, $m''(x)$, and $f_X(x)$ is only obtainable locally, implying a small effective local sample size.

A straightforward way to address the effect of the design density $f_X(x)$ is to vary the bandwidth by basing it on nearest neighbor distances, as is true for the loess estimator (although, asymptotically at least, the bandwidth is then proportional to $f_X(x)^{-1}$ rather than the optimal $f_X(x)^{-1/5}$). Figure 5.15 illustrates how this can work. The figure is a scatter plot of the change in adoption visas issued by the Immigration and Naturalization Service for the purpose of adoption by residents of the United States from 1991 to 1992 versus the change in visas issued from 1988 to 1991 for 37

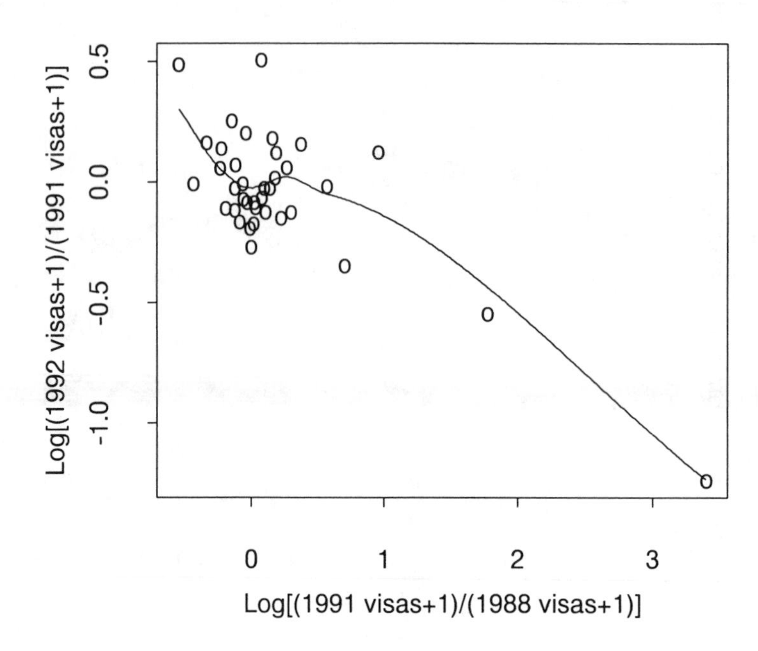


**Fig. 5.15.** Scatter plot of change in adoption visas from 1991 to 1992 versus change from 1988 to 1991, with local linear loess estimate using a 75% span superimposed.

countries (with change measured as the logarithm of the ratio of visas incremented by one, to handle zero values). The design density is long-tailed, with many countries having abscissa values in the range $[-.5, 5]$, and then a few stretching out to values greater than 3.

A local linear loess curve is superimposed (using the standard tricube kernel) where the bandwidth is chosen so that the kernel covers 75% of the observations (the *span* of the estimate). The smaller bandwidth at the left of the plot allows the nonmonotone relationship to come through, while the larger bandwidth at the right of the plot allows a smooth, reasonably straight fit to the higher values of the predictor. There is a generally inverse relationship between change in visas from 1988 to 1991 and change from 1991 to 1992, reflecting a tendency for increases in visas issued to be followed by decreases, and vice versa, but this pattern does not hold uniformly. A fixed-bandwidth estimator could not give this picture, as it would either oversmooth, and miss, the structure on the left, or undersmooth, and interpolate, the observations on the right (or be undefined for some values because of the lack of any observations with positive weights).

As useful as nearest neighbor distances can be, they generally cannot

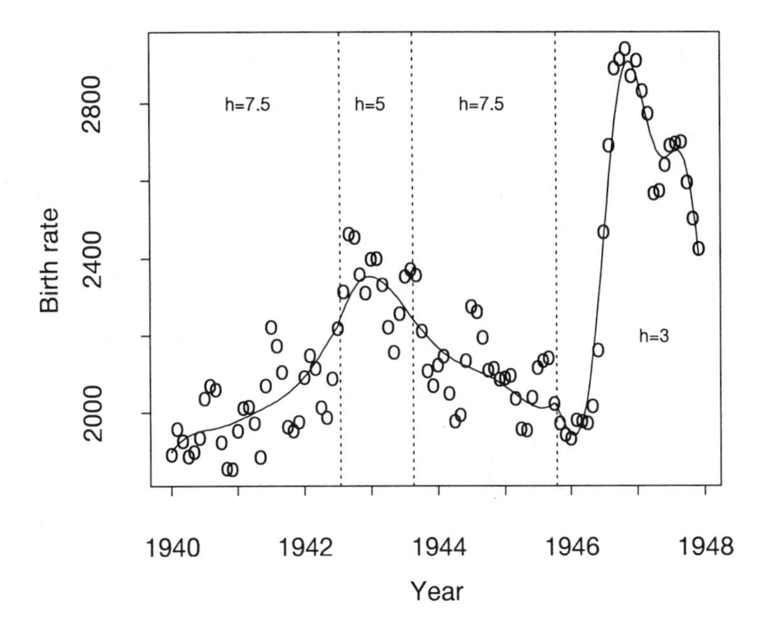


**Fig. 5.16.** Scatter plot of monthly birth rate, with local cubic estimate superimposed. The bandwidth is varied as indicated on the plot.

address all the needs for locally varying bandwidths, since they do not relate to $m''(x)$ or $\sigma^2(x)$. The designer of a well-designed experiment probably would guarantee more data (and larger $f_X(x)$) at values of $x$ that correspond to interesting structure in $m$ (larger $|m''(x)|$), and would want more data (and larger $f_X(x)$) in areas where the responses follow the regression line closely (smaller $\sigma^2(x)$). An experiment designed that way would be very amenable to analysis using local polynomials based on nearest neighbor distances, since taking $h(x)$ to be inversely related to $f_X(x)$ (as nearest neighbors do) also would result in an inverse relationship with $|m''(x)|$ and a direct relationship with $\sigma^2(x)$, as (5.18) requires.

Unfortunately, many data sets arrive with either a fixed uniform design, where nearest neighbor distances correspond exactly to fixed bandwidths (except at the boundaries), or some random design, where it is impossible to guarantee that $f_X$ has the desirable form (it might even be so perversely formed as to have low density in the places with the most structure and high density in places with little structure). Locally varying bandwidths are still useful, but are considerably more difficult to construct automatically.

Figure 5.16 illustrates the results of locally varying the bandwidth man-

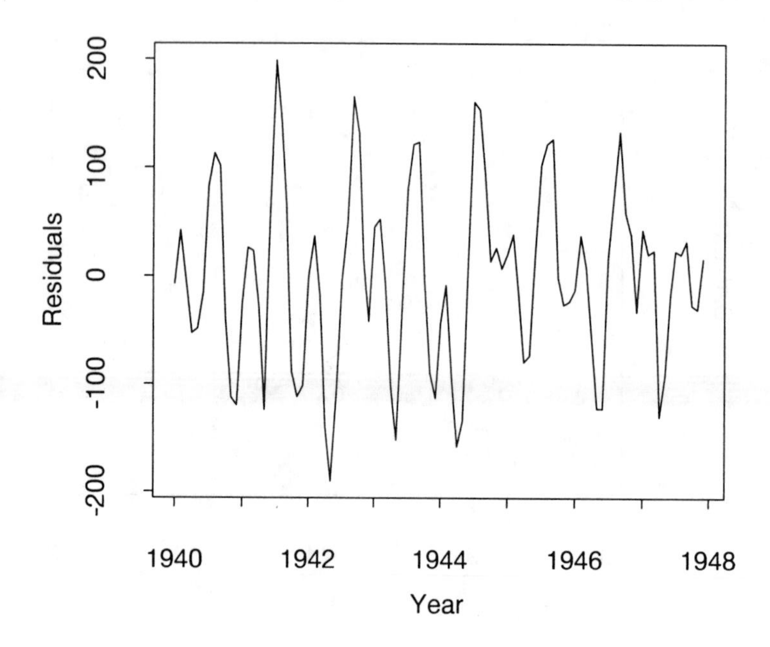


**Fig. 5.17.** Time series plot of residuals from local cubic regression estimate with locally varying bandwidth for birth rate data.

ually. The figure is a scatter plot of the monthly birth rates in the United States for the period from January 1940 through December 1947. As the observations are equispaced, nearest neighbor bandwidths are equivalent to fixed bandwidths in the interior. The curve is a local cubic regression estimate, with the bandwidth varying over four regions of the data.

The birth rate pattern provides a compelling glimpse of the dynamics of life in the United States during the 1940s, lagged by nine months. It rises smoothly from January 1940 through July 1942 (small $|m''|$), so a wide ($h = 7.5$) bandwidth is appropriate. From August 1942 through September 1943, there is a sharp peak in the birth rate (large $|m''|$), so a narrower bandwidth ($h = 5$) is appropriate. Note that this peak corresponds to a sudden increase in the birth rate from August 1942 through February 1943 (9 to 14 months after the entry of the U.S. into World War II), followed by a sharp decline in birth rate starting in March 1943 (most potential fathers being in the armed forces by nine months before this date). From October 1943 through October 1945 there is a steady decline in birth rate (corresponding to the war years), and a larger bandwidth ($h = 7.5$) shows this. Finally, starting in January 1946 (nine months after the end of the

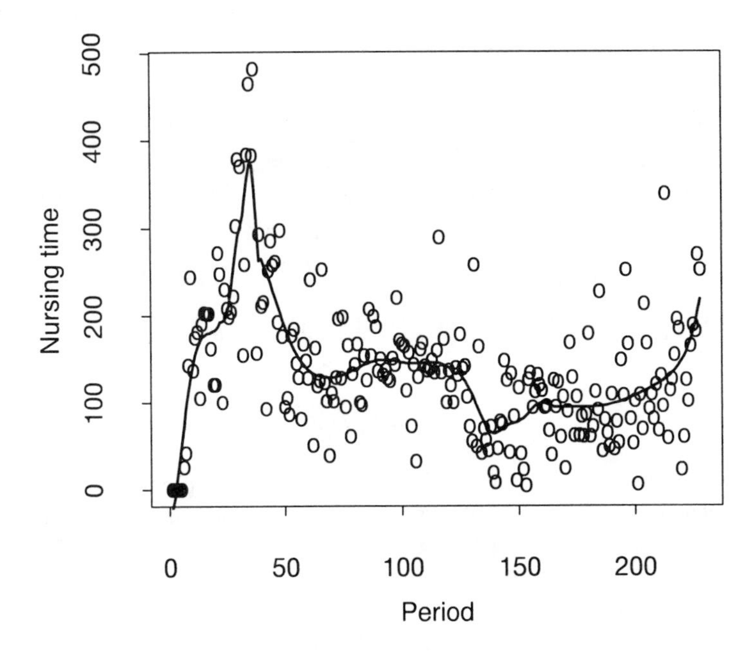


**Fig. 5.18.** Scatter plot of nursing time of beluga whale calf by time period, with local quadratic estimate superimposed. The bandwidth is varied as discussed in the text.

war in Europe), there is a very sharp rise in birth rate (to January 1947) followed by a sharp dip and rise around April 1947 (large $|m''|$), so a small bandwidth is the best choice ($h = 3$). If a fixed bandwidth had been used, either the sharp peaks and dips would have been smoothed over, or the flatter areas would have been undersmoothed.

Figure 5.17, which is a time series plot of the residuals from the fit ($y_i - \hat{m}(x_i)$), shows that the regression estimate captures the main relationship well, with no evidence of systematic over- or underestimation. There also is strong autocorrelation in the series, with the birth rate dipping at the beginning of the summer and the beginning of the winter, and peaking in midwinter and the beginning of the autumn. Issues relating to this autocorrelation are discussed further in Section 5.5.2.

Figure 5.18 gives an example where heteroscedasticity also determines the correct local bandwidth. The figure is a scatter plot that relates the nursing time (in seconds) of a newborn beluga whale calf named Hudson (born in captivity at the New York Aquarium) to the time after birth, where time is measured in six-hour time periods. Once again, the design is fixed and equispaced, so nearest neighbor distances are equivalent to fixed bandwidths in the interior. The superimposed curve is a local quadratic estimate using a locally varying bandwidth. The following table summarizes the bandwidths used and the reasons for the choices made:

| *Time periods* | *Bandwidth* | *Reason for choice* |
|---|---|---|
| 1–20 | 5.5 | Smooth rise in nursing |
| 21–40 | 2.75 | Sharp peak in nursing 7 days postpartum |
| 41–120 | 11 | Drop, then flattening out in nursing, with increase in variance |
| 121–160 | 6 | Sharp dip in nursing 35 days postpartum |
| 161–228 | 11 | Smooth rise in nursing with large increase in variance |

The dip in nursing time at 35 days postpartum is particularly noteworthy, since this was followed 10 days later by a diagnosis of bacterial infection (Hudson was administered antibiotics for 24 days and recovered). Thus, the observation of such a dip can be an "early warning sign" of potential health problems in a newborn calf.

These data sets were divided into blocks informally here, but the division can be formalized into an automatic procedure. The sharp change in degree of smoothing from block to block could be avoided by smoothing the blockwise bandwidths over adjacent blocks (although the discontinuities in bandwidths from block to block do not seem to have caused serious problems for these examples). Other, more complex, approaches to automating locally varying bandwidth choice exist as well, but it must be remembered that any attempt to vary $h(x)$ at too many places will bump up against the possibility of too little available data leading to high variability (even if asymptotically varying the bandwidth would yield a smaller MISE), unless the sample size is very large.

## 5.5 Outliers and Autocorrelation

### 5.5.1 Outliers and robust smoothing

Local polynomial regression estimates, being based on least squares estimation, can be affected by observations with unusual response values (outliers) in much the same way that ordinary (global) least squares estimates are. If an observed response is sufficiently far from the bulk of the observed responses for nearby values of $x$, $\hat{m}$ will be drawn towards the aberrant response and away from the bulk of the points.

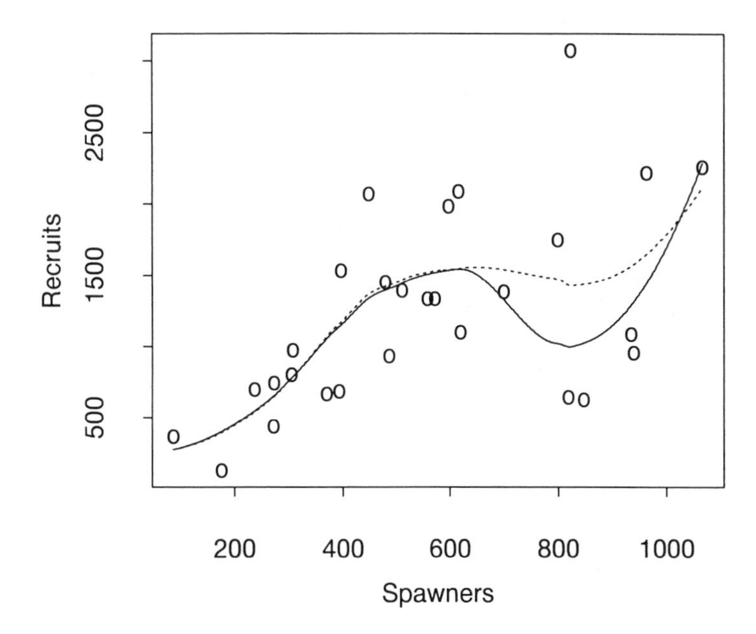


**Fig. 5.19.** Scatter plot of Skeena River salmon recruits versus spawners, with loess estimates superimposed. The dashed curve is the ordinary loess estimate, while the solid curve is the robust version.

Having noted that, it is important to recognize that it is much harder to identify unambiguously a point as outlying for a nonparametric regression model than it is when fitting a parametric model, such as a linear regression model. The reason is that the inherent flexibility of the nonparametric regression model (5.2) makes it difficult to tell the difference between an outlier and an acceptable property of the model. For example, several observations (with similar $x$) with response values that are unusually distant from the conditional mean could be outliers, or they could be the result of a larger variance $\sigma^2(x)$ off the regression line for that $x$. An isolated observation in a very sparse region of the design is virtually impossible to identify as an outlier without making further assumptions on $m$; there is no way to know if the observed response reflects the true $m$ or is an outlying response, since it is difficult to estimate $m$ in sparse regions of the design.

Still, in regions with enough data, it is possible to recognize an observation as reflecting a different conditional mean of $y$ given $x$ than that of nearby observations and to want to downweight (or eliminate) its influence on the fitted $\hat{m}$. Such a modified smoother is called a *robust* smoother. Figure 5.19 illustrates the effect of a possibly unusual observation on both a nonrobust and a robust estimate. The data relate the size of the annual spawning stock and its production of new catchable-sized fish (recruits) for 1940 through 1967 for the Skeena River sockeye salmon stock (in thousands of fish). The dashed curve is a loess (local quadratic) curve with span equaling 60% of the observations. The estimate shows a close-to-linear relationship between recruits and spawners up to about 600,000 spawners, after which the regression curve flattens out. This nonlinearity is not unexpected, since a linear relationship would imply that recruitment could be increased without limit by reducing fishing (allowing more spawners).

The solid curve is a robust version of the loess curve. The construction of the estimate is as follows:

(1) Construct the usual loess curve (the dashed curve). Let $e_i = y_i - \hat{m}(x_i)$ be the residual of the $i$th observation from the fitted curve, and let $s$ be the median of the absolute residuals.

(2) Define the robustness weight $\delta$ for each observation as

$$\delta_i = B[e_i/(6s)],$$

where $B(x)$ is the biweight function. Let $\Delta = \text{diag}(\delta_1, \ldots, \delta_n)$ be the diagonal $n \times n$ matrix of robustness weights.

(3) Construct the loess estimate, forming the weighted regression estimates using weight matrix $\Delta W$. Calculate the residuals $\mathbf{e}$ and scale measure $s$.

(4) Repeat steps (2) and (3) (three more times is standard).

This iterative process has the effect of taking observations that are far from the loess curve and repeatedly downweighting them in successive fittings, so that they have less effect on the constructed curve (although using the nonrobust loess fit in step (1) can lead to missed outliers if the nonrobust curve has been drawn too closely to them). For these data, the robust curve tracks the nonrobust one closely, except at around 800,000 spawners, where it dips markedly. The difference between the two curves is completely determined by the value for 1944, where 824,000 spawners were followed by 3,071,000 recruits, by far the largest amount over this time frame. The nonrobust curve treats the wide range of recruit values corresponding to around 800,000 spawners (from 627,000 to 3,071,000) as reflecting a change in $\sigma^2(x)$ (but not $m(x)$) and puts the fitted curve roughly through the middle of the values, while the robust version treats 1944 as an outlier and puts the fitted curve roughly through the middle of the remaining values. It is impossible to know just from the data which of the two curves is a better representation of reality, but they do offer two competing models to consider.

Figure 5.20 gives a clearer case of an outlier. The plot refers to radioimmunoassay calibration data that relates counts of radioactivity to the concentration of the dosage of the hormone TSH (in micro units per ml of incubator mixture). There is a roughly hyperbolic relationship between

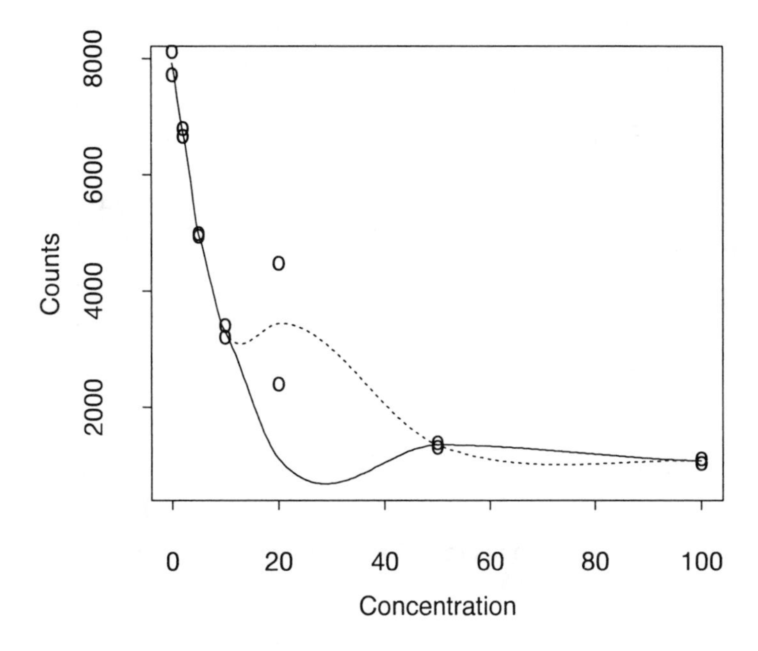


**Fig. 5.20.** Scatter plot of radioimmunoassay calibration data, with loess estimates superimposed. The dashed curve is the ordinary loess estimate, while the solid curve is the robust version.

counts and concentration, with one observation (at (20, 4478)) a clear outlier. The nonrobust loess (local linear) estimate (with span equaling 36% of the observations) follows the outlier, causing a troubling peak in the fitted curve. The robust loess curve follows the general trend more closely (though not as well as would be hoped), but it is not completely unaffected by the outlier (dipping below the nonoutlying value at $x = 20$).

Outliers also can cause problems by affecting data-based methods that control the appearance of the smoother, such as methods to choose the bandwidth. Figure 5.21 illustrates this effect. The data given relate the difference between the Democratic and Republican votes on voting machines to the difference between the Democratic and Republican votes from absentee ballots for 22 State Senate elections that took place in Philadelphia County from 1982 through 1993. The superimposed curve in Fig. 5.21(a) is a local linear estimate based on the plug-in choice for the bandwidth, $\hat{h} = 7114.4$. The curve shows the nonlinear (roughly quadratic) relationship between absentee votes and machine votes, but is undersmoothed. One reason that the plug-in estimate is too small is that it has been affected by the two outlying points at the top of the plot; the bandwidth selector tries

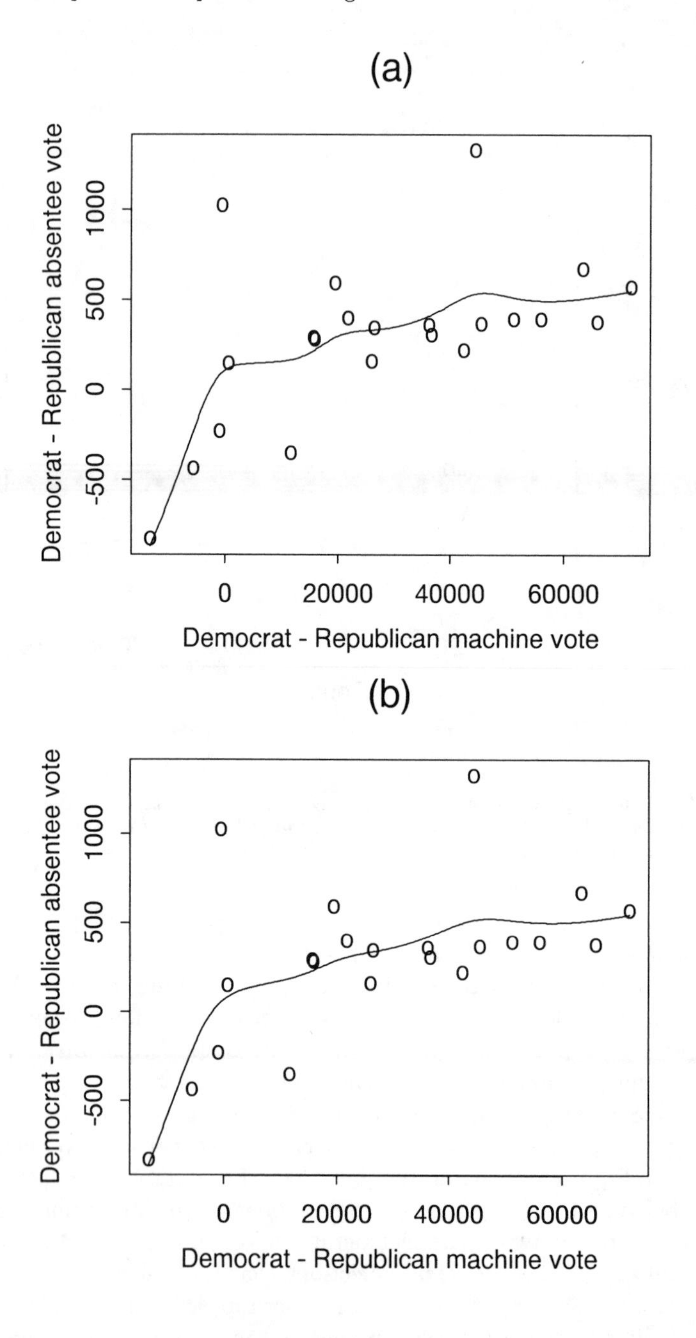


**Fig. 5.21.** Scatter plots of Democratic plurality in absentee votes versus plurality in machine votes, with local linear estimates superimposed. (a) Plug-in bandwidth choice using entire data set. (b) Plug-in bandwidth choice omitting outliers.

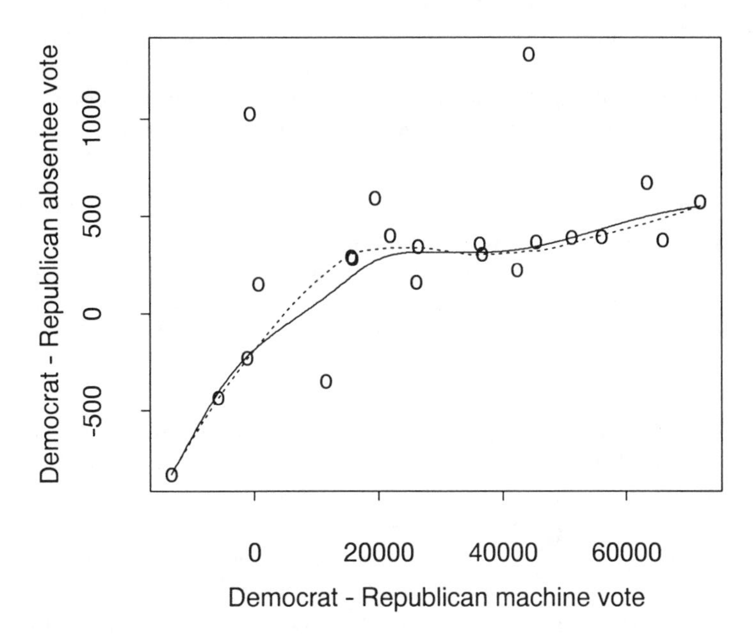


**Fig. 5.22.** Scatter plot of voting fraud data, with local linear estimate (estimate and plug-in bandwidth choice based on data omitting outliers — solid curve) and robust loess estimate (dashed curve) superimposed.

to include these points as part of the general trend by decreasing the bandwidth.

Figure 5.21(b) gives the local linear estimate (based on the entire data set) with bandwidth $\hat{h} = 8237$, the plug-in choice based on the data after omitting the two outliers. The larger bandwidth has the desired effect of reducing the spurious bumpiness in the curve. An even more sensible estimate is the solid curve in Fig. 5.22, which is the local linear estimate (with $h = 8237$) based on the data with the two outliers omitted. This eliminates the effects of the points on the estimate itself, and the fitted curve goes through the observations smoothly. A robust loess estimate with span equal to .5 (the dashed curve) is very similar to the local linear estimate, the only noteworthy difference being at around a difference of 10,000 machine votes. This is because the loess curve also downweights the 1990 election observation that falls below the estimated line.

The two previously noted outlying observations correspond to a regular election in 1992 (the right point) and a special election in 1993 (the left point). The latter observation is particularly interesting, since a Federal

District Court judge ultimately reversed the election result due to irregularities in the absentee ballot process. The nonparametric regression curves support the position that this election was very unusual, with a surprisingly large plurality of Democratic over Republican absentee ballots (consistent with possible voting fraud). The 1992 election (in which no fraud was alleged) encourages caution in this inference, however, since a surprisingly large plurality of Democratic over Republican absentee ballots also occurred then.

### 5.5.2 The effects of autocorrelation

All of the theory developed in this chapter has been based on the assumption that the error terms $\epsilon_i$ can be viewed as statistically independent of each other. This assumption is questionable if the observations have a natural ordering, such as in a time series, since then it is likely that nearby errors are correlated with each other (that is, the error process exhibits autocorrelation). Possible effects of this autocorrelation include changes in the asymptotic properties of the regression estimator $\hat{m}$, and changes in the behavior of data-based bandwidth selectors.

Consider data in the form of a time series, where the predictor falls in an equispaced fixed design (usually representing time), and the errors follow a zero-mean stationary time series with covariance function $\gamma(k) \equiv E(\epsilon_i \epsilon_{i+k}) \equiv \sigma^2 \rho(k)$ (where $\rho(\cdot)$ is the autocorrelation function) independent of the sample size that satisfies $\sum_{k=1}^{\infty} k|\gamma(k)| < \infty$ (examples of such processes include the well-known autoregressive moving average [ARMA] processes). Asymptotic analysis then shows that the MISE for $\hat{m}_1$ (the local linear estimate) satisfies

$$\text{MISE}(\hat{m}_1) = \frac{h^4 \mu_2(K)^2 \int m''(u)^2 du}{4} + \frac{R(K)[\sigma^2 + 2\sum_{k=1}^{\infty} \gamma(k)]}{nh} + o[h^4 + (nh)^{-1}]. \tag{5.19}$$

The bias contribution to MISE is identical to that for independent data, but the variance term adds $2\sum_{k=1}^{\infty} \gamma(k)$ (which is zero for independent observations) to $\sigma^2$. Equation (5.19) implies that the consistency properties of $\hat{m}$ for autocorrelated errors are broadly similar to those under independent errors, with bandwidth $h = O(n^{-1/5})$ implying the usual convergence rate $O(n^{-4/5})$.

At a more detailed level, the change in integrated variance means that the autocorrelation structure determines the optimal bandwidth, and bandwidth selectors that ignore this autocorrelation will not work correctly. Straightforward minimization of (5.19) gives the asymptotically optimal bandwidth,

$$h_{0,\gamma} = \left\{ \frac{R(K)[\sigma^2 + 2\sum_{k=1}^{\infty} \gamma(k)]}{n\mu_2(K)^2 \int m''(u)^2 \, du} \right\}^{1/5}. \tag{5.20}$$

There is a simple multiplicative relationship between $h_{0,\gamma}$ and $h_0$, the asymptotically optimal bandwidth for independent data,

$$h_{0,\gamma} = \left[1 + 2\sum_{k=1}^{\infty} \rho(k)\right]^{1/5} \times h_0.$$

Thus, if the data are positively autocorrelated ($\rho(k) \geq 0$ for all $k$), the optimal bandwidth is larger than that for independent data.

Autocorrelation in the errors often severely affects data-based bandwidth selectors. Such selectors tend to be "fooled" by the autocorrelation, interpreting it as reflecting the regression relationship and variance function. So, the cyclical pattern in positively autocorrelated errors is viewed as a high frequency regression relationship with small variance, and the bandwidth is set small enough to track the cycles (giving an undersmoothed fitted regression curve). The alternating pattern above and below $m$ for negatively autocorrelated errors is interpreted as a higher variance, and the bandwidth is set high enough to smooth over the variability (giving an oversmoothed fitted regression curve).

In a very real sense, this problem reflects an inherent identifiability problem when smoothing autocorrelated data. Without a rigid parametric structure, there is no unambiguous way to distinguish between positive autocorrelation and a high frequency response other than from general beliefs about which is more plausible for the data. It is therefore not surprising that data-based bandwidth selectors are unable to make this choice either.

Figure 5.23 illustrates this effect. The data are the U.S. monthly birth rate data given previously in Fig. 5.16. There is strong autocorrelation of the errors for these data, as the residual plot in Fig. 5.17 shows. The superimposed curve in Fig. 5.23 is the local linear estimate using the plug-in bandwidth $\hat{h} = 1.813$. The curve is very undersmoothed, as it tries to follow the cyclical pattern in the residuals rather than the dominant regression relationship.

Figure 5.24 shows that this effect also can occur even if the predictor variable is not a time index. The plots refer to (a) the "Old Faithful Geyser" eruption data, as in Fig. 5.2, and (b) the electricity usage data, as in Fig. 5.6. Each plot has superimposed on it the local linear estimate using the plug-in bandwidth ($\hat{h} = .204$ and $\hat{h} = 5.07$, respectively). Each curve exhibits some spurious waviness due to undersmoothing.

Equation (5.20) provides a way to choose the bandwidth for time series data. Besides estimating $\sigma^2$ and $\int m''(u)^2\,du$, the term $G \equiv \sum_{k=1}^{\infty} \gamma(k)$ also must be estimated. Unfortunately, the usual estimates of $\sigma^2$ and $\int m''(u)^2\,du$ are not appropriate, since they are calculated assuming independent errors, and other estimates must be used instead. Although the usual residual-based estimators of the autocovariances $\hat{\gamma}(\cdot)$ are consistent, $\sum_{k=1}^{\infty} \hat{\gamma}(k)$ is not a consistent estimator of $G$. A different approach to estimating $G$ is to fit a parametric model to the residuals (thereby obtaining

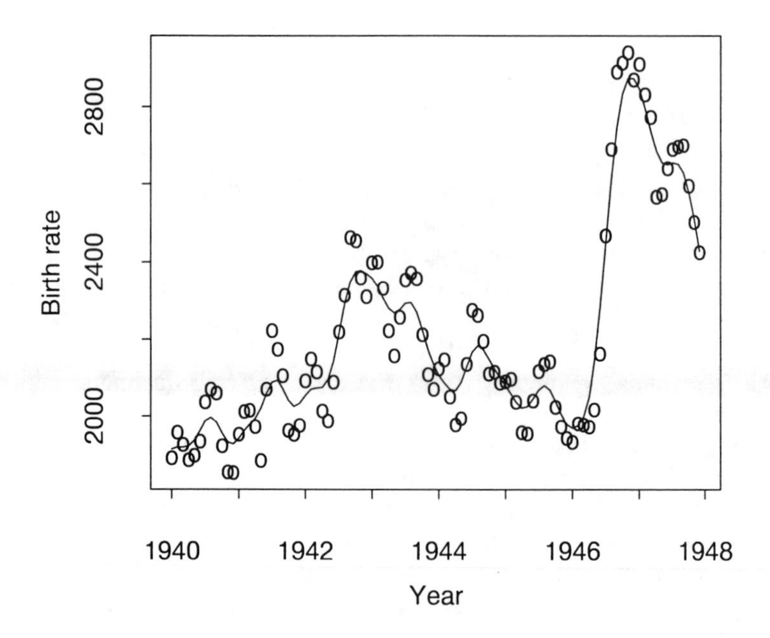


**Fig. 5.23.** Scatter plot of birth rate data, with local linear estimate based on plug-in bandwidth superimposed.

estimates of $\gamma(\cdot)$) and use these estimates in (5.20). Of course, if the assumed parametric model is incorrect, these estimates can be far from the truth, resulting in a poor choice for $h_{0,\gamma}$. In any case, such methods are unlikely to be very effective for small-to-moderate-sized data sets.

## 5.6 Spline Smoothing

### 5.6.1 Roughness penalties and nonparametric regression

Section 3.5 described the roughness penalty approach to density estimation. This methodology is easily applied to nonparametric regression as well (and, in fact, was first applied in this context). The underlying principle is identical to that described earlier: estimate the unknown smooth (regression) function by explicitly trading off fidelity to the data with smoothness of the estimate. For regression data, the residual sum of squares is a natural measure of fidelity to the data, so the roughness penalty estimator is the minimizer of

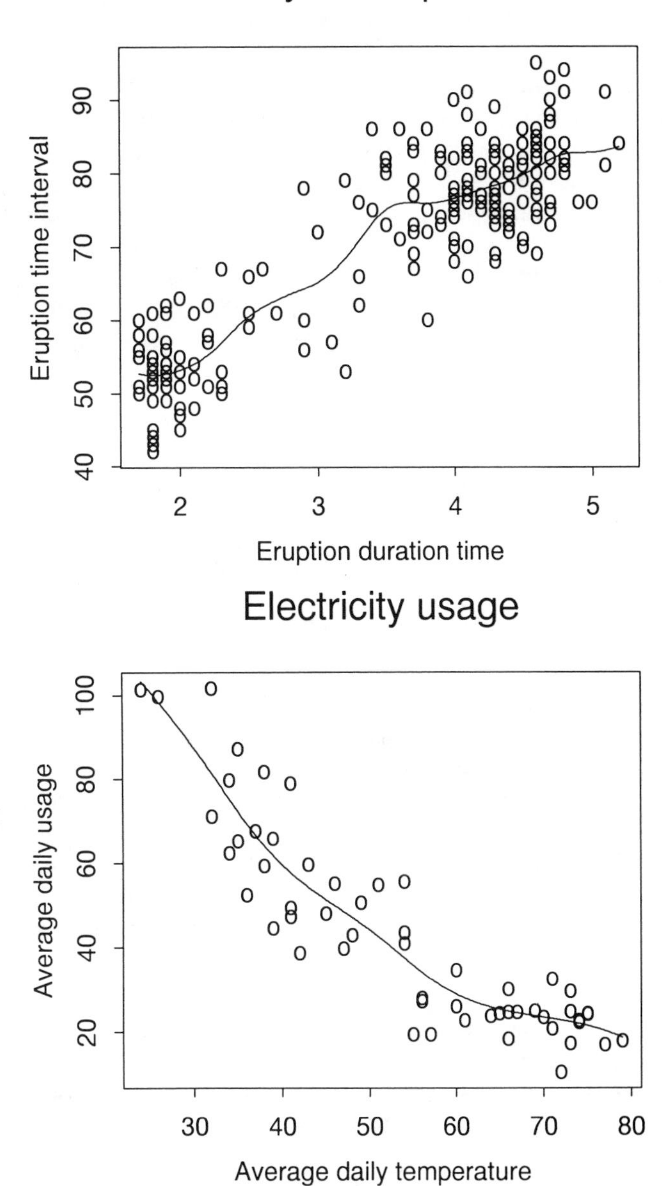


**Fig. 5.24.** Scatter plots with local linear estimate using plug-in bandwidth superimposed, for "Old Faithful Geyser" eruption data and electricity usage data.

$$L = \frac{1}{n}\sum_{i=1}^{n}[y_i - m(x_i)]^2 + \Phi(m),$$

where $\Phi(m)$ is a roughness penalty that decreases as $m$ gets smoother. In this context, it is easiest to assume that the sample is ordered over the interval $[a, b]$ with respect to the predictor values; that is, $a \le x_1 \le x_2 \le \cdots \le x_n \le b$. The class of possible minimizing functions must be restricted, or else the minimization problem has no unique solution. A natural constraint to put on $\hat{m}$ is that it be sufficiently smooth, which amounts to a condition on its derivatives. If $\hat{m}$ belongs to $W_2^\ell[a, b]$, the class of functions on $[a, b]$ with square integrable $\ell$th derivative and absolutely continuous derivatives up through order $\ell - 1$, an appropriate form for $L$ is

$$L = \frac{1}{n}\sum_{i=1}^{n}[y_i - m(x_i)]^2 + \alpha \int m^{(\ell)}(u)^2\, du. \tag{5.21}$$

Here $\alpha$ acts as a smoothing parameter analogous to the bandwidth for kernel and local polynomial estimators.

The most common version of $L$ takes $\ell = 2$, defining $\hat{m}$ as the minimizer of

$$L = \frac{1}{n}\sum_{i=1}^{n}[y_i - m(x_i)]^2 + \alpha \int m''(u)^2\, du$$

over the class of functions with $m$ and $m'$ absolutely continuous and $m''$ square integrable. The estimator is then a *cubic smoothing spline* with knots at predictor values $\{x_i\}$. Recall that a cubic spline is a function that is a piecewise cubic polynomial on any subinterval defined by adjacent knots, has two continuous derivatives, and has a third derivative that is a step function with jumps at the knots. The smoothing spline is more correctly a natural cubic spline, since it is constructed to have zero second and third derivatives at the boundaries (the so-called natural boundary conditions); that is, $\hat{m}$ is linear for $x \in [a, x_1)$ and $x \in (x_n, b]$.

### 5.6.2 Properties of cubic smoothing splines

The form of the cubic smoothing spline implies various properties of it. If $\alpha = 0$, the smoothing spline becomes an interpolating spline that goes through each of the responses $y_i$, while if $\alpha \to \infty$, $\hat{m}$ approaches the linear least squares regression line. The smoothing spline is a linear estimator, so the vector of fitted values $\hat{y}_i = \hat{m}(x_i)$ can be written as $\hat{\mathbf{y}} = A(\alpha)\mathbf{y}$. The matrix $A(\alpha)$ is called the *hat matrix* because it takes the observed data $\mathbf{y}$ to the fitted values $\hat{\mathbf{y}}$, with the $i$th row corresponding to $S_x$ with $x = x_i$ in the definition of the local polynomial estimator (5.5).

The asymptotic properties of $\hat{m}$ depend on the characteristics of the true regression function $m$ — its smoothness properties, of course, but also its behavior at the boundaries. Away from the boundary, assuming four bounded derivatives, the bias is asymptotically $O(\alpha)$, while the variance is $O[n^{-1}\alpha^{-1/4}]$ (assuming $\alpha \to 0$ and $n\alpha^{1/4} \to \infty$). The minimizer of asymptotic MSE is thus $\alpha = O(n^{-4/9})$, giving MSE $= O(n^{-8/9})$, the optimal rate. These rates hold close to $a$ and $b$ only if $m$ satisfies the natural boundary conditions $m''(a) = m''(b) = m^{(3)}(a) = m^{(3)}(b) = 0$, however, since the smoothing spline itself satisfies these conditions.

If $m$ does not satisfy the boundary conditions, the bias of the smoothing spline increases near the boundary. Specifically, if $m'' = 0$ but $m^{(3)} \neq 0$, the bias near the boundary is $O(\alpha^{3/4})$, while if $m'' \neq 0$, it is $O(\alpha^{1/2})$. These higher order biases hold within $O(\alpha^{1/4})$ of the boundaries, after which the $O(\alpha)$ rate of the interior takes over.

These results sound very much like what an even degree local polynomial estimator, or a higher order kernel estimator (without boundary correction), would imply, and in a sense that is the case. Asymptotically, in the interior $\hat{m}$ takes the form

$$\hat{m}(x) = \frac{1}{nf_X(x)h(x)} \sum_{i=1}^{n} K\left[\frac{x - x_i}{h(x)}\right] y_i, \tag{5.22}$$

with

$$K(u) = \frac{1}{2}e^{-|u|/\sqrt{2}} \sin\left(\frac{|u|}{\sqrt{2}} + \frac{\pi}{4}\right)$$

and $h(x) = \alpha^{1/4} f_X(x)^{-1/4}$. As noted in Section 3.5, this is a fourth order kernel estimator with locally varying bandwidth of order $\alpha^{1/4}$. The optimal choice of $h(x)$ is thus $O(n^{-4/9})^{1/4}$, or $O(n^{-1/9})$, and the bias is of order $h^4$, as would be expected for a fourth order kernel estimator (this also suggests a correspondence to a local polynomial estimator with $p = 2$ or 3).

In the boundary region (which (5.22) shows is of length $O(\alpha^{1/4})$), the kernel is deformed in a way that results in boundary bias. If $m'' = 0$ but $m^{(3)} \neq 0$, the bias becomes $O(h^3)$, while if $m'' \neq 0$, it is $O(h^2)$. Again, these rates are consistent with a Taylor Series expansion of the bias associated with a non-boundary corrected, renormalized kernel estimator, with the exception that the condition $m' \neq 0$ does not lead to $O(h)$ bias (as is also true for a local linear estimator). So, the smoothing spline exhibits order of magnitude asymptotic behavior similar to that of a fourth order kernel estimator (except for the avoidance of $O(h)$ boundary bias if $m' \neq 0$) or a local quadratic polynomial estimator (except for the existence of $O(h^2)$ bias if $m'' \neq 0$). In particular, the smoothing spline achieves MISE $= O(n^{-4/5})$ assuming two bounded derivatives for $m$, whatever the boundary conditions are. In this loose sense, the cubic spline estimator can be thought of as falling somewhere between the two kernel-type estimators.

The roughness penalty approach also has a straightforward Bayesian interpretation. If the errors are normally distributed with constant variance, and the prior density for $m$ is proportional to $\exp[-\alpha \int m''(u)^2\, du/2]$ over the space of all smooth functions, the smoothing spline corresponds to the posterior mode given the data.

### 5.6.3 Choosing the smoothing parameter

The parameter $\alpha$ acts as a smoothing parameter in the roughness penalty approach, with larger values of $\alpha$ penalizing roughness more, yielding a smoother estimate. As always, $\alpha$ can be chosen to minimize the cross-validation score

$$\mathrm{CV}(\alpha) = \frac{1}{n}\sum_{i=1}^{n}[y_i - \hat{m}^{(i)}(x_i)]^2,$$

where $m^{(i)}(x_i)$ is the spline estimate based on all the observations except $x_i$, evaluated at $x_i$. It can be shown that for linear smoothers, $\mathrm{CV}(\alpha)$ can be written as a function of the fitted values,

$$\mathrm{CV}(\alpha) = \frac{1}{n}\sum_{i=1}^{n}\left[\frac{y_i - \hat{m}(x_i)}{1 - A_{ii}(\alpha)}\right]^2, \tag{5.23}$$

where $A_{ii}(\alpha)$ is the $i$th diagonal element of the hat matrix. $A_{ii}(\alpha)$ is called the leverage value, since it measures the potential for the observed response at $x_i$ $(y_i)$ to exert influence on the fitted value at $x_i$ $(\hat{y}_i)$. An observation with a high leverage value (a *leverage point*) is potentially problematic, since the fitted regression curve will follow the observed response value irrespective of the general trend implied by the responses of other observations.

A variation on (5.23) is *generalized cross-validation* (GCV), which replaces each value $1 - A_{ii}(\alpha)$ with their average, $1 - n^{-1}\,\mathrm{trace}[A(\alpha)]$. The generalized cross-validation selector of $\alpha$, $\hat{\alpha}_{\mathrm{GCV}}$, is the minimizer of

$$\mathrm{GCV}(\alpha) = \frac{\sum_{i=1}^{n}[y_i - \hat{m}(x_i)]^2}{n\{1 - n^{-1}\,\mathrm{trace}[A(\alpha)]\}^2}. \tag{5.24}$$

An equivalent form of (5.24) makes clearer the connection between $\mathrm{GCV}(\alpha)$ and $\mathrm{CV}(\alpha)$, since

$$\mathrm{GCV}(\alpha) = \frac{1}{n}\sum_{i=1}^{n}\left(\left\{\frac{1 - A_{ii}(\alpha)}{1 - n^{-1}\,\mathrm{trace}[A(\alpha)]}\right\}^2 [y_i - \hat{m}^{(i)}(x_i)]^2\right).$$

If the $A_{ii}(\alpha)$ values are equal, as would be true for equispaced data (except for boundary effects), $\mathrm{GCV}(\alpha)$ and $\mathrm{CV}(\alpha)$ are identical, but otherwise GCV downweights the effects of high leverage points on the selection criterion, which is desirable. Thus, $\hat{\alpha}_{\mathrm{CV}}$ and $\hat{\alpha}_{\mathrm{GCV}}$ are very close for equispaced data, but $\hat{\alpha}_{\mathrm{GCV}}$ will generally be preferable for unequally spaced observations.

Figure 5.25 gives two examples of the use of generalized cross-validation

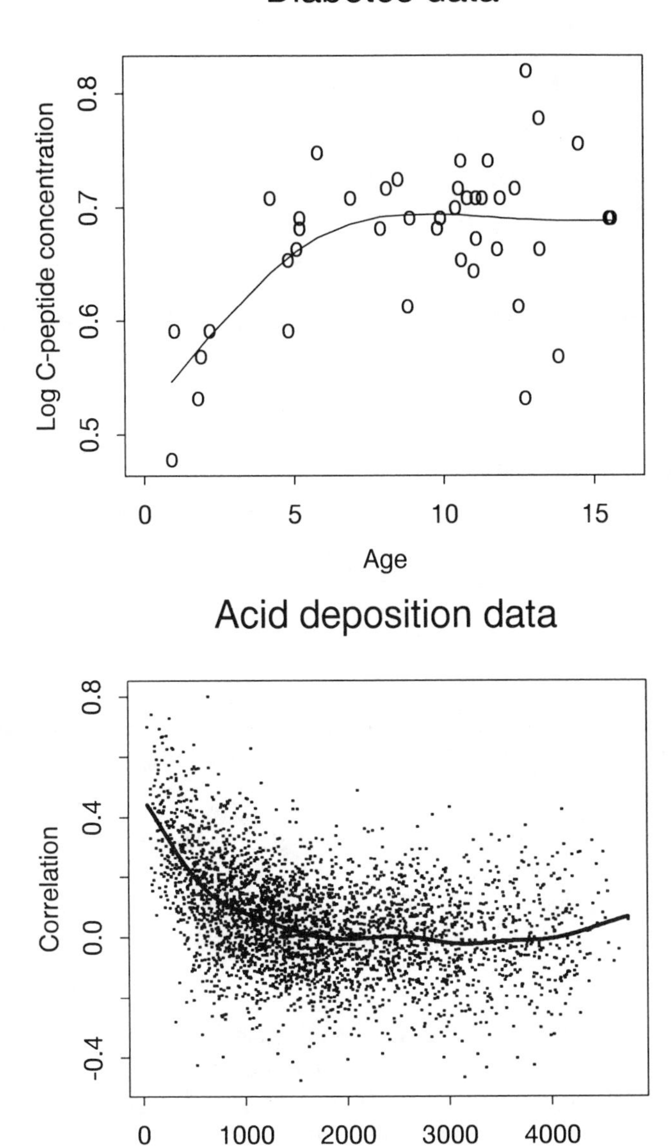


**Fig. 5.25.** Scatter plots with cubic spline estimate using generalized cross-validation smoothing parameter superimposed, for diabetes data and acid deposition data.

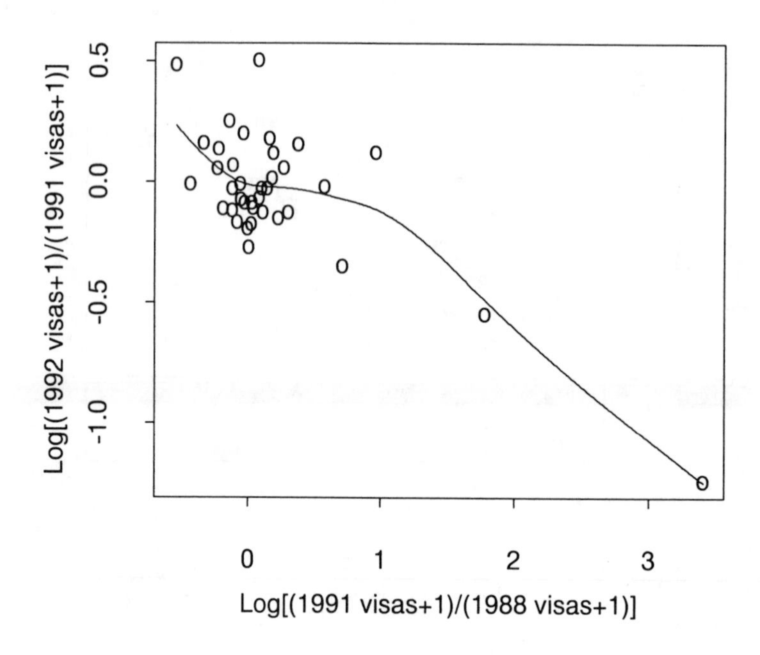


**Fig. 5.26.** Scatter plot of change in adoption visa data, with cubic smoothing spline based on generalized cross-validation smoothing parameter superimposed.

for the cubic smoothing spline. The plots give the spline estimates for the diabetes data given in Fig. 1.5 ($\hat{\alpha}_{\text{GCV}} = .938$) and the acid deposition data given in Fig. 1.6 ($\hat{\alpha}_{\text{GCV}} = 167,566$). The superimposed regression estimates are similar to the kernel and lowess estimates, respectively, given in those figures.

The approximate form (5.22) of the spline estimator as a fourth order kernel estimator with locally varying bandwidth inversely related to the design density suggests that the spline estimator should be effective when such kernel (or local polynomial) estimators are. Figure 5.26 is a scatter plot of the change in adoption visas data given in Fig. 5.15, where a nearest neighbor local polynomial estimate (loess) was used to estimate the regression. The spline estimate ($\hat{\alpha}_{\text{GCV}} = .001245$) is very similar to the loess estimate, as it also is able to use a smaller effective bandwidth to the left of the plot and a larger one to the right.

Figure 5.27 shows that the "higher order kernel" characteristic of the cubic spline allows the estimate to pick up rapid changes in the curvature of the regression curve. The figure is a scatter plot of the total lug count

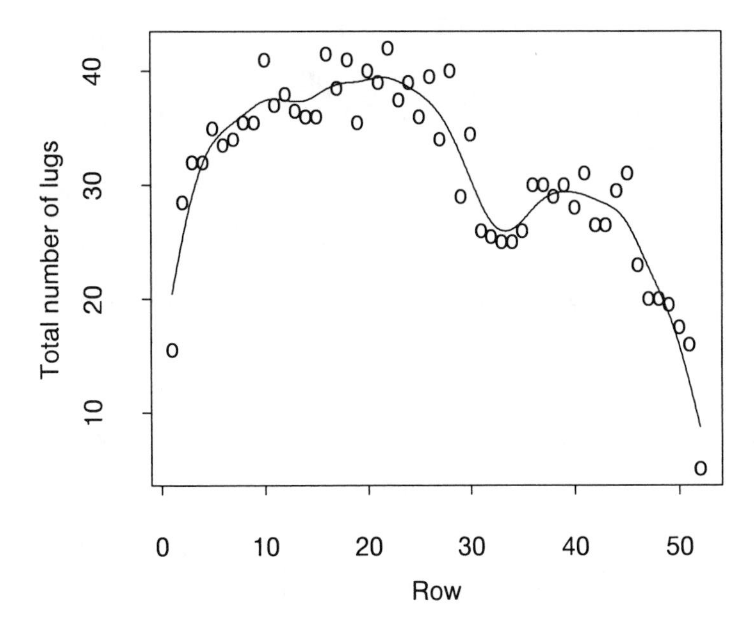


**Fig. 5.27.** Scatter plot of total lug count vineyard data, with cubic smoothing spline superimposed.

vineyard data, which Figures 5.9 and 5.10 showed to be better fit with a local cubic estimate than a local linear estimate; that is, a higher order (reduced bias) method. The cubic smoothing spline ($\alpha = .102$) similarly picks up the dip in total number of lugs in rows 30 through 40, while not undersmoothing the other rows.

Unfortunately, but not unexpectedly, $\hat{\alpha}_{\text{GCV}}$ also can suffer from the bane of cross-validated smoothing parameter selectors: undersmoothing of the regression curve. One such data set is the vineyard data of Fig. 5.27. The smoothing parameter choice based on generalized cross-validation is $\hat{\alpha}_{\text{GCV}} = .0293$, which is too small. Figure 5.28 gives two more examples of this problem. In each plot, the solid curve represents the spline based on $\hat{\alpha}_{\text{GCV}}$. For both the ethanol data (as in Fig. 5.8), where $\hat{\alpha}_{\text{GCV}} = 3.02 \times 10^{-8}$, and the newspaper circulation data (as in Fig. 5.5), where $\hat{\alpha}_{\text{GCV}} = 2032.1$, the estimates based on generalized cross-validation are very undersmoothed. The problem is not with the spline estimator itself, since larger values of $\alpha$ ($\alpha$ equaling $1.92 \times 10^{-7}$ and $245,223$, respectively) provide very reasonable representations of the regression relationships (the dashed curves).

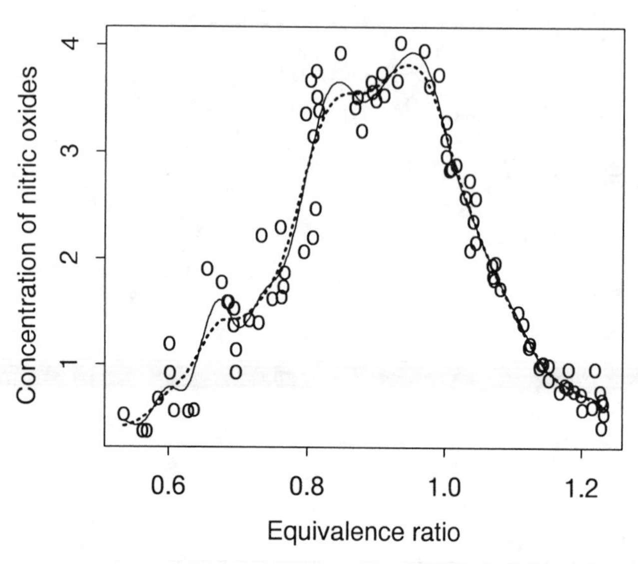


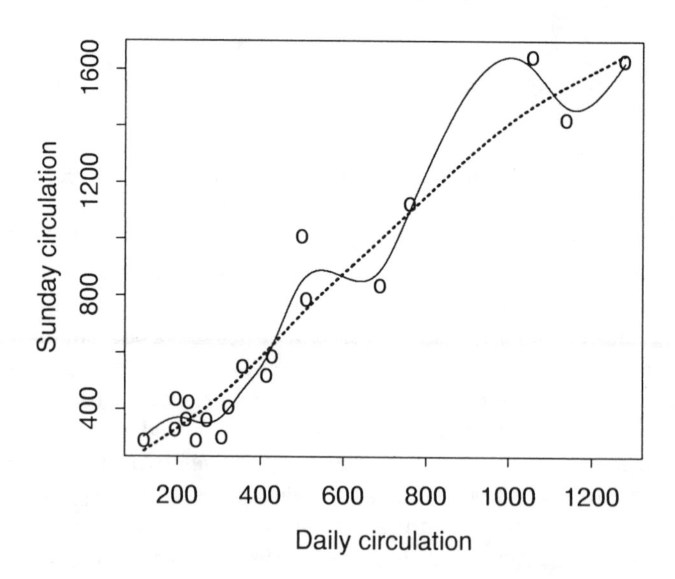


**Fig. 5.28.** Scatter plots with cubic spline estimate using generalized cross-validation (solid line) and larger smoothing parameter (dashed line) superimposed, for ethanol data and newspaper circulation data.

The recognition of $A(\alpha)$ as the hat matrix provides a motivation for an estimator of $\sigma^2$ based on spline smoothing. Since in the linear regression model the trace of the hat matrix equals the number of fitted parameters, the effective number of parameters fit by the spline can be taken to be trace$[A(\alpha)]$. Thus, a possible estimate of $\sigma^2$ is

$$\hat{\sigma}_s^2 = \nu_s^{-1} \sum_{i=1}^{n} [y_i - \hat{m}(x_i)]^2, \tag{5.25}$$

where $\nu_s$, the error degrees of freedom, equals

$$\nu_s = \text{trace}[(I - A(\alpha)] = n - \text{trace}[A(\alpha)]. \tag{5.26}$$

There are two key differences between this estimator and the local polynomial-based estimator described in (5.15) and (5.16). First, $\hat{\sigma}_s^2$ is based on the same regression estimate $\hat{m}$ as would be used for estimation of $m$, whereas the regression estimate $\tilde{m}$ used in (5.15) is asymptotically undersmoothed for estimation of $m$. Second, the form for the error degrees of freedom ($\nu$) is different. As $\alpha$ increases the two forms become identical, but for small $\alpha$ they can differ, with the value in (5.16) generally having better properties.

### 5.6.4 Violations of assumptions

Not surprisingly, the smoothing spline (being a linear estimate of the data) can be severely affected by outliers, particularly at high leverage points (which correspond to observations in sparser regions of the design density). Similarly, data-based smoothing parameter selectors can be adversely affected by outliers. Figure 5.29 is a scatter plot of the voting fraud data of Figs. 5.21 and 5.22, where the plug-in selector for the local linear estimate led to undersmoothing because of two outliers. The outliers also affect the spline generalized cross-validation selector, but in the opposite way. In Fig. 5.29, the solid curve is a cubic spline based on $\hat{\alpha}_{\text{GCV}} = 1.725 \times 10^{12}$ and is obviously oversmoothed; a better choice is $\hat{\alpha} = 8.372 \times 10^{10}$, which yields the dashed curve in the figure. One solution to these problems is to robustify the roughness penalty criterion to downweight the effect of outliers.

As would be expected, $\hat{\alpha}_{\text{GCV}}$ is also susceptible to autocorrelation effects, with positive autocorrelation leading to undersmoothing. Figure 5.30 is a scatter plot of the birth rate data with a cubic spline based on $\hat{\alpha}_{\text{GCV}} = 7.456 \times 10^{-4}$ superimposed. The regression estimate is severely undersmoothed (even compared with that in Fig. 5.23), being virtually an interpolation of the observed birth rates.

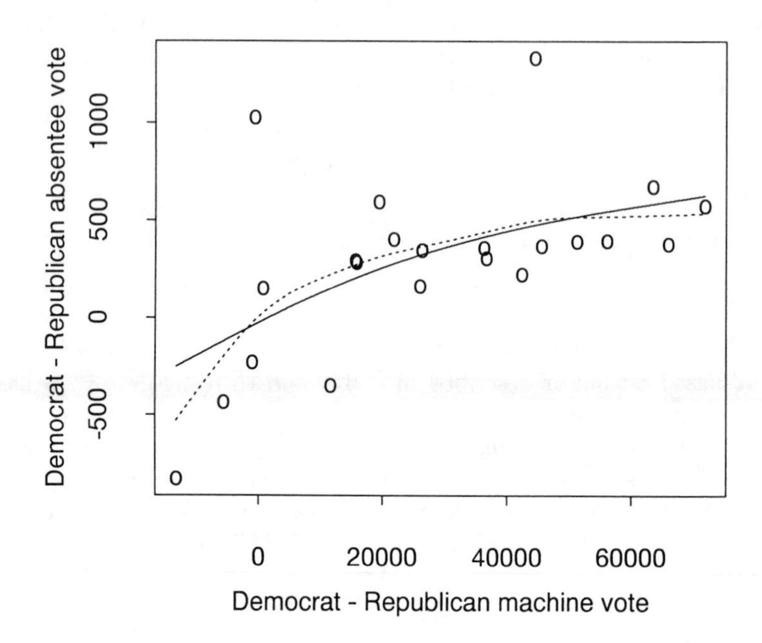


**Fig. 5.29.** Scatter plot of voting fraud data, with cubic smoothing spline based on generalized cross-validation (solid curve) and smaller smoothing parameter (dashed curve) superimposed.

## 5.7 Multiple Predictors and Additive Models

### 5.7.1 Nonparametric regression with multiple predictors

Local polynomial estimation generalizes in a straightforward way to multiple predictors, allowing easy application to the many real situations where the response variable $y$ is possibly related to more than one predictor. Consider, as the simplest case, local linear estimation. The estimator is the entry $\hat{\beta}_0 \equiv \hat{m}_1(\mathbf{x})$ of the minimizer $\hat{\boldsymbol{\beta}}$ of

$$\sum_{i=1}^{n}[y_i - \beta_0 - \beta_1(x_1 - x_{1i}) - \cdots - \beta_d(x_d - x_{di})]^2 K_d[H^{-1}(\mathbf{x} - \mathbf{x}_i)],$$

where the multivariate kernel function $K_d$ and bandwidth matrix $H$ are defined as in Section 4.2, and the $d \times 1$ vector $\mathbf{x}$ corresponds to the set of predictor variables.

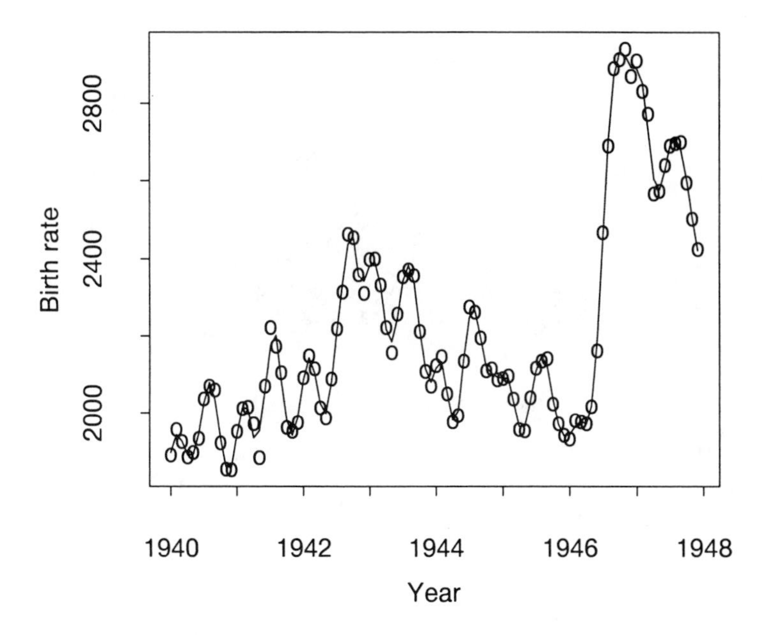


**Fig. 5.30.** Scatter plot of birth rate data, with cubic spline estimate based on generalized cross-validation superimposed.

Using the same parameterization of $K_d$ and $H$ as was used in Section 4.2, if $h \to 0$ and $nh^d \to \infty$ as $n \to \infty$, the conditional asymptotic MSE of $\hat{m}_1(\mathbf{x})$ in the interior equals

$$\text{AMSE}(\hat{m}_1(\mathbf{x})|\mathbf{x}_1, \ldots, \mathbf{x}_n) = \frac{R(K)\sigma^2(\mathbf{x})}{nh^d f(\mathbf{x})} + \frac{h^4}{4}\{\text{trace}[AA' \bigtriangledown^2 m(\mathbf{x})]\}^2.$$

The curse of dimensionality is apparent, as the variance is of order $(nh^d)^{-1}$, implying an optimal conditional AMSE $= O_p(n^{-4/(d+4)})$ taking $h = O(n^{-1/(d+4)})$. The local linear estimator for $d > 1$ achieves this conditional AMSE rate all the way to the boundary, avoiding boundary bias problems in the same way as when $d = 1$.

Figure 5.31 gives an example of the application of the local linear estimator to data with two predictors. The estimate uses nearest neighbor weights (that is, it is a loess estimate) with the span being .3 (the nearest neighbor distances use the predictors after standardizing each by its 10% trimmed standard deviation) and diagonal $H$. The fitted regression surface is an estimate of the mean points scored per minute conditional on the number of minutes played per game and height in centimeters for the 96 NBA

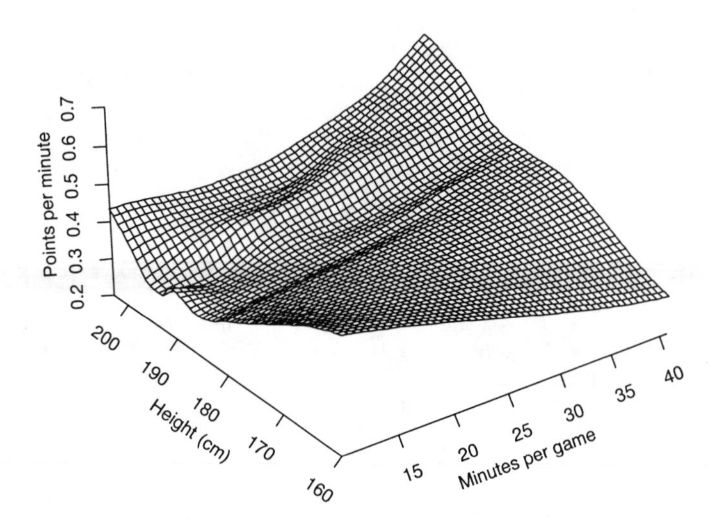


**Fig. 5.31.** Perspective plot of loess estimate of points scored per minute as a function of minutes played per game and height.

players described in Chapter 4. A trend towards higher scoring as a player's height and minutes played increase dominates the perspective plot of the regression surface, a not unexpected result. There is a plateau in the surface for heights roughly between 175 and 190 cm, which could correspond to an area of "typical" (that is, stable) performance. A surprising pattern is the negative slope of $\hat{m}$ for values of minutes per game less than 20 and height less than 180, but as this corresponds to one isolated observation (Keith Jennings, who scored .5 points per minute, but only played eight games before getting injured), its importance should not be overemphasized.

Spline (roughness penalty) estimators can be generalized to multiple predictors in various ways, with different estimators resulting from different generalizations of the roughness penalty. One possibility is to define the estimator as the minimizer (over a suitable class of smooth functions) of

$$\frac{1}{n}\sum_{i=1}^{n}[y_i - m(\mathbf{x}_i)]^2 + \alpha \int \cdots \int \sum_{\nu_1+\cdots+\nu_d=\ell} \frac{\ell!}{\nu_1!\cdots\nu_d!}\left(\frac{\partial^\ell m}{\partial u_1^{\nu_1}\cdots\partial u_d^{\nu_d}}\right)^2 d\mathbf{u}.$$

Taking $\ell = 2$ generalizes the univariate cubic spline. So, for example, for bivariate data the *thin plate smoothing spline* is the minimizer of

$$\frac{1}{n}\sum_{i=1}^{n}[y_i - m(\mathbf{x}_i)]^2 + \alpha \int\int \left[\left(\frac{\partial^2 m}{\partial u^2}\right)^2 + 2\left(\frac{\partial^2 m}{\partial u \partial v}\right)^2 + \left(\frac{\partial^2 m}{\partial v^2}\right)^2\right] du\, dv$$

(the term "thin plate" spline is used because the bending energy of an infinite elastic thin plate when deformed is, to first order, proportional to the roughness penalty above).

### 5.7.2 Additive models

Direct estimation of a multivariate regression surface is limited by the multidimensional character of the problem. It is difficult to visualize regression surfaces for more than two predictors and difficult to interpret the complex structure that can arise. Further, the curse of dimensionality implies that as the number of predictors increases, massive amounts of data will be required for accurate estimation.

A way around these difficulties is to restrict the form of the regression function $m(\mathbf{x})$. A natural possibility is to generalize ordinary (multiple) linear regression to allow arbitrary additive functions, as in

$$m(\mathbf{x}) = \alpha + \sum_{j=1}^{d} f_j(x_j). \tag{5.27}$$

Univariate smoothers can be used to estimate the functions $f_j$, thereby avoiding the curse of dimensionality, but at a crucial cost: If the additive form (5.27) is not correct, the estimator $\hat{m}$ need not even be consistent. If $m$ does satisfy (5.27), the convergence rate of $\hat{m}$ is identical to that of the one-dimensional smoother (if it does not, convergence is to the closest additive approximation to $m$ in the sense of MSE). In addition, interpretation of the model is simpler, as the structure related to any predictor is defined in a univariate way (conditional on the other smooth functions in the model).

Figure 5.32 illustrates the application of additive modeling to the basketball data of Fig. 5.31. The univariate smoothers used are local linear loess estimates with span equal to .5. The curve on each plot illustrates graphically the contribution of each predictor to $\hat{m}$ given the other predictor. Superimposed on each plot are partial residuals (the fitted term plus standardized residuals), which can help show the strength of the observed relationship. The additive model implies that scoring rate increases steadily (roughly linearly) with increasing playing time (given height), with one notably large positive residual (Michael Jordan) apparent. In contrast, scoring rate increases with height for height greater than roughly 185 cm, but is flat for smaller players (the positive slope for height less than 170 cm corresponds to only one player).

This succinct representation is not dissimilar from that implied by the loess fit in Fig. 5.31, as Fig. 5.33 shows. This figure is a perspective plot of the fitted additive model and appears to be a smoother version of the direct

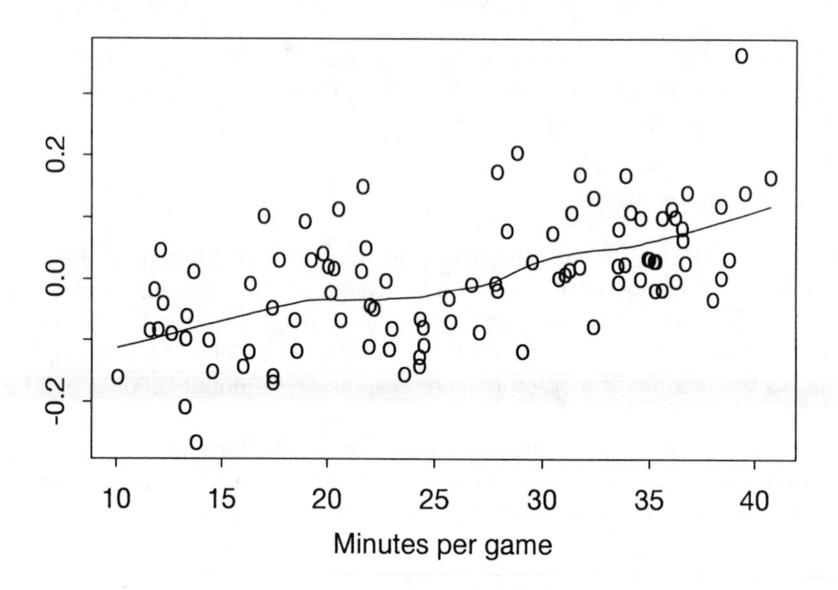


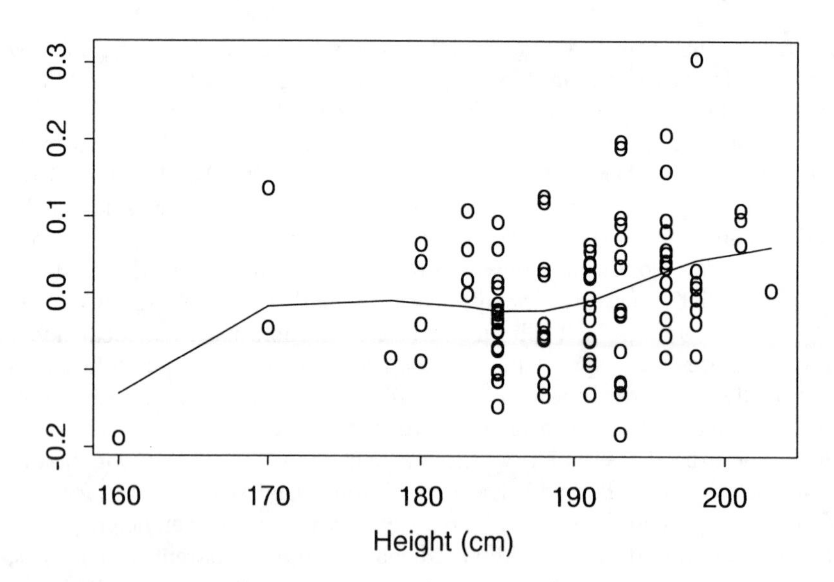


**Fig. 5.32.** Plots of contributions of minutes per game (top) and height (bottom) to additive model estimate of points scored per minute, with partial residuals superimposed.

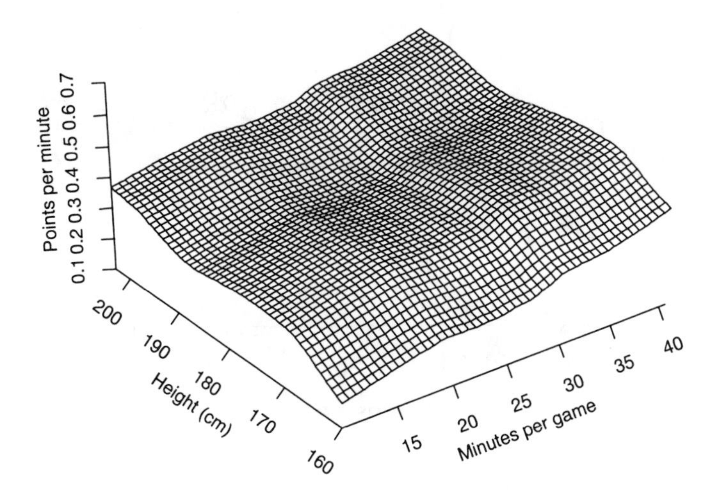


**Fig. 5.33.** Perspective plot of additive model estimate of points scored per minute as a function of minutes played per game and height.

loess surface in the earlier figure, with the only major difference being the fit for the one player with height less than 180 cm who played less than 20 minutes per game.

The real power of additive models comes from the ability to summarize relationships in an intuitive way. Figure 5.34 gives an example of this. The plots correspond to those of Fig. 5.32, except that the response variable is now assists made per minute, rather than points scored per minute, and the univariate smoother used is a cubic spline with $\text{trace}[A(\alpha)] = 5$. The contribution of minutes played per game is similar to that for points scored per minute, but the relationship with height is opposite, with increasing height being associated with lower assists per game for players over 185 cm tall (who are more likely to be shooting guards). This pattern is not surprising, and it helps explain the negative association between points per minute and assists per minute among shooting guards that was evident in Fig. 4.5.

As is true for linear regression modeling, the relationships represented in Figs. 5.32 and 5.34 are conditional ones, given the other predictors in the model. In many circumstances, this is precisely the relationship wanted,

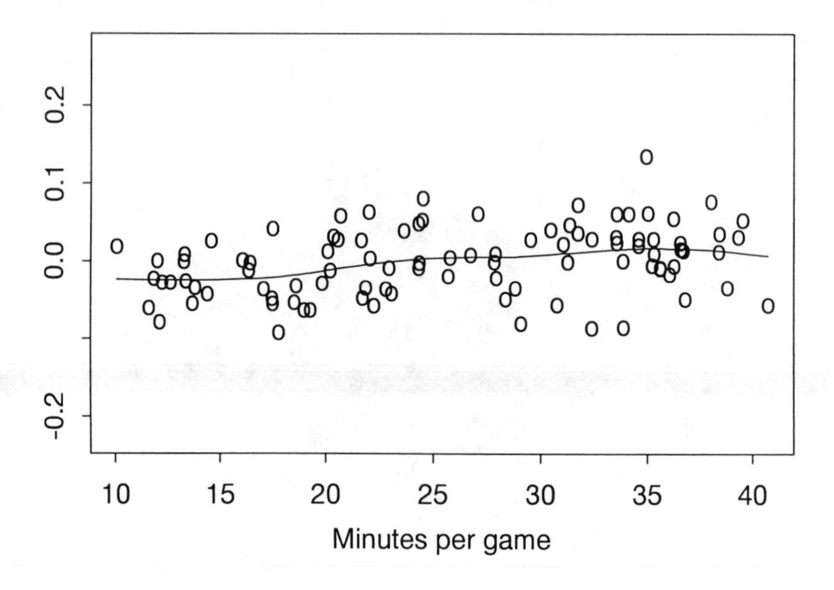


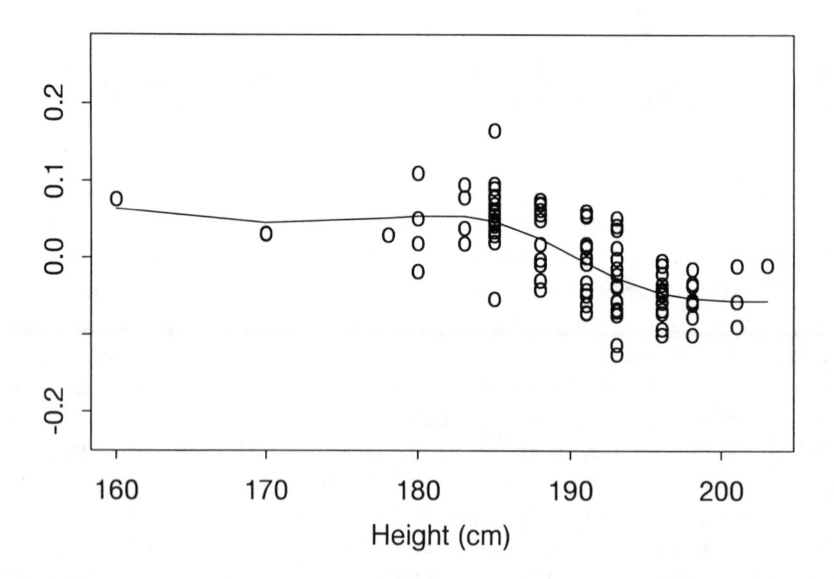


**Fig. 5.34.** Plots of contributions of minutes per game (top) and height (bottom) to additive model estimate of assists made per minute, with partial residuals superimposed.

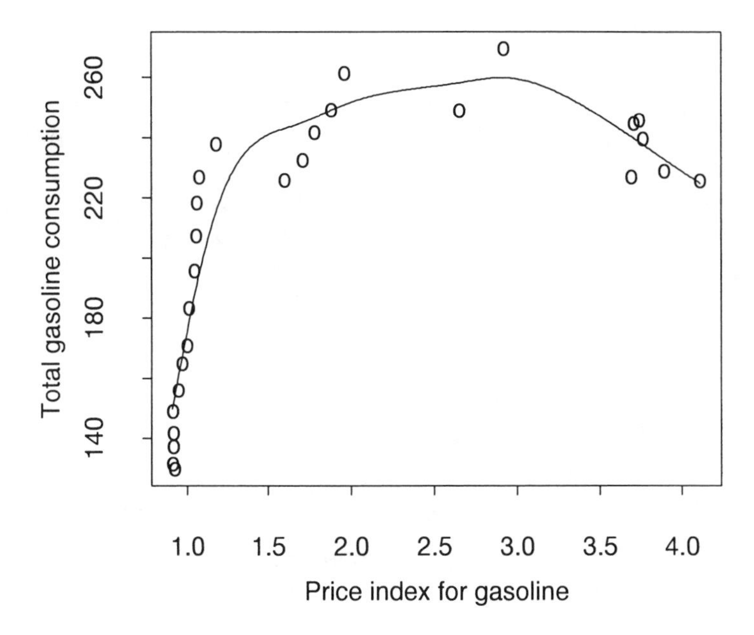


**Fig. 5.35.** Scatter plot of total gasoline consumption versus price index for gasoline, with spline estimate superimposed.

and additive modeling can help explain an otherwise puzzling relationship in the data. Consider Fig. 5.35, a scatter plot of the annual U.S. consumption of gasoline (in tens of millions of 1967 dollars) versus the price index for gasoline (in 1967 dollars) for the years 1960 through 1986. Standard economic theory states that there should be an inverse relationship between demand for a product and its price, but the observed relationship (with a spline estimate superimposed) is instead direct (and nonlinear) over most of its range. Does this mean that the economic theory fails in practice?

Additive modeling can help resolve this apparent inconsistency. Figure 5.36(a) summarizes the contributions of per capita disposable income and price index for used cars (both in 1967 dollars) to an additive model (that also includes price index for gasoline) based on splines. The curve for per capita disposable income shows a direct linear relationship given the other variables, while that for price index of used cars shows slight nonlinearity. Figure 5.36(b) gives the corresponding plot for price index for gasoline, and corrects the mistaken impression from Fig. 5.35. As economic theory suggests, given per capita disposable income and price index of used cars,

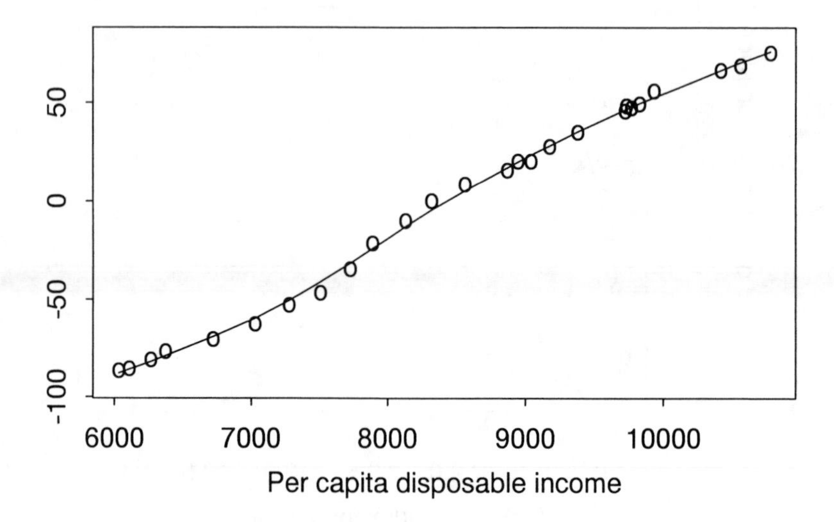


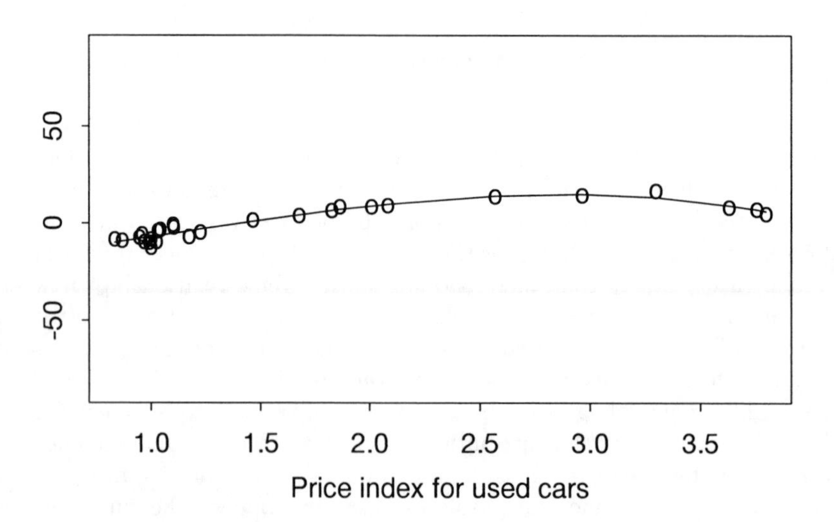


**Fig. 5.36(a).** Plots of contributions of per capita disposable income (top) and price index for used cars (bottom) to additive model estimate of total gasoline consumption, with partial residuals superimposed.

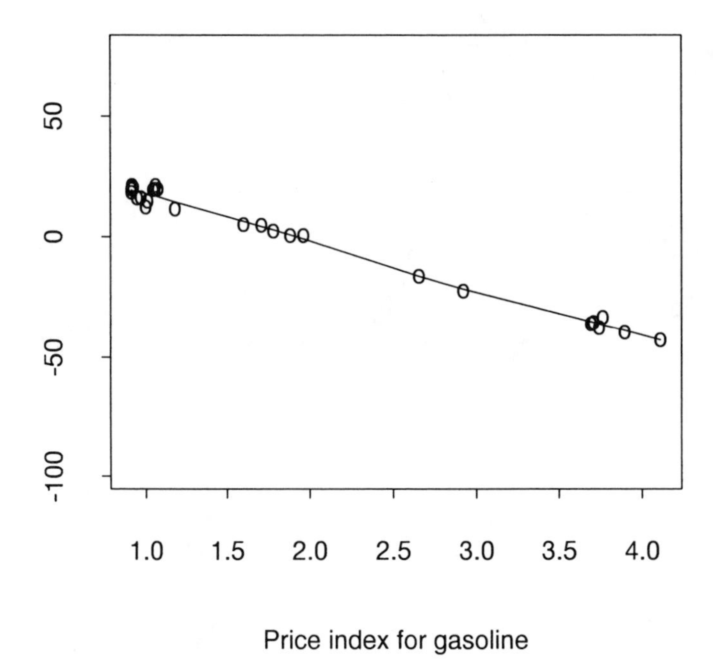


**Fig. 5.36(b).** Plot of contribution of price index for gasoline to additive model estimate of total gasoline consumption, with partial residuals superimposed.

the gasoline demand is inversely, and linearly, related to the price index of gasoline.

An appealing generalization of additive modeling is to allow alternatives to the "regression curve plus error" form (5.2) using the likelihood function. This is the essence of the *generalized additive model.* In this model, the response $y$ has a density in the exponential family,

$$f_Y(y) = \exp\left[\frac{y\theta - b(\theta)}{a(\phi)} + c(y, \phi)\right], \tag{5.28}$$

where $\theta$ is called the natural parameter, and $\phi$ is a scale parameter. In a generalized linear model, the mean of $y$, $\mu$, is related to covariates by $g(\mu) = \eta = \alpha + \beta_1 x_1 + \cdots + \beta_d x_d$, and $g(\cdot)$ is called the link function. So, for example, in a logistic regression, $y$ is binomially distributed, and the link function is the logit link

$$\eta = \log\left(\frac{\mu}{1-\mu}\right).$$

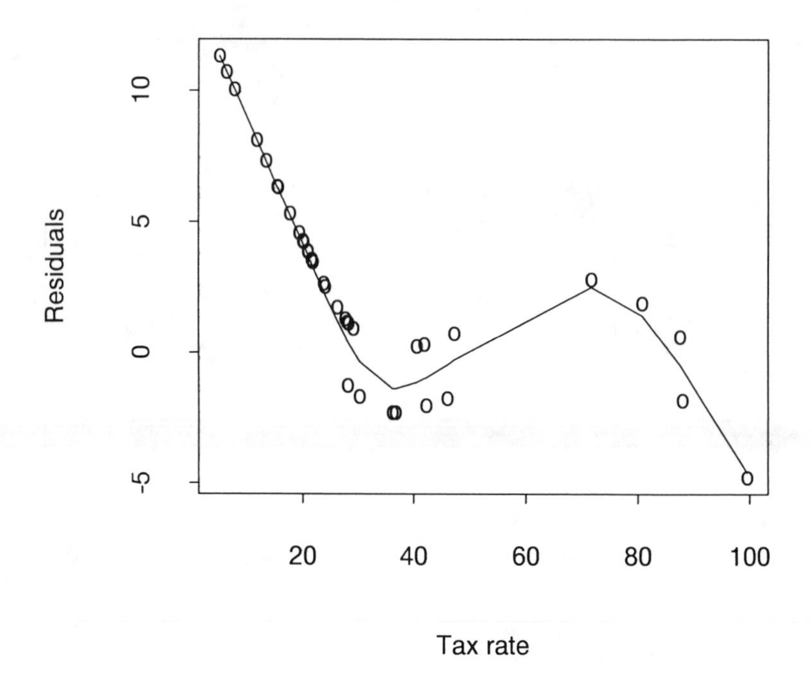


**Fig. 5.37.** Plot of spline fit to generalized additive model estimate of probability of passage of school budget from tax rate in logit scale.

The mean $\mu$ is related to the natural parameter $\theta$ by $\mu = b'(\theta)$, and the link function for which $\eta = g(\mu) = \theta$ is called the canonical link (this is the logit link for binomial data, for example).

The generalized additive model generalizes this by allowing the link function to be additive, as in $g(\mu) = \alpha + \sum_{j=1}^{d} f_j(x_j)$, rather than only linear. Generalized linear models are special cases of generalized additive models, with the functions $f_j(x_j)$ taken to be $f_j(x_j) = \beta_j x_j$. Another special case is to have both linear contributors to $\eta$ and (at least) one smooth term, which results in a generalized partially linear model, or more generally a semiparametric model.

Figures 5.37 and 5.38 illustrate application of a generalized (logistic) model. The data relate the voting result (passage or failure to pass) in 38 Long Island school district budget votes in 1993 to the average proposed equalized property tax rate (in dollars per \$100 of assessed valuation) in the budget. Figure 5.37 gives the spline fit to the probability of passage in the logit scale. A linear logistic regression fit would correspond to a straight line in this plot, but the fit here is decidedly nonlinear (the plotted data points are the partial residuals from this logistic fit). The probability

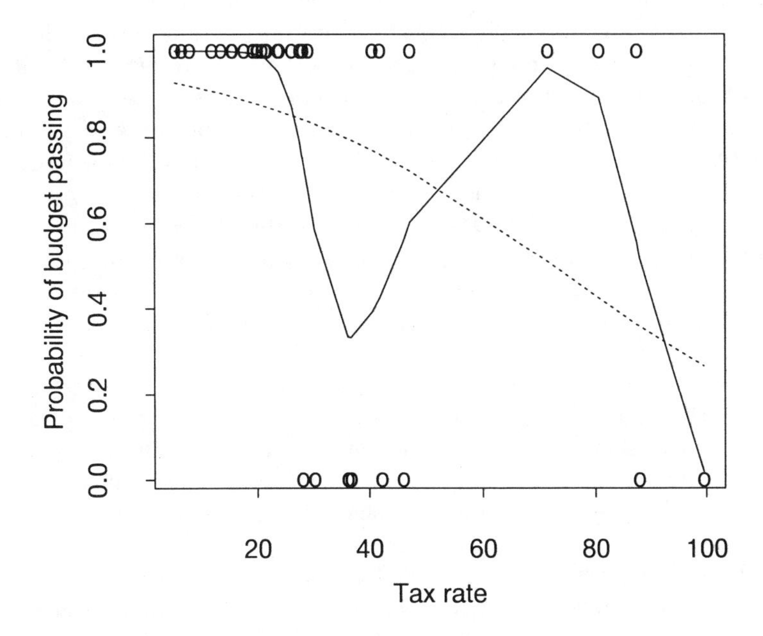


**Fig. 5.38.** Plot of generalized additive (solid curve) and generalized linear (dashed curve) estimates of probability of passage of school budget in the original scale.

of passage decreases for tax rates up to \$40, whereupon it rises for rates up to \$80 and then falls again. Thus, the model implies that besides a general pattern of increasing tax rate being associated with failure to pass the budget (an expected result), there is a "middle ground" of rates in the range of \$40–\$80 where the probability of passage is higher.

Figure 5.38 represents the data in the original scale, with the fitted estimate based on the spline model (solid curve) and linear model (dashed curve) superimposed. The spline model fits the observed pattern more accurately than the linear model does, as it picks up a sharp drop in passage rate for tax rates between \$30 and \$40, followed by an increase for rates between \$40 and \$80. The logistic regression model cannot fit such a pattern, as the fitted probability of success is a monotone function of tax rate. Of course, the evidence for nonmonotonicity is based on only a few observations, so it should be considered somewhat speculative.

## 5.8 Comparing Nonparametric Regression Methods

Local polynomial estimators and smoothing spline estimators have been the focus of this chapter, although they are not, by any means, the only possible approaches to nonparametric regression. There are good reasons for this focus. Each of these estimation schemes can be viewed as a "natural" approach to smoothing; each has reasonable properties in the sense of closeness to the true regression function (including in the boundary region); and each is accessible (in various forms) in generally available software.

Having said that, there are also considerable differences between the methods. Local polynomial estimators are easy to interpret, since they generalize the most commonly used statistical method — linear regression — to allow local nonlinearity (and easily allow local variations in smoothing). This local regression nature also allows the explicit use of least squares regression theory to derive the properties of the estimator. Spline estimators put the smoothing problem in the appealing framework of optimizing a penalized version of the likelihood, resulting in elegant theoretical developments, but it is much more difficult to understand how the smoothing spline uses the observed data in any local sense (the only method being to appeal to a kernel approximation).

Historically, a strong argument for roughness penalty methods over kernel estimators was the ease with which they could generalize from a least squares criterion (which is implicitly based on a Gaussian density for the error) to arbitrary likelihood functions. If the conditional density of $y$ given the predictors comes from the exponential family (5.28), for example, the roughness penalty estimate of the mean function $\eta(x)$ using the canonical link is the minimizer of

$$-\frac{1}{n}\sum_{i=1}^{n}\{y_i\eta(x_i) - b[\eta(x_i)]\} + \alpha\int \eta^{(\ell)}(u)^2\,du$$

(the scale parameter $\phi$ is a nuisance parameter and is ignored in the minimization).

Local polynomial estimators, however, are as easy to generalize to arbitrary likelihoods, using the idea of local likelihood. The corresponding local linear estimator of $\eta(x)$ to the roughness penalty estimator above is $\hat\beta_0$, where $\hat{\boldsymbol\beta}$ is the maximizer of

$$\sum_{i=1}^{n}\left[\frac{y_i(\beta_0+\beta_1 x_i) - b(\beta_0+\beta_1 x_i)}{a(\phi)} + c(y_i,\phi)\right] K\left(\frac{x-x_i}{h}\right)$$

(the local nature of the maximization means that the first order asymptotic properties of $\hat\eta$ do not depend on $\phi$). Link functions other than the canonical link also can be used for both the roughness penalty and local polynomial estimators.

Ultimately, the choice between the two approaches might come down to the ease of smoothing parameter selection. With the increasing use of smoothers as ingredients of more complex estimators (such as in generalized additive modeling), effective automatic choice of smoothing parameters becomes more important. On this count local polynomial estimators have a clear advantage. Their kernel-like nature allows adaptations of stable bandwidth methods (such as plug-in methods) in addition to cross-validation-based methods, while such approaches have not been developed for roughness penalty estimators.

## Background material

### Section 5.1

Weisberg (1985, pp. 231 and 234) gave the geyser eruption data. The auto exhaust data are from Brinkman (1981) and are included as part of the S–PLUS package.

Nadaraya (1964) and Watson (1964) proposed the Nadaraya–Watson kernel estimator. Greblicki (1974) introduced the kernel estimator (5.4), which has been studied by Johnston (1982), Greblicki and Krzyzak (1980) and Georgiev (1984a,b). Priestley and Chao (1972) and Benedetti (1977) proposed and studied the Priestley–Chao kernel estimator. Jones, Davies, and Park (1994) proposed the modified Priestley–Chao estimator $\hat{m}_{\mathrm{PC1}}$. That paper examined different formulations of kernel estimators, focusing on the relative benefits of using estimates of $f_X(x)$ ("external") or $f_X(x_i)$ ("internal") in the kernel weights $w_i$, with the internal choice being preferred. Gasser and Müller (1979, 1984) originated the Gasser–Müller estimator, and Cheng and Lin (1981) examined the special case $s_i = x_i$. Jennen-Steinmetz and Gasser (1988) described a generalization of the estimator. Altman (1992) discussed different kernel-type approaches to regression estimation, while Härdle (1990) gave a book-length treatment.

### Section 5.2

**5.2.1.** Hastie and Loader (1993), Cleveland and Loader (1996), and Chapter 5 of Wand and Jones (1995) gave general descriptions of local polynomial estimation. The second reference describes early applications of the idea, dating back to Spencer (1904) and Macaulay (1931). Fan and Gijbels (1996) is a book-length treatment of local polynomial estimation.

**5.2.2.** Stone (1977) examined the consistency properties of many nonparametric regression estimators, including local polynomial estimators. Cleveland (1979) was a catalyst for renewed interest in local polynomials, introducing lowess as using a tricube kernel function with bandwidth based on nearest neighbor distances (the tricube kernel is $K(x) = (1 - |x|^3)^3$ for $x \in (-1, 1)$ and zero otherwise).

Katkovnik (1979) and Cleveland and Loader (1996) gave exact expressions for the conditional bias and variance of $\hat{m}_p$. For $p = 1$, if $m$ has bounded second derivative, then

$$E(\hat{m}_1(x) - m(x)|x_1, \ldots, x_n) = \frac{1}{2}\sum_{i=1}^{n}(x - x_i)^2 \ell_i(x) m''(\theta_i),$$

where

$$\ell(x)' = (\ell_1(x), \ldots, \ell_n(x)) = \mathbf{c}(x)'(X_x' W_x X_x)^{-1} X_x' W_x,$$
$$\theta_i \text{ satisfies } (x - \theta_i)(x_i - \theta_i) \leq 0,$$

and $\mathbf{c}(x)' = (1, x, \ldots, x^p)$. That is, for any sample size and any design, the bias does not involve $m'$. Thus, the local linear estimate is unbiased if the true regression relationship between $x$ and $y$ is linear. The same is true for quadratic functions if local quadratic fitting is used, and so on. For general $p$, the conditional variance equals

$$V(\hat{m}_p(x)|x_1, \ldots, x_n) =$$
$$\sigma^2(x)\mathbf{c}(x)'(X_x' W_x X_x)^{-1} X_x' W_x^2 X_x (X_x' W_x X_x)^{-1}\mathbf{c}(x).$$

Several authors have examined the asymptotic properties of local polynomial estimators, including Lejeune (1985), Tsybakov (1986), Müller (1987), Fan (1992, 1993), Ruppert and Wand (1994), and Fan, Gasser, Gijbels, Brockmann, and Engel (1997). Fan (1992, 1993) showed that the local linear estimator is (close to) optimal, in the following sense. Let $\mathcal{C}_2$ be the set of joint densities $f$ that satisfy

$$\mathcal{C}_2 = \{f(\cdot,\cdot) : |m(x) - m(x_0) - m'(x_0)(x - x_0)| \leq C(x - x_0)^2/2,$$
$$|m(x_0)| \leq C^*\}$$
$$\cap \{f(\cdot,\cdot) : \sigma^2(x) \leq B, f_X(x_0) \geq b, |f_X(x) - f_X(y)| \leq c|x - y|^\alpha\}$$

for $C$, $C^*$, $B$, $b$, $c$, and $\alpha$ positive constants (the condition on $m$ is slightly weaker than having a bounded second derivative). Then the local linear estimator based on the Epanechnikov kernel achieves the minimum value over all linear smoothers of the maximum asymptotic conditional MSE over $\mathcal{C}_2$ (that is, it is minimax optimal over linear smoothers) and has asymptotic minimax efficiency at least 89.6% over all smoothers over $\mathcal{C}_2$. In order to guarantee that the denominator in (5.6) is not zero, Fan added $n^{-2}$ to that denominator, which results in finite unconditional variance.

Lejeune (1985), Müller (1987), Chu and Marron (1991a) (and the discussion thereon), Jones, Davies, and Park (1994), and Seifert and Gasser (1996a) compared local polynomial estimators to earlier kernel estimators. Various strengths and weaknesses of kernel estimators were catalogued and compared (referencing earlier theoretical work on these estimators), and the possible position of local polynomial estimation as a "gold standard" was noted and evaluated.

Seifert and Gasser (1996b) closely examined the conditional and unconditional properties of local polynomial estimators for finite samples. They noted that sparse regions of the design lead to unstable behavior of local polynomial estimators using compact kernels and infinite unconditional variance (and hence infinite unconditional MSE). In contrast, everywhere positive kernels yield finite unconditional variance and MSE.

Seifert and Gasser (1996b) also described ways to improve the variance properties of local polynomial estimators based on compact kernels, including increasing the bandwidth locally in sparse regions until the variance of the estimator is stabilized, and modifying the local least squares regression estimator to be a local ridge regression estimator. Hall and Turlach (1996) suggested handling sparse design regions by adding interpolated "pseudodata" to the original data set in places where there are gaps in the design.

Fan and Marron (1994) and Seifert *et al.* (1994) described fast implementations of kernel and local polynomial estimates that improve on naive application of (5.5) using binning and updating formulas. Seifert and Gasser (1996a) described improvements of both the naive and updating approaches. Cleveland and Grosse (1991) described the implementation of loess, which involves fitting the estimate at a small number of points and then interpolating the value at any desired evaluation points.

**5.2.3.** Berenson and Levine (1992, p. 66) gave the newspaper circulation data, while Chatterjee, Handcock, and Simonoff (1995, p. 292) gave the electricity usage data set and provided regression analysis of it.

Fan and Gijbels (1992) derived and described the boundary properties of the local linear estimator. The conditional variance of $\hat{m}_{\rm NW}$ at the boundary is

$$V(\hat{m}_{\rm NW}(x)|x_1,\ldots,x_n) = \frac{2R(K)\sigma^2(x)}{nhf_X(x)} + o_p[(nh)^{-1}],$$

or twice the conditional variance in the interior. Comparing this value to the value in (5.10) shows that the conditional variance of $\hat{m}_1$ is roughly 3 to 3.5 times that of $\hat{m}_{\rm NW}$ at the boundary, depending on the kernel. Ruppert and Wand (1994) extended these results to higher degree polynomials. The boundary kernel methods described in Chapter 3 also will work for kernel regression estimators (indeed, most were originally proposed in the regression rather than density estimation context).

Cheng, Fan, and Marron (1997) established the minimax optimality of $\hat{m}_1$ at the boundary. Specifically, they showed that the minimum value at the left boundary for all linear smoothers of the maximum asymptotic MSE over $\mathcal{C}_2$ is attained by a local linear estimator using the kernel

$$K(u) = \begin{cases} 1-u, & \text{for } 0 \le u \le 1, \\ 0, & \text{otherwise.} \end{cases}$$

Using a Gaussian kernel results in only a 2% loss in efficiency compared with the optimal kernel. Sidorenko and Riedel (1994) also described optimal kernels for local polynomial estimators in the boundary region.

The problem of constructing confidence bands for nonparametric regression functions is very difficult. The bias of $\hat{m}(x)$ makes the usual parametric-type approaches ineffective, unless it is accounted for in some way. Also, if $\sigma^2(x)$ is not constant, confidence regions must account for the heteroscedasticity. The distribution of the error term $\epsilon$ and the marginal density $f_X$ (for random designs) also have a potentially strong effect on the properties of confidence regions.

Several different approaches to confidence region construction have been proposed. These include ones where the interval is centered on a bias-corrected estimate of $m(x)$, using the asymptotic variance to determine the width of the region (Eubank and Speckman, 1993); ones based on bootstrapping (Härdle and Bowman, 1988; Faraway, 1990b; Härdle and Marron, 1991; Hall, 1992b); and ones based on, or related to, tube formulas for coverage probabilities (Knafl, Sacks, and Ylvisaker, 1985; Hall and Titterington, 1988; Johansen and Johnstone, 1990; Sun and Loader, 1994). Loader (1993) compared the properties of some of these methods.

A more general question is the utility of such confidence regions at all. An obvious use of such regions is to try to determine if the data are consistent with some parametric model (a linear model, for example), but this can be formulated easily as a goodness-of-fit problem. Chapter 7 includes discussion of such smoothing-based goodness-of-fit tests.

**5.2.4.** The vineyard data can be found in Chatterjee, Handcock, and Simonoff (1995, p. 304). That reference also provides an extensive description and discussion of the vineyard and its characteristics, and includes a (somewhat crude) map of the property (p. 86).

Cleveland and Loader (1995) noted that the "best" degree $p$ of local polynomial fitting might not be obvious, in that (for example) a higher value of $p$ might be needed to estimate sharp peaks and troughs in the regression curve but could lead to unacceptably large variance near the boundary. They proposed choosing the degree to be a possibly noninteger value $b$. Let $b = p + f$, where $p$ is an integer and $0 \le f < 1$. If $f = 0$, then use a local polynomial of degree $p$; otherwise, the estimated regression function is a weighted average of the estimates for local polynomials of degree $p$ and $p+1$, with the weights being $1-f$ and $f$, respectively. The mixing parameter $b$ is chosen to minimize a cross-validation sum of absolute deviations,

$$\sum_{i=1}^{n} \frac{|y_i - \hat{m}(x_i)|}{1 - h_{ii}(x_i)}, \tag{5.29}$$

where $h_{ii}(x_i)$ is the $i$th diagonal element of the prediction matrix $H_{x,b}$ for the mixed degree polynomial estimator.

## Section 5.3

Bandwidth selection for kernel regression estimators has been the subject of a good deal of research. Härdle and Marron (1986) showed that the choice of squared error criterion is asymptotically unimportant, in the sense that for kernel estimators (assuming boundedness and continuity conditions),

$$\sup_{h \in H_n} \left| \frac{d(h) - \text{MISE}(h)}{\text{MISE}(h)} \right| \stackrel{\text{a.s.}}{\rightarrow} 0$$

as $n \to \infty$, where $H_n = [n^{\delta-1}, n^{\delta}]$, $0 < \delta < .5$, and $d(h)$ represents the distances ASE$(h)$, MASE$(h)$, and ISE$(h)$. Thus, any sequence of bandwidths that is asymptotically optimal with respect to one of these distances (in the sense used earlier, as in (2.20)) is optimal with respect to MISE as well. Härdle and Marron (1985) showed that $\hat{h}_{\text{CV}}$ is such a selection method.

Kernel regression bandwidth selectors fall into three main categories. The "leave-one-out" form of CV$(h)$ is due to Stone (1974) and was first applied to kernel smoothing by Clark (1975) (other related references are Allen, 1974, and Wahba and Wold, 1975). A second approach is to base the fitting criterion on the resubstitution estimate of the prediction error,

$$p(h) = \frac{1}{n} \sum_{i=1}^{n} [y_i - \hat{m}(x_i)]^2,$$

modified to be an unbiased estimate of ASE$(h)$. The asymptotic bias of $p(h)$ is

$$-\frac{2}{n^2 h} \sum_{i=1}^{n} \frac{K\left(\frac{x - x_i}{h}\right) \sigma^2(x_i)}{\hat{f}_X(x_i)}.$$

Let $z_i = (nh)^{-1} K(0)/\hat{f}_X(x_i)$. The penalizing function approach chooses a function $\Xi(z_i)$ such that

$$G(h) = \frac{1}{n} \sum_{i=1}^{n} [y_i - \hat{m}(x_i)]^2 \Xi(z_i)$$

is an asymptotically unbiased estimator of ASE$(h)$ (plus a constant). Taylor Series expansion of $G(h)$ shows that the penalizing function should satisfy

$$\lim_{u \to 0} \Xi(u) = 1 + 2u,$$

which still leaves many possible choices of $\Xi$. Various proposals include generalized cross-validation (GCV) $[\Xi(u) = (1-u)^{-2}]$ (Craven and Wahba, 1979); Shibata's selector $[\Xi(u) = 1 + 2u]$ (Shibata, 1981); AIC $[\Xi(u) =$

$\exp(2u)$] (Akaike, 1970); finite prediction error [$\Xi(u) = \{(1+u)/(1-u)\}$] (Akaike, 1974); and Rice's $T$ [$\Xi(u) = (1-2u)^{-1}$] (Rice, 1984b). Härdle, Hall, and Marron (1988) showed that CV($h$) also has this form. They also showed that the selectors based on each of these rules are asymptotically equivalent, the relative error of each converging to the same normal distribution at the rate $O_p(n^{-1/10})$. Monte Carlo simulations showed that the selectors generally tend to undersmooth, and Chiu (1990) provided a theoretical explanation for this pattern.

The third commonly considered approach to bandwidth selection, which includes plug-in selection, allows significant improvement by changing the target from $\hat{h}_0$ to $h_0$, which can be estimated much more accurately. Chiu (1991c) described a selector for the Priestley–Chao kernel estimator on a fixed, uniform grid. The selector is based on replacing $p(h)$ in any of the penalizing function methods with a version that has much less variability, obtained by modifying the periodogram of the observations. Gasser, Kneip, and Köhler (1991) proposed an iterative plug-in selector for the Gasser–Müller kernel estimator and fixed designs. Each of these selectors can achieve a relative rate of $O_p(n^{-1/2})$ to $h_0$, assuming enough derivatives for $m$. (The $O_p(n^{-1/10})$ rate is optimal if only two derivatives are assumed.)

Härdle, Hall, and Marron (1992) proposed a selector for the Nadaraya–Watson estimator that uses two kernel smooths of the data, which they termed *double smoothing*. This approach is related to the smoothed cross-validation method for density estimation of Hall, Marron, and Park (1992) and also can achieve $O_p(n^{-1/2})$ relative convergence to $h_0$.

Herrmann (1997) gave a generalization of the Gasser, Kneip, and Köhler (1991) bandwidth selector to boundary-corrected kernel estimation. The local linear plug-in selector described in the text is due to Ruppert, Sheather, and Wand (1995). They showed that $g$ (the bandwidth used when estimating $\int m''(u)^2 f_X(u)du$) should (asymptotically) satisfy

$$g = C_2(K)\left[\frac{\sigma^2}{n|\int m''(u)m^{(4)}(u)f_X(u)du|}\right]^{1/7},$$

where $C_2$ is a function of $K$ that takes on one of two different forms depending on whether $\int m''(u)m^{(4)}(u)f_X(u)du$ is positive or negative. The bandwidth $\lambda$ (which is used when estimating $\sigma^2$) should satisfy

$$\lambda = C_3(K)\left\{\frac{\sigma^4}{[n\int m''(u)^2 f_X(u)du]^2}\right\}^{1/9},$$

where $C_3(K) = [4R(K*K-2K)/\mu_2(K)^4]^{1/9}$ and $(K*K)(x)$ is the convolution of $K$ with itself, $\int K(u)K(x-u)du$. These bandwidths are themselves functions of unknown properties of $m$ ($m''$ and $m^{(4)}$), which are estimated by dividing the range of $x$ into blocks and estimating $m$ over each block using a polynomial (see Härdle and Marron, 1995). Monte Carlo evidence confirms that the plug-in bandwidth selector tends to undersmooth for near-linear regression functions (Wand, 1995). Turlach and Wand (1996) showed how many of the necessary quantities can be calculated quickly using binning methods.

A rule-of-thumb selector that is appropriate for roughly uniform design density over $[a, b]$ and constant variance $\sigma^2$ can be based on estimating $m''(x)$ and $\sigma^2$ from a fourth order polynomial fit $m(x) = \alpha_0 + \alpha_1 x + \cdots + \alpha_4 x^4$ to the data (Fan and Müller, 1995; Fan and Gijbels, 1996). The resultant bandwidth selector from (5.14) is

$$\hat{h} = \left[ \frac{R(K)(b-a)\hat{\sigma}^2}{n\mu_2(K)^2 \int_a^b (2\hat{\alpha}_2 + 6\hat{\alpha}_3 u + 12\hat{\alpha}_4 u^2)^2 du} \right]^{1/5}.$$

Cleveland and Devlin (1988) suggested constructing regression estimates for a range of smoothing parameters in order to see how the resultant tradeoff between bias and variance affects the estimate for that data set. Fan and Müller (1995) suggested plotting regression estimates based on the set of bandwidths $\{h = 1.4^j \hat{h}_{\mathrm{CV}}, j = -2, \ldots, 2\}$ on the same plot in order to assess the sensitivity of the appearance of $\hat{m}$ to bandwidth choice.

Hall and Marron (1990) proposed the variance estimator (5.15) when using the Nadaraya–Watson kernel estimate and showed that it is optimal, in a minimax sense. Cleveland (1979) suggested it for local polynomial regression. Ruppert, Sheather, and Wand (1995) examined its asymptotic properties when estimating $m$ using $\hat{m}_p$ for $p$ odd. Fan and Gijbels (1995a) proposed a plug-in type estimator $\sigma^2(x)$ for a possibly changing variance that uses a weighted version of $p(h)$,

$$\hat{\sigma}^2(x) = \frac{1}{\mathrm{trace}[W - (X'WX)^{-1}X'W^2X]} \sum_{i=1}^{n} [y_i - \hat{m}(x_i)]^2 K\left(\frac{x_i - x}{h}\right).$$

They also derived a bandwidth selector based on minimizing $\int \mathrm{RSC}(u)du$ over $h$, where

$$\mathrm{RSC}(x) = \hat{\sigma}^2(x)[1 + (p+1)V], \tag{5.30}$$

and $V$ is the first diagonal element of $(X'WX)^{-1}(X'W^2X)(X'WX)^{-1}$.

A different approach to variance estimation comes from time series analysis and estimates $\sigma^2$ using finite differences of observed $y_i$ values. Examples include those of Rice (1984b), Gasser, Sroka, and Jennen-Steinmetz (1986), Hall, Kay, and Titterington (1990), and Seifert, Gasser, and Wolf (1993). Buckley, Eagleson, and Silverman (1988) and Ullah and Zinde-Walsh (1992) developed a general framework for variance estimation based on quadratic functions of the data and discussed optimal choices for that class (see also Carter and Eagleson, 1992).

The bandwidth $h$ can be chosen based on loss functions other than squared error. Jhun (1988) examined absolute error loss for the Nadaraya–Watson estimator and derived the minimizer of asymptotic integrated absolute error.

## Section 5.4

Chatterjee, Handcock, and Simonoff (1995, p. 19) presented the data of Fig. 5.15 and noted the generally inverse relationship between changes in visas from 1988 to 1991 and from 1991 to 1992. Cook and Weisberg (1994, p. 35) analyzed the U.S. birth rate data and noted the general patterns described here.

Chatterjee, Handcock, and Simonoff (1995, p. 306) gave the beluga whale nursing data, and analyzed them using various regression models, including ones based on differencing the data. Russell, Simonoff, and Nightingale (1997) discussed the biological implications of several of those analyses.

Several authors have discussed local variation of the bandwidth for kernel estimators. Müller and Stadtmüller (1987) examined the convolution estimator using an equispaced fixed design and showed that using a data-dependent $\hat{h}(x)$ that is a consistent estimator of the optimal $h_0(x)$ results in an $\hat{m}$ based on this $\hat{h}(x)$ that behaves asymptotically as well as $\hat{m}$ based on $h_0(x)$. Under weak conditions, this implies that the asymptotic MISE of the resultant $\hat{m}$ is less than or equal to that using the optimal global bandwidth $h_0$.

Müller and Stadtmüller (1987) proposed an estimator of $h_0(x)$ based on first estimating the global optimal bandwidth and then adapting it locally using an estimate of $m''(x)$. Staniswalis (1989a) proposed estimating the exact (finite sample) MSE of $\hat{m}$ at a point $x$ with a consistent estimator and then estimating $h_0(x)$ as the minimizer of estimated MSE. Brockmann, Gasser, and Herrmann (1993) and Herrmann (1997) adapted the iterative plug-in method for choosing a global bandwidth of Gasser, Kneip, and Köhler (1991) to allow for local adaptivity. Schucany (1995) proposed estimating the asymptotic bias of $\hat{m}(x)$ for different possible bandwidths, and then estimating $m''(x)$ using a least squares regression fit to this bias/bandwidth relationship. He showed that the estimator achieves a relative rate to $h_0(x)$ of $O_p(n^{-1/7})$, a rate shared by the method of Staniswalis. Schucany (1995) noted the theoretical difficulties of plug-in local bandwidth selection at inflection points (since $m''(x) = 0$) and gave a modification to the selector to try to handle these problems. Vieu (1991) examined the Nadaraya–Watson estimator and proposed choosing $\hat{h}(x)$ to minimize a locally weighted version of cross-validation, which he showed yields an asymptotically optimal bandwidth estimator.

Härdle and Marron (1995) argued that a simple, yet effective method to choose the bandwidth in a data-dependent way is to divide the range of the predictor into blocks and use a plug-in choice within each block based on fitting parametric models within the block. This can lead to either a global (fixed) bandwidth estimate that combines blockwise estimated variances and biases or locally varying bandwidths that are smoothed blockwise bandwidths. They specifically addressed the Nadaraya–Watson estimator, but the methods can be adapted to higher degree local polynomials as well. They estimated $f_X(x)$ (and $f'_X(x)$, which is needed for bias estimation for $\hat{m}_{\mathrm{NW}}$) using a histogram, and estimated $m'(x)$ (needed for $\hat{m}_{\mathrm{NW}}$) and $m''(x)$ by fitting blockwise parabolas to the data. The estimated local variance $\hat{\sigma}^2(x)$ comes from the residual sum of squares from these blockwise parabolas. They recommended taking the number of blocks (which corresponds to a smoothing parameter) to be between three and five.

Fan and Gijbels (1995a) used the RSC estimate of variance (5.30) to motivate a locally varying bandwidth that is also based on blocking the data. Within each block, an optimal bandwidth (for local polynomial fitting of degree $p+2$) is found based on the integral of RSC over that interval. The resultant bandwidths are smoothed, and these smoothed bandwidths are used to estimate $m^{(p+1)}$, $m^{(p+2)}$, and $\sigma^2$ locally. These pilot estimates then yield blockwise minimizers of the estimated MISE. These final bandwidths are then smoothed once more. Monte Carlo simulations showed the potential for the locally varying bandwidths to improve on a global bandwidth choice, particularly for sample sizes of at least 200. Fan and Gijbels (1995a) also applied their method to examples with very complex structure (with $n = 2048$), and found that it performed similarly to the wavelet thresholding approach of Donoho and Johnstone (1994); see also Herrmann (1997).

Cleveland and Loader (1996) proposed choosing a local bandwidth based on minimizing a local version of Mallows' (1973) $C_p$ statistic,

$$C(h) = \frac{1}{\mathrm{trace}(W)}\Big\{2\,\mathrm{trace}[(X'W^{-1}X)^{-1}(X'W^2X)] - \mathrm{trace}(W) + \frac{1}{\hat{\sigma}^2}\sum_{i=1}^{n} w_i(x)[y_i - \hat{m}_p(x_i)]^2\Big\},$$

for any $x$ over a suitably chosen interval on $h(x)$.

Results of Wang and Gasser (1996) reinforce the difficulty in determining estimators of locally optimal bandwidths. They showed that the best possible relative rate of $\hat{h}(x)$ to the minimizer of ISE$(x)$ is

$$\frac{\hat{h}(x)}{\hat{h}_0(x)} - 1 = O_p(n^{-2/45}), \tag{5.31a}$$

while the best possible relative rate to the minimizer of MISE$(x)$ is

$$\frac{\hat{h}(x)}{h_0(x)} - 1 = O_p(n^{-2/9}) \tag{5.31b}$$

assuming four derivatives for $m$. Fan, Gijbels, Hu, and Huang (1996) showed that the selector of Fan and Gijbels (1995a) achieves the rate (5.31b).

The rates (5.31) can be compared with the values for global bandwidth selection, $O_p(n^{-1/10})$ and $O_p(n^{-1/2})$, respectively. Since they converge to zero considerably more slowly, data-dependent locally varying bandwidth selection requires much larger samples to be reasonable than does data-dependent global bandwidth selection.

Another way to allow a local polynomial estimator to have different degrees of smoothing locally is to vary the degree of the polynomial, rather than vary the bandwidth for a fixed-degree local polynomial. Cleveland and Loader (1995) generalized their polynomial mixing estimator to allow varying degrees of mixing, chosen using a locally weighted version of the cross-validation criterion (5.29). Fan and Gijbels (1995b) suggested estimating the MSE of different (integer) degree polynomials for a given bandwidth at different locations and then choosing the degree for each location that minimizes the estimated MSE. Monte Carlo evidence suggests that this approach is insensitive to the bandwidth choice, often allowing a simple rule-of-thumb bandwidth selector to be used (thereby avoiding the problem of accurate bandwidth selection entirely).

**Section 5.5**

**5.5.1.** Cleveland (1979) proposed the iterative algorithm to construct a robust version of the loess estimate (his paper referred to loess's predecessor, lowess). That paper recommended two iterations of the robust downweighting operation, but the default in the S–PLUS implementation of loess is four, and it is the latter number that is given here. Mächler (1989) noted the potential nonrobustness of basing the robust estimate on the nonrobust version and proposed corrective actions.

Carroll and Ruppert (1988, p. 140) gave the Skeena River sockeye salmon data and fitted linear and nonlinear regression models to them. Tiede and Pagano (1979) gave the radioimmunoassay calibration data and noted the effect of the outlier on nonlinear regression estimation. Chatterjee, Handcock, and Simonoff (1995, p. 306) gave the voting fraud data and investigated the statistical evidence of fraud in the 1993 special election.

Several authors have discussed methods of robustifying the kernel regression estimator directly. *M-type kernel smoothers* are formed by minimizing a new local criterion; rather than the estimator being the minimizer $\hat{\beta}_0$ of

$$\sum_{i=1}^{n} [y_i - \beta_0]^2 K\left(\frac{x - x_i}{h}\right),$$

it is the minimizer of

$$\sum_{i=1}^{n} \rho[y_i - \beta_0] K\left(\frac{x - x_i}{h}\right), \tag{5.32}$$

where the function $\rho(\cdot)$ is chosen to downweight outlying observations. The minimizer solves the equation

$$\sum_{i=1}^{n} \psi[y_i - m(x)] K\left(\frac{x - x_i}{h}\right) = 0, \tag{5.33}$$

where $\psi = \rho'$, assuming differentiability of $\rho$. The usual Nadaraya–Watson estimator takes $\psi(u) = u$, while robust versions take $\psi$ to be monotone nondecreasing, antisymmetric, and bounded. A common example is the proposal of Huber (1981), $\psi(u) = \max[-c, \min(u, c)], c > 0$.

Brillinger (1977) proposed the use of $M$-type smoothers. Stuetzle and Mittal (1979) studied this estimator based on a uniform kernel. Härdle (1984) generalized the estimator to general kernels and proved consistency and asymptotic normality; see also Boente and Fraiman (1989). Härdle and Tsybakov (1988) extended the estimator to include simultaneous estimation of the scale function $\sigma(x)$. They showed that the resultant estimator of $m(x)$ has the same asymptotic bias as that of $\hat{m}_{\mathrm{NW}}$, with asymptotic variance that depends on $\psi$ and the true conditional distribution of $y|x$. Härdle (1987) described a one-step version of the $M$-smoother (which corresponds to one iteration of the solution to (5.33), using $\hat{m}_{\mathrm{NW}}$ as the starting value) and gave Fortran code to calculate it. Härdle and Gasser (1984), Antoch and Janssen (1989), and Hall and Jones (1990) discussed the construction of $M$-type convolution estimators. Kozek (1992) described and established consistency of a corresponding $M$-estimation construction for local polynomials, which is defined as in (5.32), adding higher degree polynomial terms. Fan, Hu, and Truong (1994) derived the asymptotic normal distributions for the $M$-type local linear estimator both in the interior and at the boundary. Manchester (1996) described the construction of a graphical method (termed the influence surface) to assess the effect of a single (possibly outlying) observation on a regression estimate for a given data set.

Kozek (1992) also proposed using general local nonlinear functions, replacing local polynomials, when the context of the problem suggests such functions. Jones and Hjort (1995) noted that the order of magnitude of the bias of the resultant regression estimator depends only on the number of parameters in the function, rather than its specific form, although the exact form of the bias can be simpler if the local nonlinear function is closer to the true curve than a polynomial is.

Fan and Hall (1994) examined least absolute deviation smoothing ($\rho(u) = |u|$ in (5.32)). They examined the estimator assuming one bounded derivative for $m$ and showed that the estimator is asymptotically optimal in a minimax sense (with a convergence rate of $n^{-1/3}$). Wang and Scott (1994) extended this idea to local polynomial estimation. They discussed the linear programming problems involved in determining the estimate and gave recommendations on how to calculate it, including how to choose the smoothing parameter (or span, for nearest neighbor bandwidths); see also Wang (1994). They examined the asymptotic properties of the estimator assuming a fixed, equispaced design and obtained the usual rates of convergence for the estimator (that is, MSE $= O(n^{-4/5})$ for local linear fitting, MSE $= O(n^{-8/9})$ for local quadratic fitting, and so on).

The field of robust estimation and outlier identification is very active. Book-length treatments include Hawkins (1980), Huber (1981), Hampel *et al.* (1986), and Barnett and Lewis (1994). Hadi and Simonoff (1993, 1994), and the references therein, proposed alternatives to robust estimation to handle outliers in regression models, which could potentially be adapted to local polynomial estimators as well.

**5.5.2.** There are many books on the analysis of time series data; see, for example, Brillinger and Krishnaiah (1984), Aoki (1990), and Brockwell and Davis (1991). Work on nonparametric regression with autocorrelated errors has focused on time series data, with equispaced time values as the predictor values, and the use of the convolution estimators $\hat{m}_{\mathrm{PC}}$ and $\hat{m}_{\mathrm{GM}}$. An exception is Chu and Marron (1991b), which focused on $\hat{m}_{\mathrm{NW}}$. They derived the asymptotic MISE of the estimator assuming a covariance stationary process for the errors (see also Altman, 1990, and Hart, 1991).

Several authors have noted the tendency for data-based bandwidth selectors to undersmooth for positively autocorrelated errors and oversmooth for negatively autocorrelated errors. Chiu (1989) formulated the problem in the frequency domain, and showed why the Mallows' (1973) $C_L$ criterion has this property (see also Hurvich and Zeger, 1990), while Hart (1991) worked in the time domain and showed that bandwidths based on cross-validation will severely undersmooth for positively autocorrelated errors (see also Hart and Wehrly, 1986).

Chu and Marron (1991b) proposed modifying cross-validation to be "leave $(2\ell + 1)$ out" (ordinary cross-validation taking $\ell = 0$), with observations $(x_{j+i}, y_{j+i}), -\ell \leq i \leq \ell$ being omitted in the calculation of $\hat{m}(x_j)$. They showed that if the errors follow an ARMA process, the modified cross-validated bandwidth choice converges in distribution to a normal distribution centered at $h_{0,\gamma}$ at the rate $n^{-1/10}$ as long as $\ell$ is large enough (without the need for any knowledge of the true covariance structure). This rate, of course, is the same (slow) rate achieved by ordinary cross-validation for independent errors.

Chiu (1989), Altman (1990), and Hart (1991) examined the modification of penalizing functions (such as GCV and $C_L$) for bandwidth selection with autocorrelated errors, based on a plug-in approach where the penalizing function criterion is modified to reflect the autocorrelation. These methods require specification of the true error process; if the specification is correct, and the covariance estimates are $\sqrt{n}$-consistent, then the resultant bandwidths have the same asymptotic distribution as that of modified cross-validation for large enough $\ell$ (Chu and Marron, 1991b). Hart (1991), Truong (1991), and Altman (1993) discussed the estimation of $\gamma(\cdot)$ in the nonparametric regression context.

Herrmann, Gasser, and Kneip (1992) and Quintela del Río (1994) proposed direct plug-in methods based on (5.20) under certain models for the error process. They established the consistency of these selectors assuming correct specification of the error process and gave some Monte Carlo evidence of their ability to address observed autocorrelation. Hart (1994) proposed a prediction-based version of cross-validation that he termed *time series cross-validation* (TSCV). The method attempts to find a good one-step-ahead predictor based on past data, using a specified parametric model for the error process. Hart showed that for a correctly chosen error process, minimizing TSCV is asymptotically equivalent to minimizing MASE.

The results of this section assume that the error process has a finitely summable covariance function. Many real-world series have error terms with autocorrelation that decays so slowly that this condition is violated (equivalently, the spectrum is infinite at frequency zero). Such processes are termed *long memory processes.* The behavior of nonparametric regression estimators is very different from the previously noted results for processes of this type. Hall and Hart (1990b) showed that if the errors have an autocorrelation function of the form $\rho(k) = C_3 k^{-\alpha}$, for $C_3 > 0$ and $0 < \alpha < 1$, then the optimal bandwidth that minimizes the asymptotic MISE is

$$h_0' = \left[\frac{C_3\alpha \int\int |u-v|^{-\alpha} K(u)K(v)du\,dv}{n^\alpha \mu_2(K)^2 \int m''(u)^2 f_X(u)du}\right]^{1/(4+\alpha)}.$$

The optimal rate for MISE is $O(n^{-4\alpha/(4+\alpha)})$, which is slower than the usual convergence rate. Ray and Tsay (1996) proposed a plug-in estimator for $h_0'$ and showed that if $\alpha$ and $C_3$ are estimated consistently, the estimated bandwidth is asymptotically equivalent to $h_0'$. Their simulations suggested that the proposed long memory method produces bandwidths closer to $h_0'$ than the method of Herrmann, Gasser, and Kneip (1992) under long memory processes, although the actual ISE and relative ISE values were not significantly different. See also Csörgő and Mielniczuk (1995b).

## Section 5.6

**5.6.1.** Several book-length treatments focusing on roughness penalty methods have appeared in recent years. Green and Silverman (1994) gave a very accessible account, while Eubank (1988) and particularly Wahba (1990) provided more theoretical treatments. See also Silverman (1985). Eilers and Marx (1996) defined the penalty function to be based on finite differences of the coefficients of adjacent B-splines (a P-spline); see also O'Sullivan (1986, 1988).

**5.6.2.** Whittaker (1923) originated the idea of spline smoothing, terming the smoothing process a graduation of the observations. The modern formulation of spline smoothing as a nonparametric regression estimator is due to Schoenberg (1964) and Reinsch (1967), who proved that the minimizer of (5.21) with $\ell = 2$ is a cubic smoothing spline. Wahba (1975b) showed that the estimator is linear in the observations, and she derived asymptotic MISE properties of it. Analysis based on the general form of the roughness penalty (5.21) generalizes the case $\ell = 2$. Kimeldorf and Wahba (1970a,b) showed that the minimizer is a polynomial spline of degree $2\ell-1$ with knots at the data points. If $m \in W_2^\ell$, MISE $= O(n^{-2\ell/(2\ell+1)})$, the optimal rate (Wahba, 1978; Craven and Wahba, 1979).

Silverman (1984) derived the approximate kernel form for the cubic smoothing spline. Messer (1991) and Messer and Goldstein (1993) investigated the connections between the smoothing spline and approximate kernel further, and the latter paper formulated asymptotically equivalent, closed form, boundary corrected kernels for this problem. See also Nychka (1995).

Rice and Rosenblatt (1983) proved the existence of boundary bias problems for the cubic smoothing spline if three or four derivatives of $m$ are bounded but the second or third derivatives are nonzero at the boundaries. Utreras (1988) extended these results to $\ell > 2$. Oehlert (1992) derived what he termed "relaxed boundary splines" that correct the boundary bias problem for equispaced data design for $\ell \geq 4$ (that is, if $m \in W_2^{2\ell}$, MISE $= O(n^{-4\ell/(4\ell+1)})$) without any boundary conditions on $m$.

Kimeldorf and Wahba (1970a,b) investigated the Bayesian formulation of smoothing splines. Wahba (1978) provided a Bayesian justification based on Gaussian processes for smoothing splines. If the prior has the distribution that is the same as the distribution of the stochastic process

$$X(t) = \sum_{j=1}^{\ell} \theta_j \phi_j(t) + \sigma(n\alpha)^{-1/2} Z(t),$$

where $\boldsymbol{\theta} = (\theta_1, \ldots, \theta_\ell)' \sim N(\mathbf{0}, \xi I_{\ell\times\ell})$, $\phi_j(t) = t^{j-1}/(j-1)!$, $j = 1, \ldots, \ell$, and $Z(\cdot)$ is the $\ell$-fold Wiener process,

$$Z(t) = \int_0^t \frac{(t-u)^{\ell-1}}{(\ell-1)!}\, dW(u),$$

$W(u)$ being the Wiener process, then the smoothing spline is the mean of the posterior process if $\xi \to \infty$ (a diffuse prior on $\boldsymbol{\theta}$).

Wahba (1983) used this Bayesian formulation to justify the construction of confidence intervals for the true function $m$ based on the smoothing spline. The intervals are constructed based on the posterior distribution of $m$ given the data, but Monte Carlo evidence suggested that they have good frequentist properties as well, in the following sense: in repeated sampling from the regression model, roughly 95% of the simulated observations are covered by each newly constructed 95% confidence region (Wahba, 1985). The reason for these good frequentist properties is that the average posterior variance is close to the MASE. See Silverman (1985), Hall and Titterington (1987a), and Nychka (1988, 1990). Wang and Wahba (1995) compared the properties of these Bayesian confidence intervals to bootstrap-based intervals.

**5.6.3.** Wahba and Wold (1975) first proposed the use of cross-validation to choose $\alpha$. Craven and Wahba (1979) introduced generalized cross-validation and argued for its superior performance over cross-validation for unequally spaced data. Marron (1985b) pointed out that the approximate kernel form for the spline implies that $\hat{\alpha}_{\text{CV}}$ and $\hat{\alpha}_{\text{GCV}}$ have properties similar to those in the kernel regression context.

The Bayesian formulation of the smoothing spline estimator given in Wahba (1978) allows the smoothing parameter selection problem to be addressed using the (marginal) likelihood function. The resultant generalized maximum likelihood choice is the minimizer of

$$\text{GML}(\alpha) = \frac{\mathbf{y}'[I - A(\alpha)]\mathbf{y}}{\{\det^+[I - A(\alpha)]\}^{1/(n-\ell)}},$$

where $\det^+[I - A(\alpha)]$ is the product of the $n - \ell$ nonzero eigenvalues of $I - A(\alpha)$ (Anderssen and Bloomfield, 1974; Barry, 1983; Wecker and Ansley, 1983; Wahba, 1985). Wahba (1985) showed that for a certain class of functions $m$, the MASE based on $\hat{\alpha}_{\text{GML}}$ converges to zero at a rate slower than that when using $\hat{\alpha}_{\text{GCV}}$, while for other classes of smooth functions, the MASE properties are similar. Kohn, Ansley, and Tharm (1991) used Monte Carlo simulations to compare $\hat{\alpha}_{\text{CV}}$, $\hat{\alpha}_{\text{GCV}}$, and $\hat{\alpha}_{\text{GML}}$. They found that the other methods outperformed $\hat{\alpha}_{\text{CV}}$ for unequally spaced data, while $\hat{\alpha}_{\text{GCV}}$ and $\hat{\alpha}_{\text{GML}}$ had similar properties for cubic smoothing splines. They also found that quintic splines ($\ell = 3$) based on $\hat{\alpha}_{\text{GML}}$ outperformed cubic splines for the situations studied. Barry (1995) used a Bayesian analysis to motivate choosing $\alpha$ using Jeffreys' prior, and his Monte Carlo simulations (using $\ell = 1$) suggested that this choice outperforms GCV for larger $\sigma$, smaller $n$, and less smooth $m$.

Wahba (1978) proposed the variance estimator (5.25) for smoothing splines. Carter and Eagleson (1992) compared variance estimators using the two forms of error degrees of freedom (5.16) and (5.26) and found that for small $\alpha$, (5.26) leads to too small an estimate of $\sigma^2$. Since generalized cross-validation sometimes undersmooths, variance estimates using splines based on $\hat{\alpha}_{\text{GCV}}$ can be unacceptably negatively biased when using (5.26), and they recommended using (5.16) instead. Hastie and Tibshirani (1990, p. 305) suggested the approximation $\nu = \nu_s - \text{trace}[A(\alpha)]/4 + .5$ relating (5.16) to (5.26) for the cubic smoothing spline.

**5.6.4.** The smoothing spline objective function (5.21) is based on assuming constant variance off the true regression line. It can be adjusted if heteroscedasticity is present, in the same way that the linear least squares estimation criterion (which is just (5.21) with $\alpha = 0$) becomes a weighted least squares criterion under nonconstant variance.

Say the true variances satisfy the relation $\sigma_i^2 = \sigma^2/w_i$. Then, the adjusted roughness penalty criterion is

$$L = \frac{1}{n}\sum_{i=1}^{n} w_i[y_i - m(x_i)]^2 + \alpha \int m^{(\ell)}(u)^2\, du, \qquad (5.34)$$

yielding a weighted smoothing spline. Just as is true in linear least squares regression, some guess for the weight vector $\mathbf{w}$ must be made, which will often be based on a preliminary unweighted spline estimate.

Silverman (1985) discussed incorporating weighting into the roughness penalty objective function to account for heteroscedasticity and suggested fitting an unweighted spline and then estimating the local variance with a local moving average of squared generalized residuals (the generalized residuals being residuals divided by $\sqrt{\sum[y_i - \hat{m}(x_i)]^2/n}$) to determine the weights. Eubank and Thomas (1993) described simple tests and diagnostic plots based on the unweighted smoothing spline estimate. The perceived need to weight the global error sum of squares in the spline formulation to account for nonconstant variance can be contrasted with the situation for local polynomial estimation, where no such weighting of the local fitting criterion is warranted (Jones, 1993c); rather, locally varying the bandwidth under heteroscedasticity is the appropriate remedy. Since the spline estimator is asymptotically a kernel estimator, this casts some doubt on the necessity of weighting the spline roughness penalty at all.

Eubank (1984) examined the hat matrix for smoothing splines closely and described how they can be used to identify potentially influential observations. Eubank (1985) extended this work to define other diagnostics analogous to those used for linear least squares regression. So, for example, the studentized residuals are defined as

$$r_j = \frac{e_j}{\hat{\sigma}_s\sqrt{1 - A_{jj}(\alpha)}}, \qquad j = 1, \ldots, n, \tag{5.35}$$

and can be used to identify outliers (see also Silverman, 1985). Diagnostics for the $j$th observation based on omitting that observation also can be defined, using the identity

$$\hat{m}^{(j)}(x_i) = \hat{m}(x_i) + \frac{A_{ij}(\alpha)e_j}{1 - A_{jj}(\alpha)}.$$

Thus, the deleted studentized residual $r_j^{(j)}$ replaces $\hat{\sigma}_s$ in (5.35) with $\hat{\sigma}_s^{(j)}$, where

$$\hat{\sigma}_s^{(j)} = \sqrt{\frac{1}{n - 1 - \operatorname{trace}(A^{(j)})}\sum_{i=1}^{n}\left[e_i + \frac{A_{ij}(\alpha)e_j}{1 - A_{jj}(\alpha)}\right]^2}$$

and

$$\operatorname{trace}(A^{(j)}) = \sum_{i \neq j}\left[A_{ij}(\alpha) + \frac{A_{ij}(\alpha)^2}{1 - A_{jj}(\alpha)}\right].$$

Different applications of $\hat{m}^{(j)}(x_i)$ yield diagnostics that correspond to commonly used diagnostics in least squares regression analysis (see Chatterjee and Hadi, 1988, for a thorough discussion of such diagnostics). Thomas (1991) described how the local influence approach of Cook (1986) can be used to construct diagnostics that can identify observations that locally influence $\hat{\alpha}_{GCV}$.

The smoothing spline estimation criterion itself can be robustified, in much the same way as kernel and local polynomial estimators can, to downweight the influence of outliers. An $M$-type smoothing spline is the minimizer over $W_2^{\ell}[a,b]$ of

$$\frac{1}{n}\sum_{i=1}^{n}\rho[y_i - m(x_i)] + \alpha \int m^{(\ell)}(u)^2\,du,$$

where $\rho(\cdot)$ is a function as in (5.32) (Anderssen, Bloomfield, and McNeil, 1974; Huber, 1979; Utreras, 1981). Cox (1983) examined the asymptotic properties of $M$-type smoothing splines.

Diggle (1985b) and Diggle and Hutchinson (1989) investigated the effects of autocorrelation on generalized cross-validation and found that positive autocorrelation tends to lead to undersmoothing. Diggle and Hutchinson proposed a modification to the generalized cross-validation criterion that assumes knowledge of the true autocovariance process. Van der Linde (1994) noted that this criterion is actually a different penalizing function criterion that does not address the autocorrelation problem, and proposed a different criterion that also depends on knowledge of the true autocovariances.

Schimek (1988) and Schimek and Schmaranz (1994) generalized the weighted objective function (5.34) to address autocorrelation for equispaced time series data. This spline-related estimate is the minimizer of

$$\sum_{i=1}^{n}\sum_{j=1}^{n}[y_i - m(x_i)]d_{ij}[y_j - m(x_j)] + \alpha\sum_{i=2}^{n-1}[m(t_{i+1}) - 2m(t_i) + m(t_{i-1})]^2,$$

where the $d_{ij}$ are the elements of $D = \Sigma^{-1}$, $\Sigma$ being the covariance matrix of the error process. This is a discrete generalized least squares formulation, just as (5.34) is a weighted least squares formulation. Kohn, Ansley, and Wong (1992) proposed a similar construction using state space methods.

Schimek (1992) used Monte Carlo simulations to investigate the effects of autocorrelation on the ordinary spline estimator and the discrete generalized least squares-based estimator. The simulations showed that while autoregressive-type autocorrelation can sometimes affect the estimation accuracy of the ordinary spline (with the generalized least squares version being an improvement), moving average-type autocorrelation has little effect, with the ordinary spline providing smaller MASE.

## Section 5.7

**5.7.1.** Cleveland and Devlin (1988) described local polynomial estimation for multiple predictors using nearest neighbor weights (loess) and described the use of approximate hypothesis tests based on analogy with analysis of variance to evaluate the fitted surface. Cleveland, Grosse, and Shyu (1992) gave a detailed discussion of application of loess to several real data sets. Ruppert and Wand (1994) derived the asymptotic properties of the multivariate local linear and local quadratic estimators. No systematic study of bandwidth matrix choice has been made for these estimators, although Herrmann *et al.* (1995) proposed plug-in approaches for bivariate convolution kernel estimators. Zhang (1991) and Bickel and Zhang (1992) examined choosing the number of predictors in a multivariate kernel regression in a data-dependent way.

Cleveland and Grosse (1991) discussed computational issues in the implementation of loess to multivariate data sets. The methods described in Wand (1994b) also can be applied to multivariate local polynomial regression estimation.

Eubank (1988, pp. 286–292), Wahba (1990, pp. 30–39), and Green and Silverman (1994, Chapter 7) described thin plate smoothing splines and gave earlier references to the method. Carmody (1988) generalized regression diagnostics based on the hat matrix to thin plate splines. It is also possible to construct multivariate regression estimators based on univariate splines using tensor products; see Green and Silverman (1994, pp. 155–159) for a discussion.

**5.7.2.** Chatterjee, Handcock, and Simonoff (1995) gave the basketball data (p. 299) and gasoline demand data (p. 291). The school budget vote data originally appeared in the May 6, 1993, issue of *Newsday* (Newsday, 1993b).

Stone (1985b) established that additive models overcome the curse of dimensionality, with convergence rates identical to those in one dimension to the closest additive approximation to the true regression curve, using spline smoothers for the univariate smoothing. Stone (1986) generalized these results to generalized additive models. Hastie and Tibshirani (1986, 1987) and Hastie (1992) provided an overview of generalized additive model fitting, including a discussion of computational issues and application to real data, while Hastie and Tibshirani (1990) gave a book-length treatment. See also Buja, Hastie, and Tibshirani (1989). McCullagh and Nelder (1989) gave a book-length treatment of the generalized linear model.

Other approaches to additive modeling have also been proposed in the literature. Wahba (1986), Chen (1993), and Gu and Wahba (1993b) allowed for the possibility of interaction among the predictors by using interaction spline models, which they called smoothing spline ANOVA. Chen (1987) and Barry (1993) studied model fitting procedures for these models. Friedman and Silverman (1989) proposed fitting additive models using piecewise polynomials with data-dependent knot choice (which they termed TURBO). Hastie (1989) adapted the method to incorporate a different algorithmic scheme (BRUTO), while Marx and Eilers (1994) proposed using P-splines.

Friedman (1991) proposed multivariate adaptive regression splines (MARS), a generalization of the TURBO method of Friedman and Silverman. Stone *et al.* (1997) described the general use of polynomial splines and their tensor products in many additive modeling situations. Smith and Kohn (1996) proposed a Bayesian approach to additive modeling that is also based on regression splines, with the Gibbs sampler being used to select knots and transformations and to identify outliers. Projection pursuit regression (Friedman and Stuetzle, 1981; Diaconis and Shahshahani, 1984; Donoho and Johnstone, 1986, 1989; Hall, 1989b; Chen, 1991; Roosen and Hastie, 1994) constructs a regression estimate that is a sum of smooth functions of linear combinations (with unknown coefficients) of the predictors. The alternating conditional expectations (ACE) method of Breiman and Friedman (1985) allows transformation of the response variable to help achieve additivity, while the additivity and variance stabilization (AVAS) method of Tibshirani (1988) adds a variance-stabilization step to the fitting of the model. Gu (1992) described diagnostics to help analysts build parsimonious versions of additive models.

The partially linear model has been studied closely. Examples include Heckman (1986, 1988), Chen (1988), Speckman (1988), Chen and Shiau (1991), and Young and Bowman (1995). A variation on this model (related to projection pursuit) is the partially linear single index model, where a smooth term is included that is a function of a linear combination of several predictors (with coefficients to be estimated), rather than a single predictor; see Härdle, Hall, and Ichimura (1993) and Carroll *et al.* (1997).

### Section 5.8

Marron (1996) gave a very cogent account of the kinds of issues that go into a comparison of nonparametric regression estimators (and smoothing methods in general). He pointed out that any such comparison should take into account many different properties of a "good" method, including general availability, interpretability, statistical efficiency, quick computability, integrability into general computer frameworks (such as S–PLUS or XLISP–STAT), and ease of mathematical analysis. Since it is unlikely that any method will dominate all others on all these criteria, the search for the "best" method is probably futile. The results of Banks, Maxion, and Olszewski (1995) support this; they compared many multivariate nonparametric regression methods, including MARS, loess, ACE, AVAS, generalized additive models, and projection pursuit regression, and found that the best method depends on many factors, including the dimensionality of the data, the dimensionality of the true regression relationship, the sample size, and the strength and complexity of the relationship.

Other approaches to nonparametric regression can be designed to optimize some criterion other than sums of squares. Mammen (1991) developed a method based on controlling the qualitative smoothness of the estimate, as measured by inflection points. Related methods include those of Mächler (1995b) and Riedel (1995). McDonald and Owen (1986) used split smoothed linear fits to allow estimation of curves with discontinuities or discontinuous lower order derivatives. Parametric regression models that are rich enough to represent smooth curves, such as polynomial and trigonometric series, can be used to estimate such curves through estimation of the coefficients of the regression model; see, for example, Cox (1988), Eubank (1988, Chapter 3), Eubank and Speckman (1990), and Tarter and Lock (1993). Hurvich and Tsai (1995) discussed the properties of using AIC to choose the number of terms in the series, while Kneip (1994) examined the use of $C_L$ for a wide class of smoothers. Antoniadis (1994) and Antoniadis, Gregoire, and McKeague (1994) discussed the application of wavelets to curve estimation, with the latter paper describing wavelet versions of the Gasser–Müller and Nadaraya–Watson kernel estimators. See also Donoho and Johnstone (1994, 1995) and Donoho *et al.* (1995).

O'Sullivan, Yandell, and Raynor (1986), Green (1987), Cox and O'Sullivan (1990), and Gu and Qiu (1994) discussed the properties of roughness penalty estimators for general likelihood functions. Eilers and Marx (1992) proposed a version of penalized likelihood using P-splines. Wahba *et al.* (1995) generalized smoothing spline ANOVA models to accommodate exponential families. Brillinger (1977) and Tibshirani and Hastie (1987) described the use of local likelihood estimation. Staniswalis (1989b) examined local constant (Nadaraya–Watson kernel) estimation for local likelihoods, while Fan, Heckman, and Wand (1995) provided asymptotic theory for local polynomials, showing that they retain the same types of properties in the local likelihood context as in ordinary nonparametric regression. See also Jones and Hjort (1995).

## Computational issues

Kernel regression estimation is available in several statistical packages, including SAS/INSIGHT, S–PLUS, XLISP–STAT, and XploRe. The `haerdle` collection in the `S` directory of `statlib` also contains S–PLUS code, based on Härdle (1991).

XploRe provides local linear regression estimation. The collection `fanmarron` in the `jcgs` directory of `statlib` contains code to calculate local linear estimates based on the algorithms in Fan and Marron (1994). Fortran code for local polynomial estimation, along with S–PLUS interface, based on Seifert *et al.* (1994), can be obtained by anonymous `ftp` at the address `biostat1.unizh.ch` in the file `pub/lpepa`. The URL address `http://cm.bell-labs.com/stat/project/locfit` allows World Wide Web access to S–PLUS code to calculate local polynomial estimates, including locally adaptive estimates and different likelihood families, as described in Loader (1995).

Many statistical packages provide lowess estimation (local linear regression with nearest neighbor bandwidth). These include Data Desk, NCSS, S–PLUS, SPSS, Stata, Systat, XLISP–STAT, and XploRe. The Smoother's Workbench (described in Manchester and Trueman, 1993) also includes it (and can be obtained as the collection `smoothwb` in the `general` directory of `statlib`), as does APL2STAT, which is available via the World Wide Web at the address `http://www.math.yorku.ca/SCS/friendly.html`. JMP, SAS/INSIGHT, and S–PLUS provide the more modern estimator loess, which allows local quadratic fitting and multiple predictors. Instructions on how to obtain Fortran and C code for loess can be found in the `loess` entry of the `general` directory of `statlib`.

S–PLUS code to calculate the local polynomial ridge regression estimator of Seifert and Gasser (1996b) is available by anonymous `ftp` at the address `biostat1.unizh.ch` in the file `pub/lpridge`.

The URL `http://www.unizh.ch/biostat/software.html` provides access to code for the global and locally adaptive local polynomial bandwidth selectors discussed in Herrmann (1997).

The Fortran code of Härdle (1987) is algorithm 222 in the `apstat` collection of `statlib`. The Smoother's Workbench code also includes this one-step $M$-smoother.

Spline smoothing estimation, with various smoothing parameter selectors, is available in various statistical packages, including JMP, SAS/INSIGHT, S–PLUS, and XploRe. Several Fortran and Ratfor implementations of spline smoothing are available via anonymous `ftp` at the address `www.netlib.org` in the directory `gcv`. These include GCVPACK (Bates *et al.*, 1987), which also fits multivariate data using thin plate splines, BART (O'Sullivan, 1985), GCVSPL (Woltring, 1986), and RKPACK (Gu and Wahba, 1991), which also includes thin plate and tensor product splines and can be used to fit smoothing spline ANOVA models. The latter set of routines is also available as the collection `rkpk` in the `general` directory of `statlib`.

Paul Eilers contributed S–PLUS code to calculate P-spline estimates for scatter plot smoothing and logistic regression to the S–news electronic mailing list, which can be found in the collection `digest153` in the `S-news` directory of `statlib` (June 21, 1994, with correction June 23, 1994).

The first author of Schimek and Schmaranz (1994) has made available code to calculate the discrete generalized least squares-based spline estimate of Schimek (1988).

GLIM, S–PLUS, and XploRe provide generalized additive model fitting as part of the package. Fortran code to fit generalized additive models is available in the collection `gamfit` in the `general` directory of `statlib`. The collection `cox_ph` in the `S` directory of `statlib` gives C, Ratfor, and S–PLUS code that allows fitting of proportional hazards models using the generalized additive modeling techniques of S–PLUS.

Fortran and S–PLUS code that implements the Bayesian additive modeling approach of Smith and Kohn (1996) can be obtained using a WWW browser at `http://www.agsm.unsw.edu.au/~mikes/software.html`.

S–PLUS and XploRe include projection pursuit regression functionality. The collection `roosen-hastie` in the `jcgs` directory of `statlib` contains S–PLUS code to calculate projection pursuit regression estimators based on the automatic smoothing spline approach of Roosen and Hastie (1994). Fortran code to carry out ACE is available as the collection `ace` in the `general` directory of `statlib`. S–PLUS and XploRe include ACE as part of the package. S–PLUS code to determine ACE estimators for logistic regression can be found in `ace.logit` of the `S` directory of `statlib`. Fortran code and S–PLUS drivers to carry out AVAS constitute the `avas` collection in the `S` directory of `statlib`. Fortran code to implement MARS is in the collection `mars3.5` in the `general` directory of `statlib`. C code with S–PLUS drivers that determine wavelet transforms and thresholding in one and two dimensions can be found in the `wavethresh` collection in the `S` directory of `statlib`. The S + WAVELETS toolkit, by Bruce and Gao (1995), performs wavelet calculations within S–PLUS. Wavelet regression estimation is also provided in XploRe.

GRKPACK, a collection of Ratfor/Fortran programs that are a generalization of RKPACK, fits smoothing spline ANOVA models to exponential families. It is available as the collection `grkpack` in the `general` directory of `statlib`.

## Exercises

**Exercise 5.1.** Calculate local constant, linear, quadratic, and cubic estimates based on a fixed bandwidth for the acid deposition data of Figs. 1.6 and 5.25. How do the estimates compare with the lowess and cubic smoothing spline estimates given in the earlier figures? Which gives the best impression of the observed regression relationship?

**Exercise 5.2.** Construct fixed-bandwidth local quadratic and cubic estimates of the newspaper circulation data of Fig. 5.5. Do these estimates suggest a near-linear relationship between daily and Sunday circulation?

**Exercise 5.3.** Construct variability plots for each of the estimates in Exercises 5.1 and 5.2. Do the higher order polynomial estimates suffer from increased variability compared to the lower order polynomial estimates?

**Exercise 5.4.** Construct variability plots for the annual lug count data of Fig. 5.14 for the local linear estimate given there and a local quadratic estimate. Which estimate would you prefer based on these plots?

**Exercise 5.5.** Determine the plug-in bandwidth choices for local linear estimation for the acid deposition data, newspaper circulation data, diabetes data (Figs. 1.5 and 5.25), and beluga whale data (Fig. 5.18). Graph the resultant regression estimates. Are you satisfied with the impressions from the fitted curves? If not, why do you think that the plug-in-based local linear estimator failed?

**Exercise 5.6.** Calculate bandwidths for the local linear estimate based on cross-validation and generalized cross-validation for the ethanol data (Fig. 5.8), total lug count data (Fig. 5.13), annual lug count data, acid deposition data, newspaper circulation data, diabetes data, and beluga whale data. How do the chosen bandwidths compare with the plug-in bandwidths? Do the cross-validation rules tend to lead to undersmoothing for these data sets?

**Exercise 5.7.** Construct loess estimates (linear and quadratic) for the newspaper circulation data. Are they better representations of the regression relationship than the fixed-bandwidth estimates?

**Exercise 5.8.** Rows 1 and 52 in the total lug count data appear to be possible outliers. Refit the regression estimates of Figs. 5.9, 5.10, and 5.12 with these points omitted. Do the fitted curves change very much? What does the robust version of the loess estimate look like?

**Exercise 5.9.** The beluga whale data set also includes corresponding information for a second calf (Casey) born at the New York Aquarium around the same time as Hudson. Fit a local polynomial estimate to Casey's data. Is his nursing pattern similar to Hudson's? Does varying the bandwidth locally give a better regression estimate?

**Exercise 5.10.** Fit cubic smoothing splines to the annual lug count data using different smoothing parameter selectors. Which selectors result in a reasonable estimate?

**Exercise 5.11.** The true regression curve being estimated for the acid deposition data is a correlation function and as such, must satisfy certain properties. More informally, its nature as a correlation function suggests likely properties for the curve. Based on this, which of the estimates constructed in Figs. 1.6 and 5.25, and in Exercises 5.1, 5.5, and 5.6, are most likely to be accurate reflections of the true curve?

**Exercise 5.12.** Fit cubic smoothing splines to the geyser data (Figs. 5.1 and 5.24), electricity usage data (Figs. 5.6 and 5.24), birth rate data (Figures 5.16 and 5.23), and beluga whale data using generalized cross-validation to choose the smoothing parameter. Does autocorrelation of the errors seem to cause difficulties for the selector for these data sets?

**Exercise 5.13.** Fit cubic smoothing splines to the Skeena River salmon data (Fig. 5.19) and immunoassay calibration data (Fig. 5.20) using generalized cross-validation to choose the smoothing parameter. Does the presence of possible outliers cause difficulties for the selector for these data sets?

**Exercise 5.14.** Construct a perspective plot for a loess fit to the basketball data of Fig. 5.31 with the observation corresponding to Keith Jennings omitted from the data set. Does this change the fitted surface for small values of height and minutes per game? What about fitting the robust version of loess to the entire data set? How do you account for the observed patterns?

**Exercise 5.15.** The basketball data also include the player's age. Fit additive models using height, minutes per game, and age as predictors of points scored per minute and assists made per minute, respectively. Is age related to either of these response variables? Does it add anything to either fit?

**Exercise 5.16.** Fit a loess estimate to the gasoline demand data of Figs. 5.35 and 5.36, and plot two-dimensional slices of the estimated surface using perspective plots. Are your impressions similar to those from the additive model fit?

**Exercise 5.17.** The school budget vote data of Fig. 5.38 also include information on the total proposed budget, percentage change in the budget, percentage change in the tax rate, and average full-value property wealth per student for each school district. Construct the best generalized additive model you can using these variables. Which variables add significantly to the fit, and which can be omitted? Do the results make intuitive sense for these data?

Chapter 6

# Smoothing Ordered Categorical Data

## 6.1 Smoothing and Ordered Categorical Data

All the examples described in Chapters 2 – 4 referred to continuous data and estimation of a smooth density function $f$. This is reasonable, since smoothness and continuity would seem to be naturally linked to each other.

Consider now a one-dimensional categorical variable, where the sample space consists of $K$ cells, and $n_i$ observations fall in the $i$th cell, $i = 1, \ldots, K$, with $\sum n_i = n$ the sample size. The vector $\mathbf{p} = \{p_i\}$ represents the probability of an observation falling in a given cell. There are two obvious arguments against smoothing data of this type:

(1) The discrete nature of the data means that no sensible concept of "smoothness" exists (since the idea of "nearby" values being similar to each other is essential to the idea of smoothness).

(2) The observed cell frequencies $\overline{p}_i = n_i/n$ are already adequate estimates of the cell probabilities.

Neither of these arguments completely removes the possible benefits of smoothing. A categorical variable where the categories do not have any natural ordering is called a *nominal* variable (examples would be the religion or nationality of the respondent to a survey). For such data, smoothing is not very helpful, since it is very difficult to define how "close" two categories are (since their ordering is arbitrary).

A categorical variable where the categories do have a natural ordering, called an *ordinal* variable, is a very different matter. Such a variable can arise as a discretization of an underlying continuous variable (for example, $0 < x \le 10$; $10 < x \le 20$; $20 < x \le 30$, and so on) or as an inherently discrete, but ordered, set of categories (for example, strongly disagree; disagree; no opinion; agree; strongly agree). For such a variable, smoothing makes sense, as it is likely that the number of observations that fall in a particular cell provides information about the probability of falling in nearby cells as well. For example, if the variable represents a discretization of a continuous variable with smooth density $f$, the probability vector $\mathbf{p}$ also will reflect that smoothness, with $p_i$ being close to $p_j$ for $i$ close to $j$.

The second argument against smoothing is valid, but only when $\overline{\mathbf{p}}$ is a good estimator of $\mathbf{p}$. The Law of Large Numbers states that $\overline{p}_i$ is a consistent estimator of $p_i$ as long as $n_i \to \infty$ as $n \to \infty$. From a practical point of view, this corresponds to the sample size $n$ being large compared with the number of cells $K$. In many situations, however, such as for multidimensional tables, the number of cells is close to (or even greater than) the number of observations, resulting in many small (or zero) cell counts. Such a table of counts is called a *sparse* table. For such tables, $\overline{p}_i$ is not a good estimator of $p_i$, as the usual asymptotic approximations do not apply.

Consider Table 6.1. The table gives three discretizations of data representing the monthly salary of 147 nonsupervisory female employees holding the Bachelors (but no higher) degree who were practicing mathematics or statistics in 1981. Despite their inherently continuous nature, these data were given in discretized form in the original data source. Figure 6.1 gives the corresponding frequency estimates as solid lines connecting the estimated cell probabilities (the circles). The 6-cell table is not at all sparse, and the cell frequencies highlight a smooth pattern of the salaries, peaking in the third cell (\$1751–\$2150), with a noticeable bulge around the fifth cell (\$2551–\$2950).

The cell frequencies in the 12-cell table reflect the same general pattern, but sparseness now causes roughness in the probability estimates. In particular, there are four modes in $\overline{\mathbf{p}}$, and it is unlikely for them to all be genuine.

In the 28-cell discretization (the resolution given in the original source), sparseness has obscured almost all the structure, except for the peak around \$1900. There are many spurious modes, making it difficult to get an impression of the general pattern of probabilities.

**Table 6.1.** Discretizations of salary data (6 cells and 12 cells).

| 6 cells | | | 12 cells | | | | | |
|---|---|---|---|---|---|---|---|---|
| $i$ | Salary | $n_i$ | $i$ | Salary | $n_i$ | $i$ | Salary | $n_i$ |
| *1* | 951–1350 | 11 | *1* | 951–1150 | 6 | *7* | 2151–2350 | 17 |
| *2* | 1351–1750 | 27 | *2* | 1151–1350 | 5 | *8* | 2351–2550 | 10 |
| *3* | 1751–2150 | 45 | *3* | 1351–1550 | 12 | *9* | 2551–2750 | 14 |
| *4* | 2151–2550 | 27 | *4* | 1551–1750 | 15 | *10* | 2751–2950 | 11 |
| *5* | 2551–2950 | 25 | *5* | 1751–1950 | 30 | *11* | 2951–3150 | 6 |
| *6* | 2951–3750 | 12 | *6* | 1951–2150 | 15 | *12* | 3151–3750 | 6 |

**Table 6.1(cont.).** Discretization of salary data (28 cells).

| | | | | **28 cells** | | | | |
|---|---|---|---|---|---|---|---|---|
| $i$ | Salary | $n_i$ | $i$ | Salary | $n_i$ | $i$ | Salary | $n_i$ |
| *1* | 951–1050 | 5 | *11* | 1951–2050 | 6 | *20* | 2851–2950 | 5 |
| *2* | 1051–1150 | 1 | *12* | 2051–2150 | 9 | *21* | 2951–3050 | 4 |
| *3* | 1151–1250 | 0 | *13* | 2151–2250 | 5 | *22* | 3051–3150 | 2 |
| *4* | 1251–1350 | 5 | *14* | 2251–2350 | 12 | *23* | 3151–3250 | 1 |
| *5* | 1351–1450 | 2 | *15* | 2351–2450 | 7 | *24* | 3251–3350 | 2 |
| *6* | 1451–1550 | 10 | *16* | 2451–2550 | 3 | *25* | 3351–3450 | 0 |
| *7* | 1551–1650 | 5 | *17* | 2551–2650 | 10 | *26* | 3451–3550 | 1 |
| *8* | 1651–1750 | 10 | *18* | 2651–2750 | 4 | *27* | 3551–3650 | 1 |
| *9* | 1751–1850 | 10 | *19* | 2751–2850 | 6 | *28* | 3651–3750 | 1 |
| *10* | 1851–1950 | 20 | | | | | | |

Smoothing methods provide a way around this problem. Since it is reasonable to assume that $\mathbf{p}$ changes smoothly as salary increases, information in nearby cells can be "borrowed" to help provide more accurate estimation in any given cell (discretizing continuous data has already done a crude form of smoothing). Not surprisingly, modifications of the methods of earlier chapters can be used to accomplish this.

## 6.2 Smoothing Sparse Multinomials

Consider a one-dimensional table $\{n_i\}$. The standard model for this random variable is a multinomial distribution with sample size $n$ and probability vector $\mathbf{p}$, with log-likelihood

$$\sum_{i=1}^{K} n_i \log p_i, \quad \text{subject to the constraint } \sum_{i=1}^{K} p_i = 1$$

(ignoring constants). For this distribution $E(n_i) = np_i$, and hence $E(\overline{p}_i) = p_i$. It is helpful to think of the vector $\mathbf{p}$ as being generated from an underlying smooth density $f$ on $[0, 1]$ through the relation

$$p_i = \int_{(i-1)/K}^{i/K} f(u)du;$$

in this way, the existence of derivatives of $f$ corresponds to smoothness of $\mathbf{p}$. Note that the Mean Value Theorem implies that

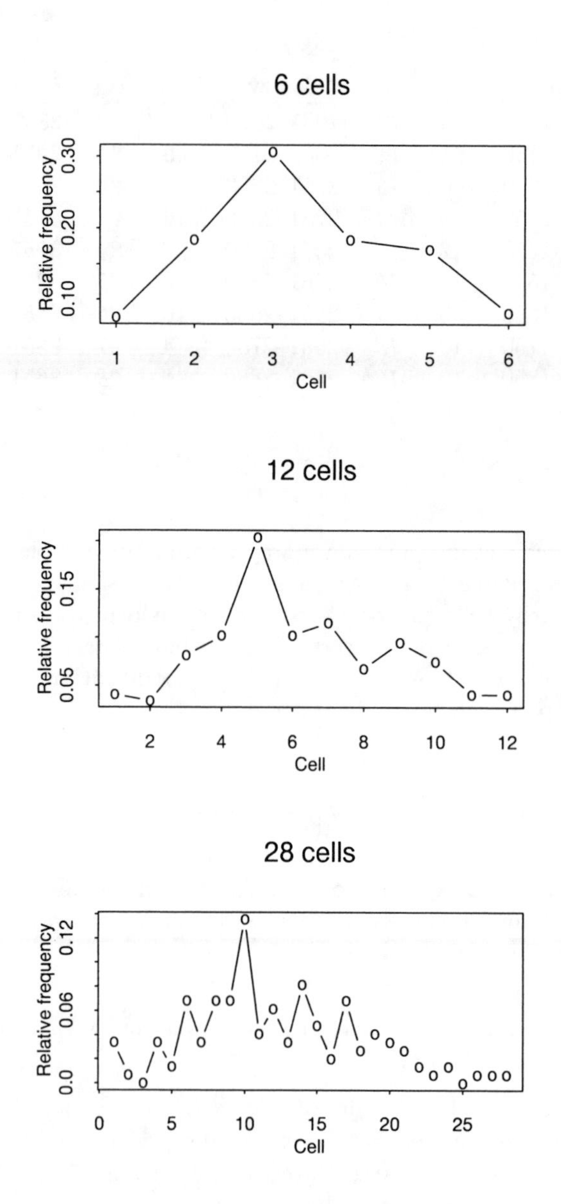


**Fig. 6.1.** Frequency estimates of cell probabilities for salary data based on discretizations into 6, 12, and 28 cells, respectively.

$$p_i = f(x_i)/K, \quad \text{for some } x_i \in [(i-1)/K, i/K], \tag{6.1}$$

which is $O(K^{-1})$ for bounded $f$.

A natural way to define a smooth estimator $\hat{\mathbf{p}}$ is by analogy with a regression of response values $\overline{p}_i$ on the equispaced design $i/K, i = 1, \ldots, K$, since in a regression model $E(y_i) = m(x_i)$, the value to be estimated. So, a Nadaraya–Watson kernel estimator of $p_i$, as in Section 5.1, would be

$$\hat{p}_i = \frac{\sum_{j=1}^{K} W\left(\frac{i/K - j/K}{h}\right) \overline{p}_j}{\sum_{j=1}^{K} W\left(\frac{i/K - j/K}{h}\right)}, \tag{6.2}$$

where $W(\cdot)$ is the kernel function. Here, the bandwidth $h$ is in units corresponding to the proportion of cells in the table; so, for example, for a kernel that is positive on $(-1, 1)$, if $h = .1$, 20% of the cells are used in the smoothing in the interior, centered at cell $i$.

Figure 6.2 gives kernel-based estimates for the three salary tables of Table 6.1, superimposed on dashed lines giving the frequency estimates. The kernel estimates are based on $h = .075$ (6 cells), $h = .067$ (12 cells), and $h = .057$ (28 cells), respectively. The 6-cell table is not sparse, and no smoothing is needed, so $\hat{\mathbf{p}}$ is close to $\overline{\mathbf{p}}$. For 12 and 28 cells, however, the smooth curves remove the spurious bumpiness in the frequency estimates and lead to the same general impressions for all three versions of the data.

As would be expected, however, the kernel estimator (6.2) is subject to boundary bias. Table 6.2 gives counts that correspond to a discretization into 55 cells of 109 time intervals between explosions in mines involving more than ten men killed in Great Britain from December 8, 1875 to May 29, 1951. Figure 6.3 is a plot of $\overline{\mathbf{p}}$, where the sparseness of the table is evident. While the estimated probabilities have a vaguely exponential shape (not surprisingly for time interval data), there is a jumble of structure around the tenth cell (300 days between explosions) that is difficult to resolve. The many zero probability estimates also are obviously not an accurate reflection of the true cell probabilities.

The top plot of Fig. 6.4 is a kernel estimate for these data using $h = .029$. The estimate resolves the structure around the tenth cell as a clear bulge, and it is apparent that the underlying density is more long-tailed than an exponential density.

Unfortunately, the kernel estimate exhibits boundary bias at the left boundary. The obvious corrective action is to move from the (kernel) local constant estimator (6.2) to higher order polynomials, so that $\hat{p}_i$ is the constant term of the minimizer $\hat{\boldsymbol{\beta}}$ of

$$\sum_{j=1}^{K} \left[\overline{p}_j - \beta_0 - \cdots - \beta_t \left(\frac{i}{K} - \frac{j}{K}\right)^t\right]^2 W\left(\frac{i/K - j/K}{h}\right).$$

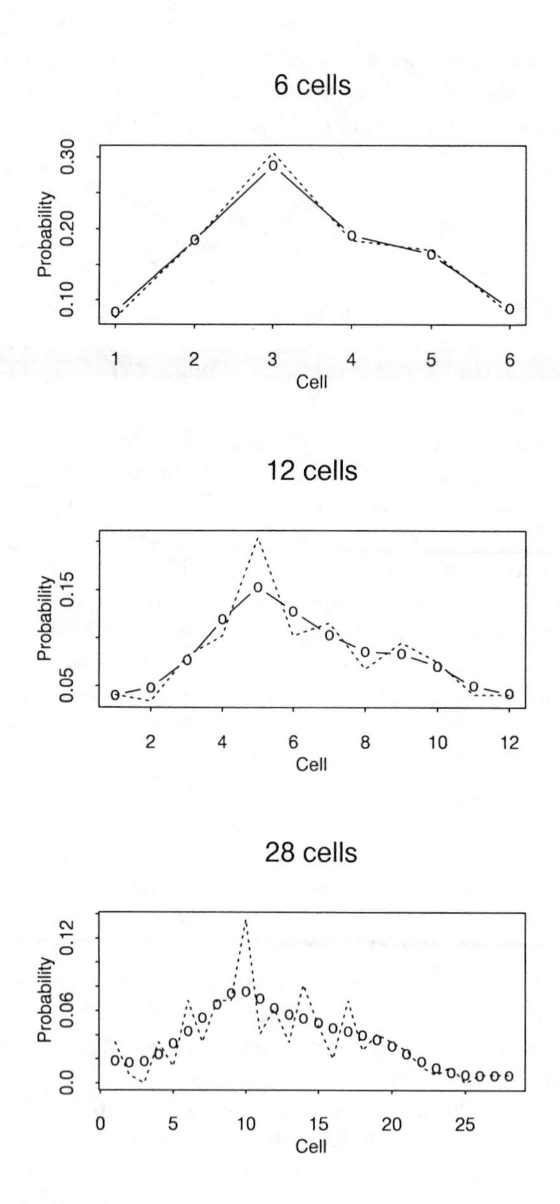


**Fig. 6.2.** Kernel estimates of cell probabilities for salary data based on discretizations into 6, 12, and 28 cells, respectively.

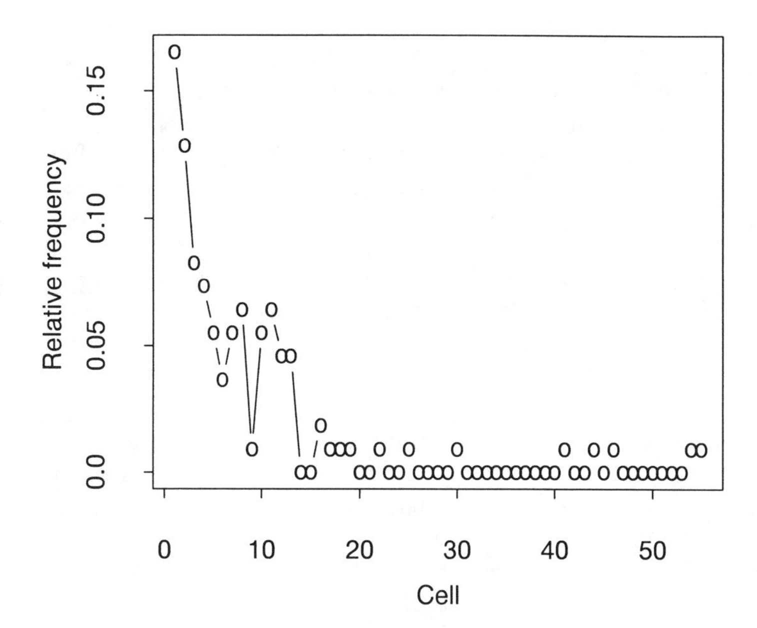


**Fig. 6.3.** Frequency estimates of cell probabilities for mine explosion data.

The gains of $\hat{\mathbf{p}}$ over $\overline{\mathbf{p}}$ show up in asymptotics that represent sparse tables. These *sparse asymptotics* require the number of cells to become infinite as the sample size becomes infinite, thereby modeling the occurrence of large tables with relatively few observations in each cell. Even though the cell frequencies $\overline{\mathbf{p}}$ are accurate when the table size is small compared with the sample size, under sparse asymptotics they are generally not even consistent, in the sense that if $K$ and $n$ both become infinite at the same rate,

$$\sup_{1 \leq i \leq K} \left| \frac{\overline{p}_i}{p_i} - 1 \right| \neq o_p(1)$$

(assuming $f$ is bounded below, so that $\inf_i p_i > 0$). Squared error analysis is based on the mean sum of squared errors MSSE ($\text{MSSE}(\tilde{\mathbf{p}}) = E[\sum_i (\tilde{p}_i - p_i)^2]$), and (6.1) shows that $K \times$ MSSE corresponds to the MISE measure used in continuous density estimation. The frequency estimator satisfies $\text{MSSE}(\overline{\mathbf{p}}) = O(N^{-1})$, so under sparse asymptotics, $K \times \text{MSSE} \not\to 0$.

In contrast, the local polynomial estimator $\hat{\mathbf{p}}$ is consistent, with properties very similar to those for nonparametric regression estimation. Say $f$

**Table 6.2.** Time intervals between mine explosions data.

| $i$ | Days | $n_i$ | $i$ | Days | $n_i$ | $i$ | Days | $n_i$ |
|---|---|---|---|---|---|---|---|---|
| *1* | 0–30 | 18 | *20* | 571–600 | 1 | *38* | 1111–1140 | 0 |
| *2* | 31–60 | 14 | *21* | 601–630 | 0 | *39* | 1141–1170 | 0 |
| *3* | 61–90 | 9 | *22* | 631–660 | 0 | *40* | 1171–1200 | 0 |
| *4* | 91–120 | 8 | *23* | 661–690 | 1 | *41* | 1201–1230 | 1 |
| *5* | 121–150 | 6 | *24* | 691–720 | 0 | *42* | 1231–1260 | 0 |
| *6* | 151–180 | 4 | *25* | 721–750 | 0 | *43* | 1261–1290 | 0 |
| *7* | 181–210 | 6 | *26* | 751–780 | 1 | *44* | 1291–1320 | 1 |
| *8* | 211–240 | 7 | *27* | 781–810 | 0 | *45* | 1321–1350 | 0 |
| *9* | 241–270 | 1 | *28* | 811–840 | 0 | *46* | 1351–1380 | 1 |
| *10* | 271–300 | 6 | *29* | 841–870 | 0 | *47* | 1381–1410 | 0 |
| *11* | 301–330 | 7 | *30* | 871–900 | 0 | *48* | 1411–1440 | 0 |
| *12* | 331–360 | 5 | *31* | 901–930 | 1 | *49* | 1441–1470 | 0 |
| *13* | 361–390 | 5 | *32* | 931–960 | 0 | *50* | 1471–1500 | 0 |
| *14* | 391–420 | 0 | *33* | 961–990 | 0 | *51* | 1501–1530 | 0 |
| *15* | 421–450 | 0 | *34* | 991–1020 | 0 | *52* | 1531–1560 | 0 |
| *16* | 451–480 | 2 | *35* | 1021–1050 | 0 | *53* | 1561–1590 | 0 |
| *17* | 481–510 | 1 | *36* | 1051–1080 | 0 | *54* | 1591–1620 | 1 |
| *18* | 511–540 | 1 | *37* | 1081–1110 | 0 | *55* | 1621–1650 | 1 |
| *19* | 541–570 | 1 | | | | | | |

has $t+1$ uniformly continuous derivatives, and $h \to 0$ with $hK \to \infty$. Then, for $t$ odd, the MSSE for the $t$th-degree local polynomial estimator satisfies

$$\begin{aligned}\text{MSSE}(\hat{\mathbf{p}}) =& \frac{1}{K}\left[\frac{h^{t+1}\mu_{t+1}(W_{(t)})}{(t+1)!}\right]^2 \int f^{(t+1)}(u)^2 du \\ &+ \frac{R(W_{(t)})}{nhK} + o[h^{2t+2}K^{-1} + (nhK)^{-1}] \qquad (6.3)\end{aligned}$$

($t$ is used here to avoid confusion with the probability vector $\mathbf{p}$).

The correspondence to the asymptotic MISE for a $t$th degree local polynomial regression estimator given in (5.12) is obvious, as $K \times$ MSSE equals MISE, with $f$ taking the place of $m$ (recalling that here $W$ is the kernel function and $t$ is the degree of the polynomial). A similar correspondence occurs for even $t$, with even degree estimators (such as the kernel estimator (6.2)) suffering from boundary bias compared with the next highest (odd) degree (the boundary properties correspond to those given for nonparametric regression estimation in Section 5.2.3). The bandwidth that minimizes the leading terms of (6.3) is

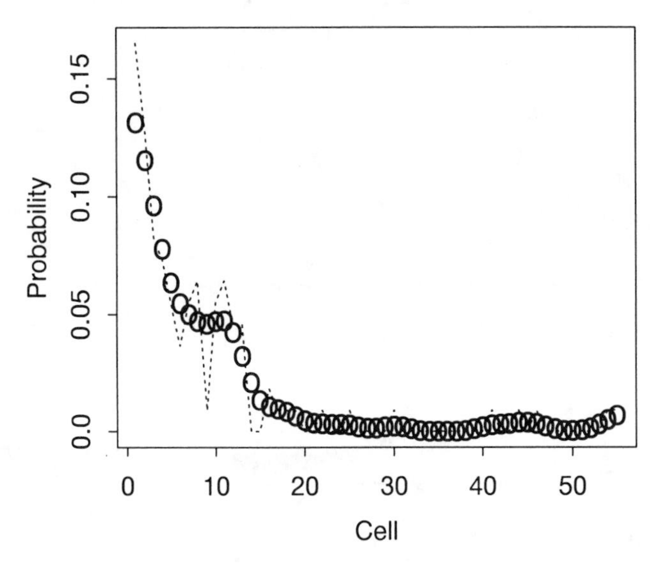


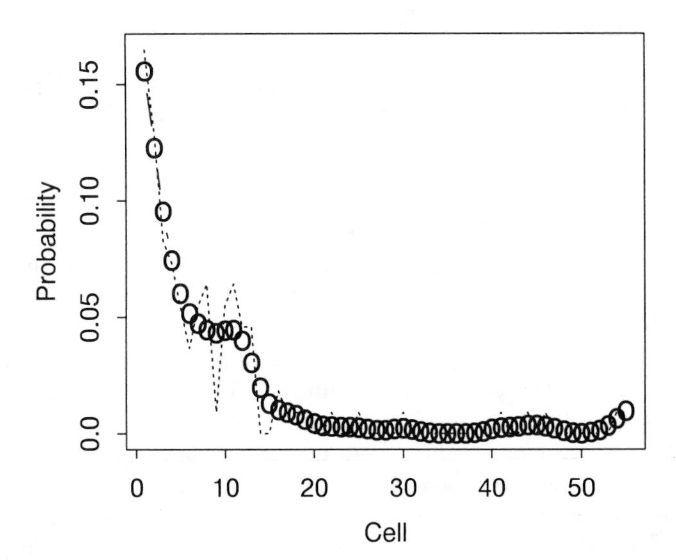


**Fig. 6.4.** Local polynomial estimates for mine explosion data. The dashed line represents the frequency estimates. Top: local constant (Nadaraya–Watson-type) estimate. Bottom: local linear estimate.

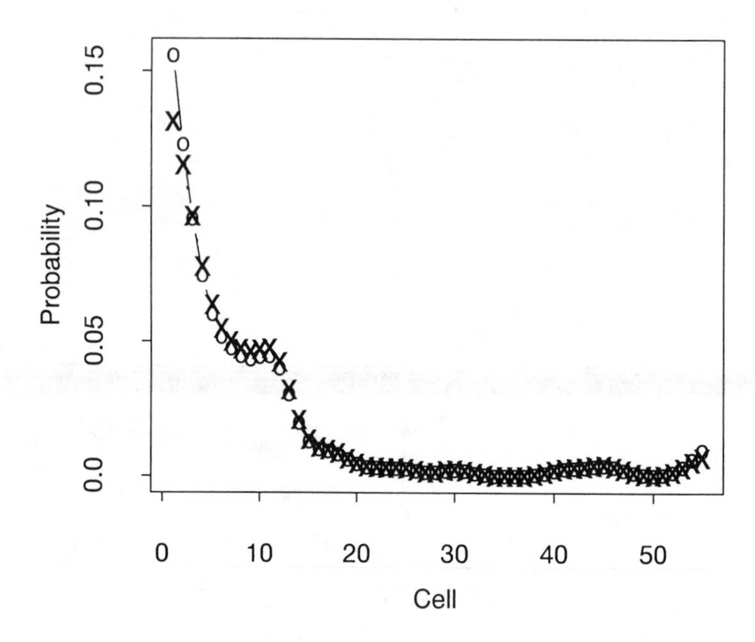


**Fig. 6.5.** Local constant (×) and local linear (∘) estimates of cell probabilities for mine explosion data superimposed on same plot.

$$h_0 = \left[ \frac{(t+1)(t!)^2 R(W_{(t)})}{2n\mu_{t+1}(W_{(t)})^2 \int f^{(t+1)}(u)^2\, du} \right]^{1/(2t+3)}, \tag{6.4}$$

which shows that the optimal MSSE converges to zero at the optimal rate of $O(n^{-(2t+2)/(2t+3)}K^{-1})$ (so, for local linear estimation, the rate is $O(n^{-4/5}K^{-1})$).

The bottom plot in Fig. 6.4 illustrates the benefits of moving from the local constant to the local linear estimator. The estimators are virtually identical over all the cells, except at the left boundary, where the local linear estimate no longer suffers from boundary bias. Figure 6.5 makes the relationship between the two estimators clearer. Here the local constant estimates are marked with the symbol ×, while the local linear estimates are marked with the symbol ∘. The close correspondence over all cells but the left boundary cells is obvious, as is the boundary bias of the local constant estimates in the left boundary region.

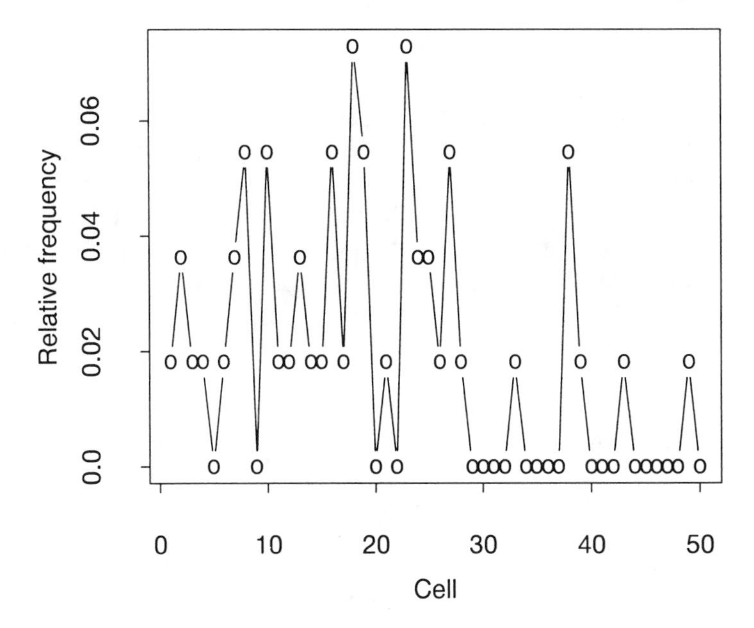


**Fig. 6.6.** Frequency estimates of cell probabilities for calcium carbonate data.

The form of the MSSE (6.3) shows that higher order local polynomials can achieve reduced bias if $f$ is smooth enough (although at the cost of increased variance in the boundary region). Table 6.3 and Fig. 6.6 give a 50-cell multinomial that represents the range of percentage concentrations of calcium carbonate for 52 sets of 5 samples each, taken from a mixing plant of raw metal. The cell frequencies are virtually impossible to interpret, being far too rough. Figure 6.7 gives local linear and local quadratic estimates for these data, using $h = .128$. The estimates are similar for cells 1 through 20, and both reflect an asymmetry in the cell probabilities, but the local quadratic estimate levels out for cells 35 through 45, which (based on Fig. 6.6) seems more reasonable than the steady decrease in the local linear estimate.

As was noted in Chapter 5, higher order local polynomial regression estimators are subject to spurious bumpiness in the tails because of increased variability. Local polynomial estimators for multinomials behave similarly. Figure 6.8 is a plot of local cubic cell probability estimates for the calcium carbonate data, again using $h = .128$. The general pattern is the same as for the local quadratic estimates, but there are now dips at both ends, presumably because of increased variability. For these data, the

**Table 6.3.** Calcium carbonate concentration data.

| $i$ | Concent. | $n_i$ | $i$ | Concent. | $n_i$ | $i$ | Concent. | $n_i$ |
|---|---|---|---|---|---|---|---|---|
| *1* | .100–.112 | 1 | *18* | .305–.316 | 4 | *35* | .509–.520 | 0 |
| *2* | .113–.124 | 2 | *19* | .317–.328 | 3 | *36* | .521–.532 | 0 |
| *3* | .125–.136 | 1 | *20* | .329–.340 | 0 | *37* | .533–.544 | 0 |
| *4* | .137–.148 | 1 | *21* | .341–.352 | 1 | *38* | .545–.556 | 3 |
| *5* | .149–.160 | 0 | *22* | .353–.364 | 0 | *39* | .557–.568 | 1 |
| *6* | .161–.172 | 1 | *23* | .365–.376 | 4 | *40* | .569–.580 | 0 |
| *7* | .173–.184 | 2 | *24* | .377–.388 | 2 | *41* | .581–.592 | 0 |
| *8* | .185–.196 | 3 | *25* | .389–.400 | 2 | *42* | .593–.604 | 0 |
| *9* | .197–.208 | 0 | *26* | .401–.412 | 1 | *43* | .605–.616 | 1 |
| *10* | .209–.220 | 3 | *27* | .413–.424 | 3 | *44* | .617–.628 | 0 |
| *11* | .221–.232 | 1 | *28* | .425–.436 | 1 | *45* | .629–.640 | 0 |
| *12* | .233–.244 | 1 | *29* | .437–.448 | 0 | *46* | .641–.652 | 0 |
| *13* | .245–.256 | 2 | *30* | .449–.460 | 0 | *47* | .653–.664 | 0 |
| *14* | .257–.268 | 1 | *31* | .461–.472 | 0 | *48* | .665–.676 | 0 |
| *15* | .269–.280 | 1 | *32* | .473–.484 | 0 | *49* | .677–.688 | 1 |
| *16* | .281–.292 | 3 | *33* | .485–.496 | 1 | *50* | .689–.700 | 0 |
| *17* | .293–.304 | 1 | *34* | .497–.508 | 0 | | | |

local quadratic estimate seems to be the best choice.

Equation (6.4) provides one way to try to choose the bandwidth $h$. A plug-in method would substitute an estimate for $\int f^{(t+1)}(u)^2\,du$ into (6.4). An alternative that is likely to work reasonably well is to take advantage of the close connection of the estimator to a regression estimator and use a regression-based method on the data $\{i/K, \overline{p}_i\}$. The bandwidths for the local linear estimates in Figs. 6.4 and 6.7 were chosen this way, using the local linear plug-in method described in Chapter 5.

## 6.3 Smoothing Sparse Contingency Tables

The problems associated with sparse multinomials are magnified when moving to higher dimensions (contingency tables). Since the number of cells in a table increases multiplicatively with the dimension of the table, higher dimensional tables are more likely to be sparse. Fortunately, smoothing methods extend directly to higher dimensions as well. Not surprisingly, however, so does the curse of dimensionality.

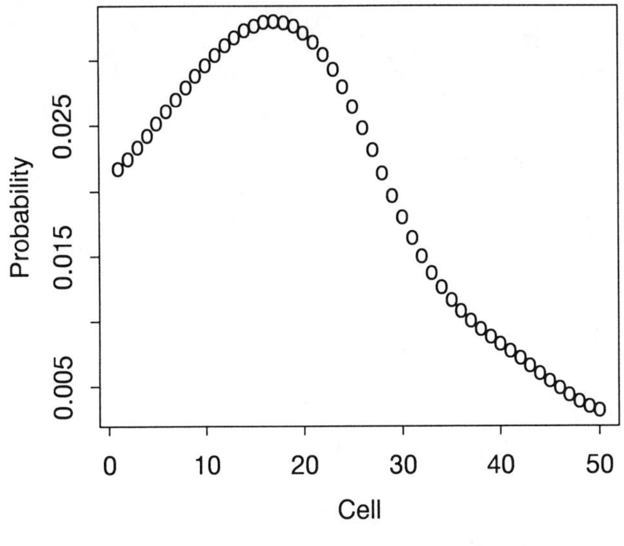


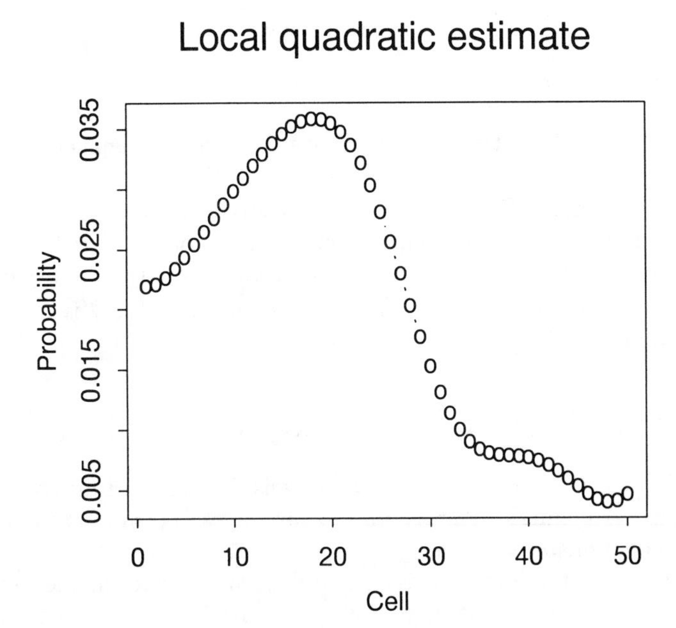


**Fig. 6.7.** Local linear (top) and local quadratic (bottom) estimates of cell probabilities for calcium carbonate data.

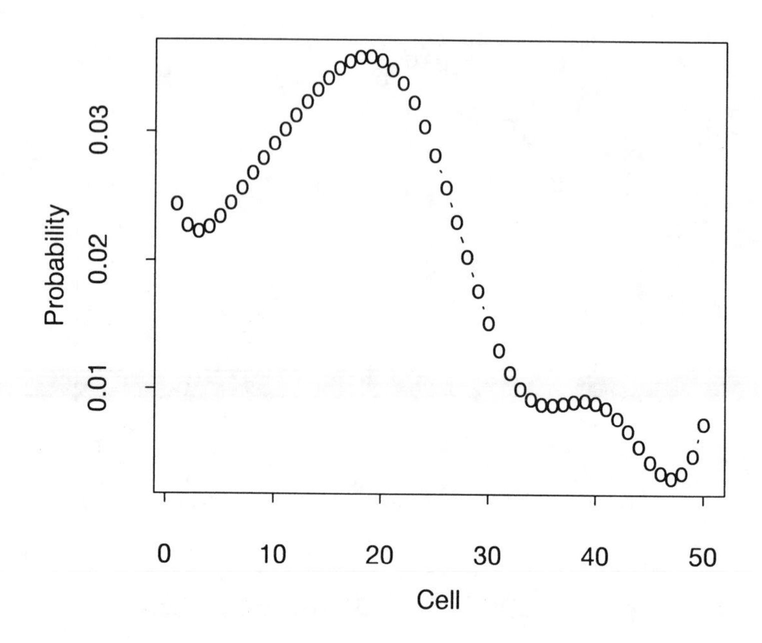


**Fig. 6.8.** Local cubic estimates of cell probabilities for calcium carbonate data.

Local polynomial estimators generalize to $d$-dimensional tables in the same way that they do for nonparametric regression with multiple predictors. For example, the local linear estimator $\hat{p}_{ij}$ for the probability of falling in the $(i,j)$th cell of an $R \times C$ two-dimensional table $\{n_{ij}\}$ is $\hat{\beta}_0$, where $\hat{\boldsymbol{\beta}}$ is the minimizer of

$$\sum_{k=1}^{R}\sum_{\ell=1}^{C}\left[\overline{p}_{k\ell} - \beta_0 - \beta_1\left(\frac{i}{R} - \frac{k}{R}\right) - \beta_2\left(\frac{j}{C} - \frac{\ell}{C}\right)\right]^2 W_{h_R,h_C}(i,j,k,\ell,R,C),$$

where $W_{h_R,h_C}(i,j,k,\ell,R,C)$ is a two-dimensional kernel function (a product kernel, for example) and $h_R$ and $h_C$ are the smoothing parameters for rows and columns, respectively.

Just as was true for regression and density estimation, the higher the dimension of the table, the more slowly the variance converges to zero, because of the relative lack of local information in higher dimensions. The variance is now $O[(nh_1 \cdots h_d K)^{-1}]$, where $K$ is the total number of cells in the table, reflecting the increasing difficulty in estimation in high dimensions (that is, the curse of dimensionality). So, for example, for a two-dimensional table where $R$ and $C$ both become infinite at a rate proportional to $\sqrt{n}$ (so

**Table 6.4.** MBA survey data. Rows represent opinion of importance of statistics in business education from least to most important; columns represent opinion of importance of economics. The numbers outside the table represent the scale used to define the rows and columns (1 = completely useless; 2 = useless; 3 = little importance; 4 = neutral; 5 = somewhat important; 6 = very important; 7 = absolutely crucial).

| Statistics \ Economics | 1 | 2 | 3 | 4 | 5 | 6 | 7 |
|---|---|---|---|---|---|---|---|
| 2 | 0 | 1 | 0 | 0 | 0 | 1 | 0 |
| 3 | 0 | 0 | 0 | 1 | 0 | 0 | 0 |
| 4 | 0 | 0 | 3 | 6 | 4 | 0 | 0 |
| 5 | 0 | 0 | 1 | 4 | 7 | 4 | 0 |
| 6 | 1 | 0 | 0 | 2 | 6 | 10 | 1 |
| 7 | 0 | 0 | 0 | 0 | 0 | 2 | 1 |

that the total number of cells in the table increases at the same rate as the sample size), the optimal MSSE rate is $O(n^{-2/3}K^{-1})$, which is attained when $h = O(n^{-1/6})$ for both rows and columns, which can be compared to the optimal $O(n^{-4/5}K^{-1})$ rate in one dimension.

Table 6.4 is a $6 \times 7$ cross-classification of the responses of 55 first year MBA students at New York University's Stern School of Business in 1991 to questions about the importance of statistics (rows) and economics (columns) in business education. Responses were coded on a 7-point scale from "completely useless" to "absolutely crucial" (no students rated statistics "completely useless").

This table is moderately sparse, the results of which can be seen in the top plot of Fig. 6.9. This is a *shade plot* (sometimes called an image plot), where greyscale shading represents the counts in the contingency table, with higher counts corresponding to a darker shade. The shade plot highlights a generally positive association between ratings of statistics and economics, with the highest counts associated with ratings of 5 or 6 on both scales.

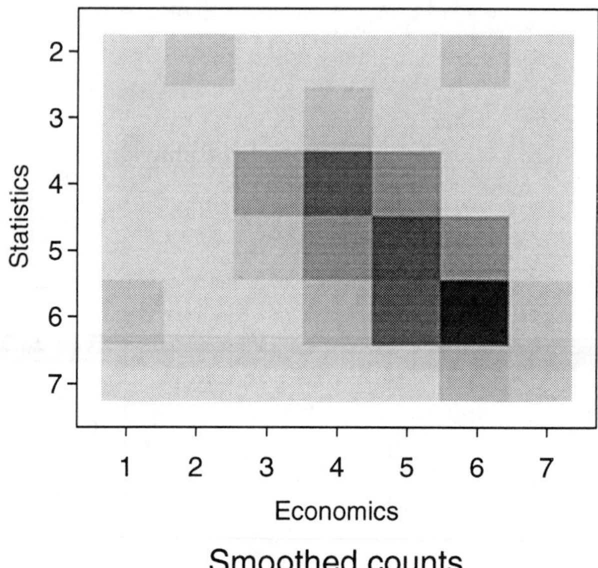


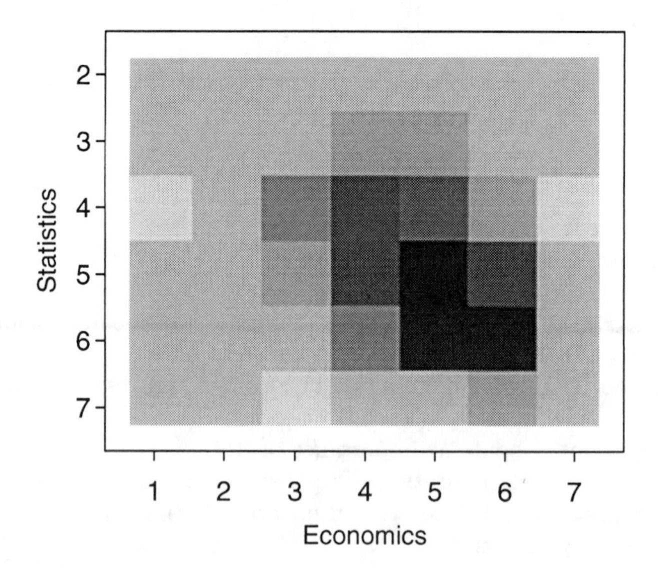


**Fig. 6.9.** Shade plots of unsmoothed counts (top) and local linear smoothed counts (bottom) for MBA survey data.

Still, the progression from light to dark in the plot is not completely smooth. There are also three cells that are outlying compared with the others (the darker shadings along the left and top boundaries), corresponding to a student with a low opinion of both statistics and economics, a student with a low opinion of statistics but a high opinion of economics, and a student with a low opinion of economics but a high opinion of statistics, respectively.

The shade plot in the bottom of Fig. 6.9 represents the smoothed (truncated) counts (that is, $\lfloor n \times \hat{p}_{ij} \rfloor$, where $\lfloor \cdot \rfloor$ is the greatest integer function) based on a local linear loess estimate with span equal to .25 (that is, the tricube-based kernel covers 25% of the cells). The estimated probability matrix is smoother than before (reflecting the not unexpected positive association between the variables), and the counts in the three unusual cells have been smoothed over to be more consistent with the cells around them.

One unfortunate property of the local polynomial estimator that is also noticeable in the shade plot is that the estimated cell probabilities can be negative (for the local linear estimator, because of boundary bias correction). The three cells with the lightest shading have probability estimates less than zero, which is meaningless.

For this table, boundary bias correction is apparently not needed, as the probabilities decrease smoothly towards the boundaries. Thus, a smoothed fit with only nonnegative probabilities can be achieved by using a local constant (kernel) estimator. Figure 6.10 gives a shade plot of a local constant loess fit with span equal to .2. The high probability region around answers $(6, 6)$ is evident, with a slow drop in probabilities in the direction of positive association (with a slightly faster dropoff to $(7, 7)$, due, no doubt, to that being the highest rating possible) and a fast drop in probabilities in the direction of negative association.

More complex structure in the probability matrix might require the use of higher degree polynomials. Table 6.5 refers again to the salary data of Table 6.1, except that now the 147 respondents have been cross-classified into a $12 \times 10$ contingency table, with the columns representing the number of years since receiving the Bachelor's degree. It is very difficult to see the relationship between the variables from the table, as it is very sparse. Figure 6.11, which gives shade plots for this table, reinforces this impression. The top plot represents the unsmoothed counts, and other than a somewhat weak positive association between salary and years since degree, little else is apparent from the table.

The bottom plot in Fig. 6.11 is a shade plot of the smoothed counts based on a local quadratic loess fit with span equal to .35. The smoothed counts are far more evocative of the pattern in the table. There is a high probability region centered at 0–5 years since degree and \$1351–\$2150 monthly salary and a generally positive association between the two variables (as would be expected). The smoothed counts also suggest bimodality in the probability matrix, with a secondary mode (representing more ex-

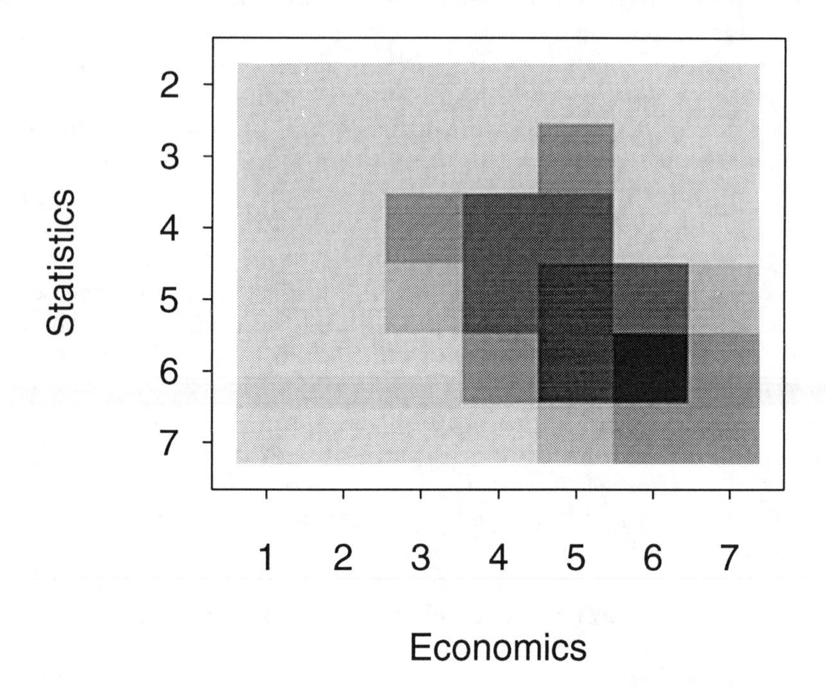


**Fig. 6.10.** Shade plot of local constant smooth counts for MBA survey data.

perienced, and higher paid, workers) centered at 15–17 years since degree and \$2351–\$2750 monthly salary.

Unfortunately, this estimate also includes negative values (the eight cells with the lightest degree of shading along the left and upper boundaries). This could be avoided by using a local constant estimator, but that estimator cannot achieve the improved bias properties of the local quadratic estimator (squared bias being $O(h^8 K^{-1})$ in the interior). It is possible, however, to achieve this rate while still guaranteeing that all cell probability estimates are nonnegative, by using a geometric combination estimator, as in (3.17). That is, for (possibly multidimensional) cell $I$, the probability estimator equals

$$p^*(I|\mathbf{h}) = \hat{p}(I|\mathbf{h})^{4/3}\hat{p}(I|2\mathbf{h})^{-1/3}, \tag{6.5}$$

where $\hat{p}(I|\mathbf{h})$ is a (possibly boundary bias-corrected) kernel estimator using bandwidth (vector) $\mathbf{h}$. This estimator improves on the local constant and linear estimators, giving sum of squared errors $\text{SSE}(\mathbf{p}^*) = O_p(n^{-8/(d+8)}K^{-1})$. For $d = 2$, this is $O_p(n^{-4/5}K^{-1})$, an improvement over the usual $O(n^{-2/3}K^{-1})$ rate and the same as the rate for the local cubic estimator (and interior rate for the local quadratic).

**Table 6.5.** Salary data. Rows represent salary; columns represent the number of years since receiving the Bachelor's degree.

| Salary | Years since degree | | | | | | | | | |
|---|---|---|---|---|---|---|---|---|---|---|
| | 0–2 | 3–5 | 6–8 | 9–11 | 12–14 | 15–17 | 18–23 | 24–29 | 30–35 | $> 35$ |
| 951-1150 | 5 | 0 | 1 | 0 | 0 | 0 | 0 | 0 | 0 | 0 |
| 1151-1350 | 2 | 1 | 0 | 0 | 0 | 0 | 0 | 2 | 0 | 0 |
| 1351-1550 | 5 | 1 | 3 | 2 | 0 | 0 | 1 | 0 | 0 | 0 |
| 1551-1750 | 5 | 5 | 2 | 1 | 0 | 1 | 0 | 1 | 0 | 0 |
| 1751-1950 | 9 | 9 | 5 | 0 | 2 | 2 | 1 | 1 | 1 | 0 |
| 1951-2150 | 3 | 5 | 2 | 1 | 2 | 0 | 1 | 0 | 0 | 1 |
| 2151-2350 | 0 | 1 | 4 | 3 | 2 | 1 | 3 | 0 | 2 | 1 |
| 2351-2550 | 0 | 0 | 4 | 0 | 1 | 2 | 2 | 0 | 0 | 1 |
| 2551-2750 | 0 | 0 | 2 | 2 | 0 | 5 | 1 | 2 | 1 | 1 |
| 2751-2950 | 0 | 0 | 1 | 0 | 0 | 1 | 4 | 0 | 2 | 3 |
| 2951-3150 | 1 | 0 | 1 | 0 | 1 | 1 | 1 | 0 | 1 | 0 |
| 3151-3750 | 0 | 0 | 0 | 0 | 0 | 0 | 5 | 0 | 0 | 1 |

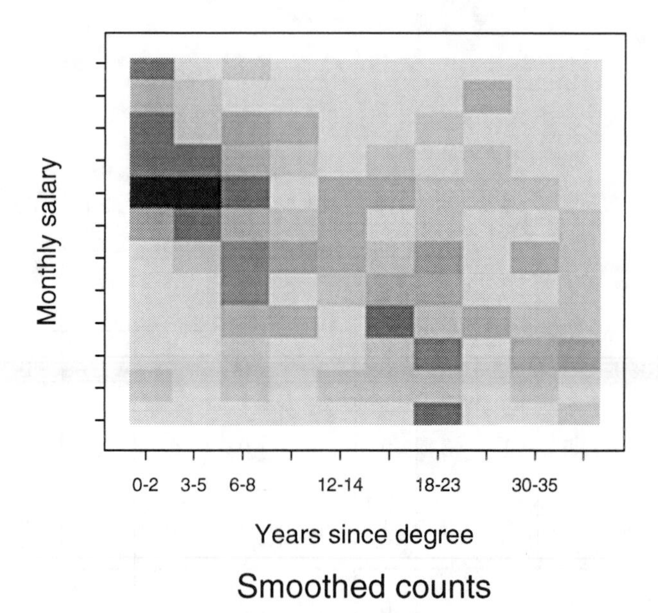


**Fig. 6.11.** Shade plots of unsmoothed counts (top) and local quadratic smoothed counts (bottom) for salary data.

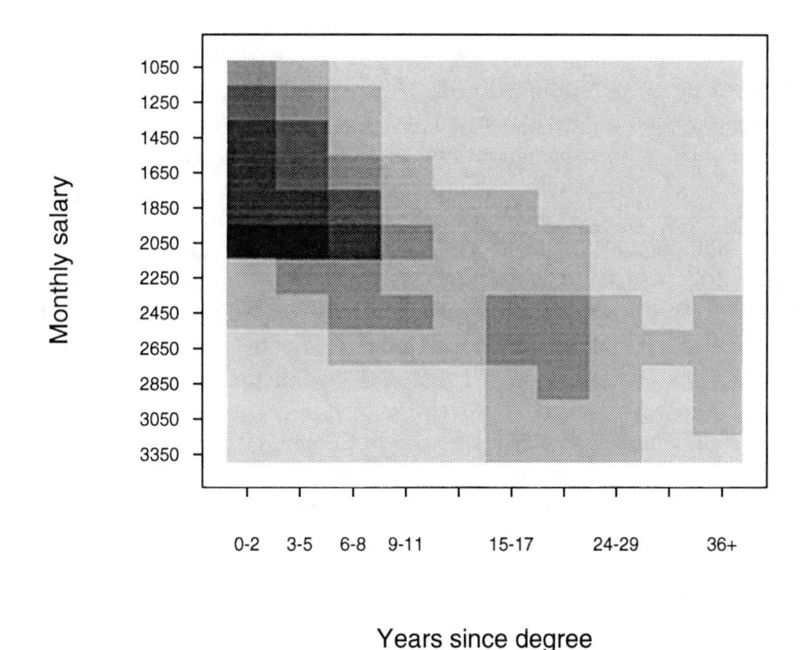


**Fig. 6.12.** Shade plot of smoothed counts using geometric combination estimate for salary data.

Figure 6.12 gives a shade plot for an application of this estimator to the salary data. The underlying estimates are local constant loess estimates with spans of .13 and .52, respectively (doubling the bandwidth in both dimensions corresponds to multiplying the span by four). The smoothed counts are similar to those of the local quadratic estimate in Fig. 6.11, with the same indications of bimodality and positive association but now with no negative cell probability estimates.

The informal exploratory analysis given here is not typically the focus of contingency table analysis. Rather, such tables are usually analyzed through the fitting of models (often log-linear) that reflect different association patterns between the rows and columns. Standard analysis, however, is based on the usual asymptotics where the number of observations in each cell is large and is thus inappropriate for sparse tables. Smoothing-based model fitting for such tables will be discussed briefly in Chapter 7.

## 6.4 Categorical Data, Regression, and Density Estimation

Categorical data smoothing provides a natural bridge between nonparametric regression and density estimation. As was noted in the two previous sections, contingency table smoothing is operationally very similar to nonparametric regression, with the cell relative frequencies being the response values and the cell indices being the predictor values. Multinomial and contingency table smoothing is also closely connected to density estimation, through (6.1), with an estimate of a cell probability $p_i$ giving a density estimate for $f(x_i)$ (it is easiest to take $f$ to be supported on $[0,1]$, so that $x_i = i/K$, but any arbitrary interval can be treated by a simple translation, with the estimated density on $[a,b]$ being the estimated density on $[0,1]$ divided by $b-a$). This is the essence of the computational efficiencies that arise from calculating density estimates on binned data.

Thus, in a sense all these smoothing problems can be treated as special cases of a general regression problem, with the response variable possibly being a true response $y$ or a set of cell relative frequencies $\overline{\mathbf{p}}$. If the bins are so narrow that only one observation can fall into any bin (except for exact duplications to the resolution of the data), the resultant regression estimator corresponds exactly to a density estimator. For example, the univariate (multinomial) local linear regression-based density estimator that is the constant term of the minimizer of

$$\sum_{j=1}^{K}\left[\overline{p}_j - \beta_0 - \beta_1\left(\frac{i}{K}-\frac{j}{K}\right)\right]^2 W\left(\frac{i/K - j/K}{h}\right)$$

becomes equivalent to the constant term of the minimizer of (3.24),

$$\int\left[n^{-1}\sum_{i=1}^{n}\delta(u-x_i) - \beta_0 - \beta_1(x-u)\right]^2 W\left(\frac{x-u}{h}\right)du,$$

as the bins narrow, substituting $i/K$ for $x$ and $j/K$ for $u$. This local polynomial estimator is equivalent to the generalized jackknifing boundary kernel given by (3.12), so higher odd order local polynomials correspond to boundary bias correction for higher order kernel estimators, and multivariate local polynomial regression-based density estimators correspond to boundary bias-corrected multivariate density estimators, which can be calculated using available local polynomial regression software.

Figure 6.13 gives an illustration of the regression-based density estimator for the mine accident data previously examined in Figs. 3.7, 3.10, 3.11, and 3.18 – 3.20. The data are binned to the integer level, which is the resolution of the data. The circles in the plot are the frequency estimates for the bins (which correspond to the response values in the regression), and the density estimate is a local linear loess estimate with span equal to .4.

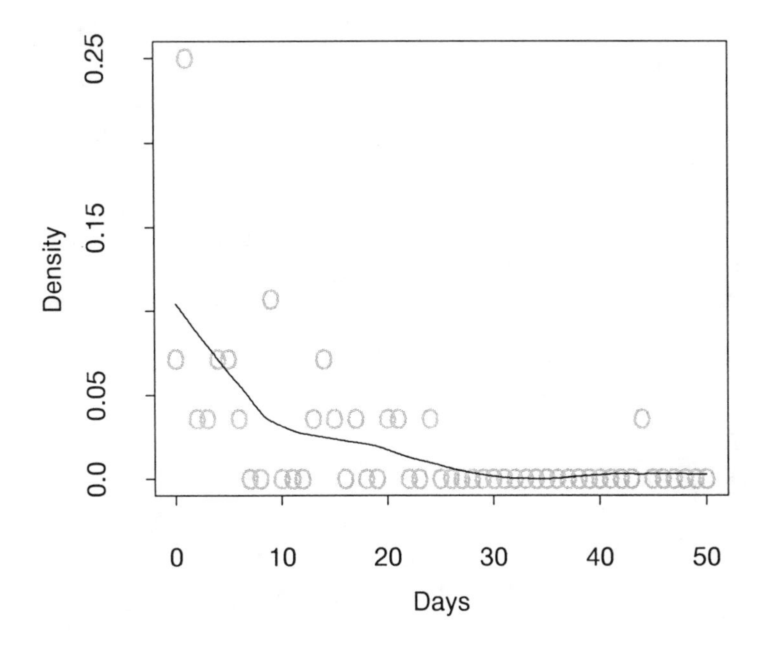


**Fig. 6.13.** Local linear loess density estimate for mine accident data.

The form of the estimate is very similar to the previous versions, with the roughly exponential shape and bulge around 10–20 days between accidents apparent.

Regression-based density estimation is probably even more useful for multivariate data, where boundary bias correction results have lagged behind those of univariate estimation. The top plot of Fig. 6.14 is a scatter plot of the percentage of people in rural areas with access to safe water in 1985 (horizontal axis) and in 1990 (vertical axis) for 70 countries reporting the values. It is apparent that there is a positive association between the two variables, but the scatter plot does not reveal any structure past that.

The bottom plot in Fig. 6.14 is a contour plot for a regression-based local quadratic loess estimate with span equal to .3. The basis of the estimate is a $10 \times 10$ table of counts, so that each bin has width 10, or ten times the resolution of the data. Despite the coarseness of the grid, the density estimate shows clear trimodality in the density, corresponding to countries with very little rural access to safe water in both years (this includes countries such as Afghanistan, Ethiopia, and Nicaragua), countries with a moderate level of access in both years (including Algeria, Honduras, and the Philippines), and countries with high access in both years (such as

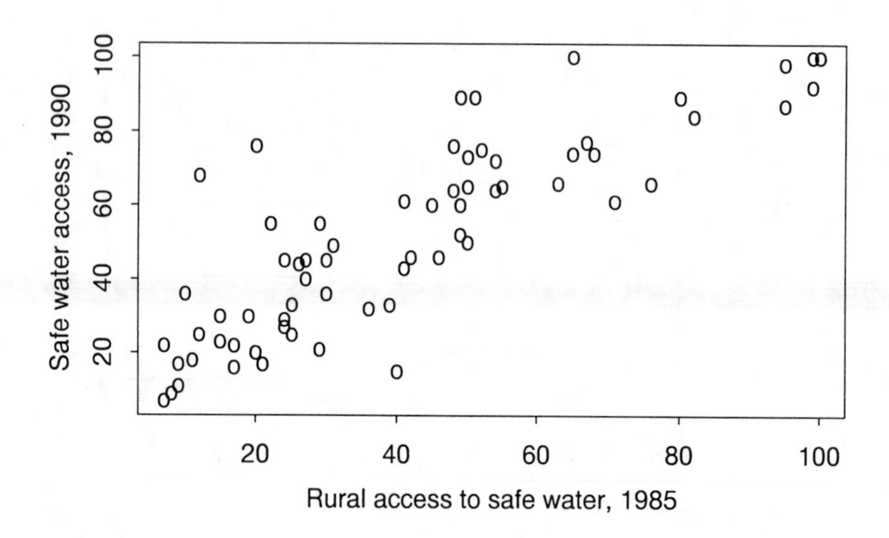


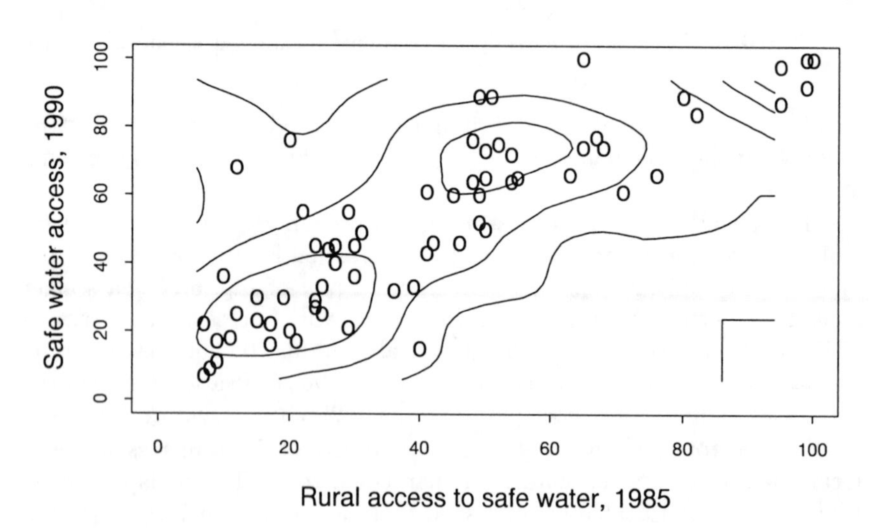


**Fig. 6.14.** Rural access to safe water data. Top: Scatter plot of rural access in 1990 versus rural access in 1985. Bottom: Contour plot of local quadratic loess density estimate.

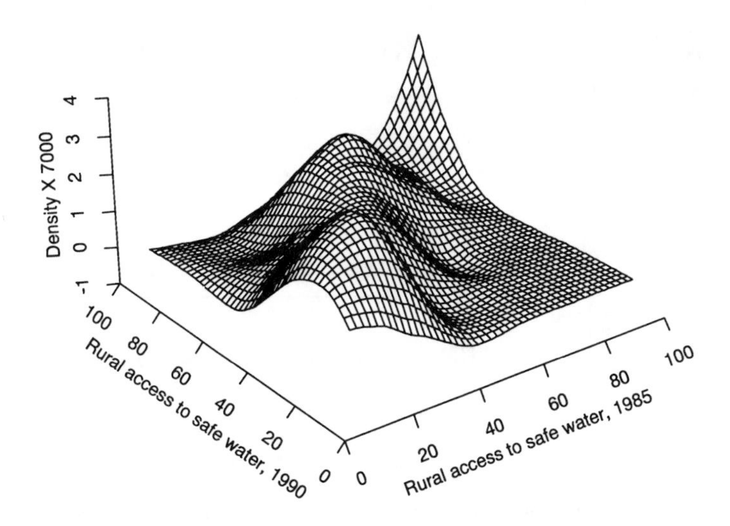


**Fig. 6.15.** Perspective plot of local quadratic loess density estimate for water access data.

Bahrain, Cyprus, and Tonga). Another noteworthy property of the density is that the modes corresponding to low and moderate access are centered at higher values for 1990 than for 1985, suggesting an encouraging pattern of improved access over the five years.

The high access mode is difficult to assess in the contour plot, since it is at the edge of the support of the data. Figure 6.15 gives a corresponding perspective plot, which shows the trimodality, including the high density "point" at the upper boundaries of the two variables.

A drawback to the regression-based density estimator, and the closely related local polynomial multinomial and contingency table estimator, is that density and probability estimates can be negative. Figure 6.16 is a local quadratic loess estimate for the mine accident data with span equal to .7. The form of the estimate is similar to that of the local linear loess estimate in Fig. 6.13, but the density estimate is now negative for values between 30 and 40 days. The bivariate local quadratic loess density estimate for the water access data also has negative regions, which correspond to the

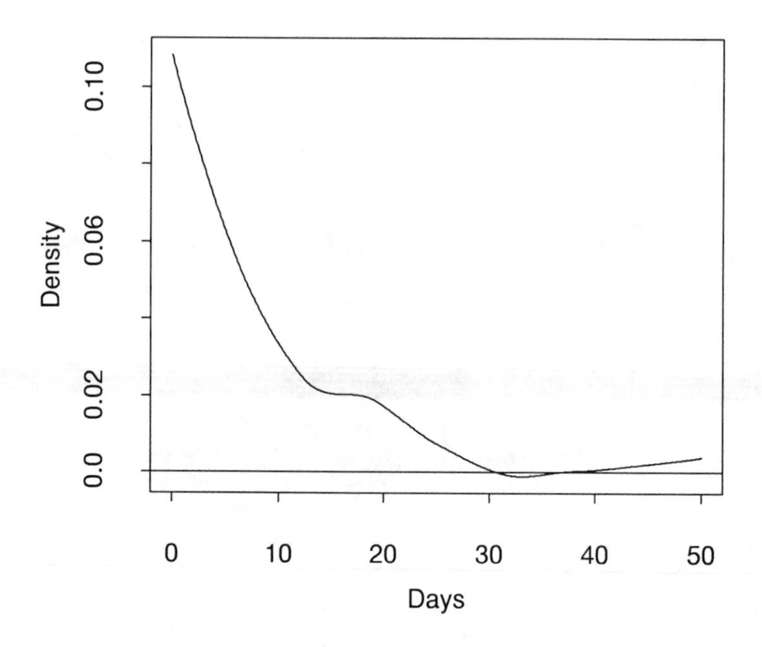


**Fig. 6.16.** Local quadratic loess density estimate for mine accident data.

lowest density contours in Fig. 6.14 and are also apparent in the perspective plot of Fig. 6.15.

One way to avoid these negative values is to use a geometric combination estimator, as in (6.5) and Fig. 6.12. A more general approach is to recognize that the main justification for the use of local least squares in the local polynomial regression objective function is the assumption that regression errors are normally distributed, since least squares then corresponds to maximum likelihood. This is not true for multinomial data, and using an objective function tied to the proper likelihood function is more sensible. That is, the appropriate regression analogy is not with least squares regression, but with generalized regression, in the sense of generalized linear and additive models, as in Section 5.7.

It is helpful to formulate this problem in terms of a Poisson model, rather than a multinomial model (the multinomial vector can be viewed as a set of Poisson random variables, conditional on $\sum n_i = n$). The log-likelihood is then

$$\sum_{j=1}^{K} [n_j \log(np_j) - np_j] \tag{6.6}$$

(ignoring constants). The canonical link for the Poisson distribution is the log link, so the local log-likelihood has the form

$$L' = \sum_{j=1}^{K} \left\{ n_j \left[ \beta_0 + \cdots + \beta_t \left( \frac{i}{K} - \frac{j}{K} \right)^t \right] - \exp\left[ \beta_0 + \cdots + \beta_t \left( \frac{i}{K} - \frac{j}{K} \right)^t \right] \right\} W\left( \frac{i/K - j/K}{h} \right) \tag{6.7}$$

for cell $i$. The density estimate is then $\exp(\hat{\beta}_0)$, where $\hat{\boldsymbol{\beta}}$ is the maximizer of (6.7). This guarantees that the estimate will be nonnegative.

For $K \to \infty$, $np_j \approx nf(x_j)/K$ as in (6.1), so (6.6) is approximately

$$\sum_{j=1}^{K} \left\{ n_j \log \left[ \frac{nf(x_j)}{K} \right] - \frac{nf(x_j)}{K} \right\}.$$

The generalized local log-likelihood (6.7) is thus approximately

$$\begin{aligned} L' &\approx \sum_{j=1}^{K} W\left( \frac{i/K - j/K}{h} \right) n_j \log \left[ \frac{nf(x_j)}{K} \right] \\ &\quad - n \sum_{j=1}^{K} W\left( \frac{i/K - j/K}{h} \right) \frac{f(x_j)}{K} \\ &\approx \sum_{i=1}^{n} W\left( \frac{x - x_i}{h} \right) \log[f(x_i)] - n \int W\left( \frac{x-u}{h} \right) f(u) du \end{aligned} \tag{6.8}$$

(ignoring constants and changing to a continuous scale by substituting $x$ for $i/K$ and approximating the second sum with an integral). Equation (6.8) is identical to the local likelihood (3.18) for density estimation, so a Poisson-based nonparametric regression on cell frequencies for a fine grid is an approximate local likelihood density estimator. Besides sharing the good properties of that estimator, the approximate estimator effectively is always defined, since the regression estimator on the equispaced grid exists as long as $h$ is large enough to include the minimal number of grid points needed for the local regression estimate to be defined (for local linear estimation, for example, the span of the estimate must be at least two grid points).

Figure 6.17 gives the approximate local (quadratic) likelihood estimate for the mine accident data that corresponds to the local (quadratic) least squares estimate of Fig. 6.16. Not surprisingly, the two estimates have almost identical forms, but the approximate local likelihood estimate is nonnegative, as it must necessarily be. Figure 6.18 is a variability plot for this estimate, which has the expected properties of widening at the boundary and in the region of the bulge, without the disquieting jaggedness of the nearest neighbor-based local quadratic variability plot given in Fig. 3.18.

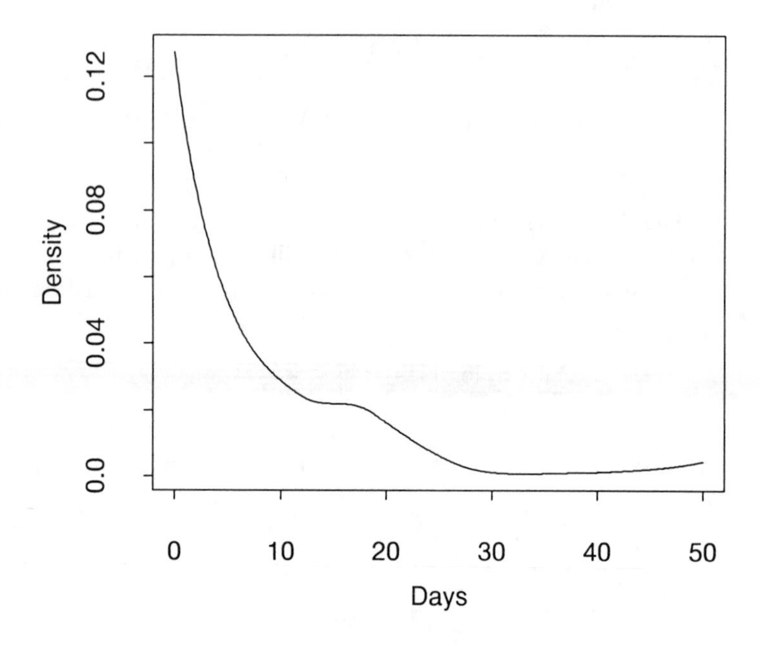


**Fig. 6.17.** Approximate local (quadratic) likelihood density estimate for mine accident data.

The local likelihood density estimator discussed in Chapter 3 allows the local degree of smoothing to vary by using nearest neighbor weights, thereby smoothing less in high density regions and more in low density regions. Nearest neighbor weights do not do this for the approximate local likelihood estimator, since the "observations" are equispaced grid points, but the same goal can be accomplished by locally varying the bandwidth in an appropriate way.

Figure 6.19 shows how this can be done. The density estimate given is an approximate local (quadratic) likelihood estimate for the racial distribution data. Figure 3.17 gave the local quadratic estimate for these data using a 70% nearest neighbor span, which allowed the peak on the right side to be highlighted without spurious bumpiness in the low density region to the left. The approximate estimate in Fig. 6.19 is virtually identical to that in Fig. 3.17 and is based on an approximate local quadratic likelihood loess fit with spans equal to .8 for $0 \leq x \leq .461$, .5 for $.461 < x \leq .81$, and .2 for $x > .81$, respectively. That is, by locally varying the bandwidth, there is more smoothing at the lower density regions and less smoothing at the high density regions.

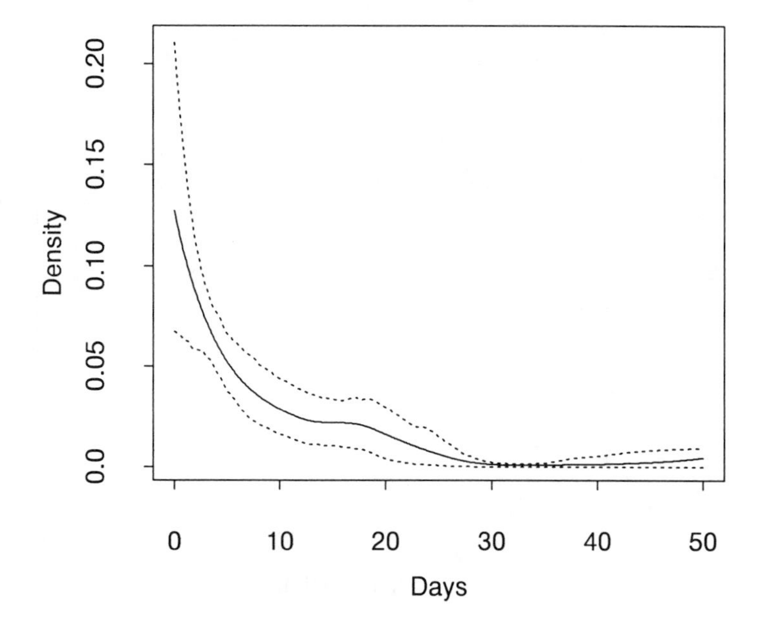


**Fig. 6.18.** Variability plot of approximate local (quadratic) likelihood density estimate for mine accident data.

The approximate local likelihood density estimator is of local polynomial form, but estimators of the same type that are not based on local polynomials can be constructed. Any nonparametric regression estimate that can be adapted to the generalized likelihood framework could be used to fit a smooth curve to the Poisson mean function in (6.6), yielding a density estimate. In particular, fitting a generalized regression model using smoothing splines instead of local polynomials is conceptually similar to the logspline estimator discussed in Section 3.5, although the details of implementation are considerably different.

## Background material

### Section 6.1

Simonoff (1995c) gave a detailed account of smoothing methods for categorical data. He showed that suggested approaches have close ties to other areas of statistical methodology, including shrinkage estimation, Bayes methods, penalized likelihood, splines, and kernel density and regression estimation.

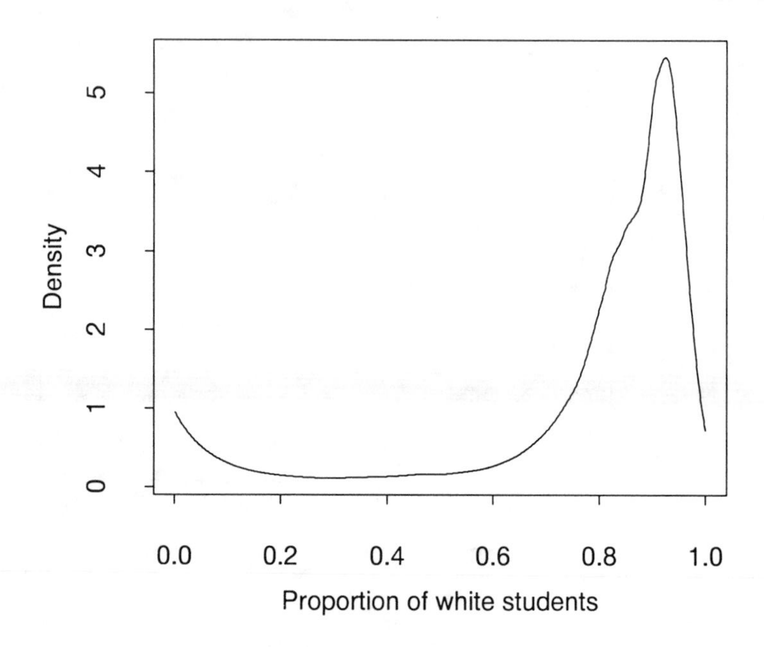


**Fig. 6.19.** Approximate local (quadratic) likelihood density estimate with varying bandwidth for racial distribution data.

The salary data, in its 28-cell form, came from Department of Energy (1982). Simonoff and Tsai (1991a) gave the 6-cell version, while Simonoff (1987) gave the 12-cell version.

### Section 6.2

Simonoff (1983) gave the 55-cell discretization of the mine explosion data. Simonoff (1985) included the 50-cell discretization of the calcium carbonate data.

Aitchison and Aitken (1976) proposed the use of kernel estimators for multinomials with unordered categories (that is, a nominal categorical variable). The kernel gives higher weight to $\overline{p}_i$ in estimation of $p_i$ and lower, constant, weight to all other cell frequency estimates. This has the effect of shrinking the frequency estimates toward a uniform distribution, and it can be shown that a shrinkage factor (which is a function of the unknown probabilities) exists such that the MSSE of the kernel estimator is smaller than that of the frequency estimator (Brown and Rundell, 1985).

When the categories are ordered, a kernel function that takes the ordering into account is sensible. Such a kernel should have the property that

$W(x)$ decreases smoothly as $|x|$ increases. Aitchison and Aitken (1976) proposed such weights for the case $K = 3$, which Aitken (1983) extended to arbitrary $K$ cells. Habbema, Hermans, and Remme (1978), Titterington (1980), and Wang and Van Ryzin (1981) gave other specific suggestions. Such an estimator can be viewed as a discretization of a continuous density estimator (as well as through analogy with regression), and Solow (1995) defined a categorical data smoother by first using a kernel estimator on continuous data and then integrating the resultant $\hat{f}$ over the range of a given cell to estimate the cell probability.

Aitchison and Aitken (1976) proposed choosing the smoothing parameter using likelihood cross-validation, which leads to a consistent estimator of $\mathbf{p}$ under standard asymptotics (Bowman, 1980) but can behave erratically when there are small cell counts in the table (Hall, 1981). For this reason, Titterington (1980), Hall (1981), Bowman, Hall, and Titterington (1984), and Brown and Rundell (1985) proposed squared-error rules, such as least squares cross-validation. Titterington and Bowman (1985) performed a small Monte Carlo study comparing several different kernels and smoothing parameter rules.

Fienberg and Holland (1973) and Bishop, Fienberg, and Holland (1975, Chapter 12) introduced the idea of sparse asymptotics, where the number of cells becomes infinite at the same rate as the sample size. They studied the properties of shrinkage estimators $\hat{p}_i = (n_i + \alpha_i)/[\sum_j (n_j + \alpha_j)]$ and showed that certain choices of $\boldsymbol{\alpha}$ lead to smaller sparse asymptotic MSSE than that of $\overline{\mathbf{p}}$. These estimators are still not sparse asymptotic consistent, however.

Burman (1987a) and Hall and Titterington (1987b) first examined the sparse asymptotic properties of kernel estimators. Burman's proposed kernel estimator was of the Nadaraya–Watson (local constant) form, and he derived the asymptotic form of its MSSE assuming $h \to 0$ with $hK \to \infty$. Burman avoided boundary bias problems by assuming that $f'(0) = f'(1) = 0$, and proposed choosing $h$ using $AIC$.

Hall and Titterington (1987b) proposed a kernel estimator of the Priestley–Chao form,

$$\hat{p}_i = \sum_{i=1}^{K} W\left(\frac{i/K - j/K}{h}\right) \overline{p}_j.$$

Note that the $\hat{p}_i$ do not necessarily sum to 1 (and usually sum to less than 1). Hall and Titterington derived the optimal convergence rate of any estimator of $\mathbf{p}$ and showed that the kernel estimator achieves it. Specifically, say $f$ has $s$ bounded continuous derivatives on $(-\infty, \infty)$ and vanishes outside a bounded interval. This condition imposes very strong boundary conditions on the true density; for example, for $s = 2$, it requires that $f$, $f'$, and $f''$ all become zero at the boundary. Let $\delta$ satisfy $\sup_i p_i \leq C\delta$ for some constant $C$ (roughly speaking, $\delta$ is equivalent to $K^{-1}$). Then, if $\delta \to 0$ as $n \to \infty$, the optimal rate of MSSE of any estimator $\hat{\mathbf{p}}$ equals

$$\text{MSSE}(\hat{\mathbf{p}}) = \begin{cases} O(n^{-1}), & \text{if } n^{1/(2s+1)}\delta \text{ is bounded away from zero,} \\ O(n^{-2s/(2s+1)}\delta), & \text{if } n^{1/(2s+1)}\delta \to 0. \end{cases}$$

Thus, if the multinomial is not overly sparse, with $K$ increasing at a rate no faster than $n^{1/(2s+1)}$, the optimal convergence rate of MSSE is $O(n^{-1})$, which $\overline{\mathbf{p}}$ achieves. Any greater sparseness, however, requires estimators that smooth by borrowing information from nearby cells. If $f$ has two bounded continuous derivatives, the optimal rate is $O(n^{-4/5}K^{-1})$, as the local constant (assuming boundary conditions) and local linear estimators achieve. Hall and Titterington proposed using least squares cross-validation to choose $h$ and proved its asymptotic optimality.

The boundary conditions needed for the kernel estimators to achieve consistency (Hall and Titterington) or optimal MSSE (Burman) are often not satisfied. Dong and Simonoff (1994) derived boundary kernels (based on the Hall and Titterington kernel estimator) that avoid the need for the boundary conditions. At the left boundary, the kernel weights are constructed from integration of a boundary kernel function that satisfies

$$\int_{-1}^{q} \omega_q(u) = 1; \int_{-1}^{q} u\omega_q(u)du = 0;$$
$$\int_{-1}^{q} u^2\omega_q(u)du = \alpha_q \neq 0; \int_{-1}^{q} \omega_q(u)^2\, du < \infty;$$
$$\omega_q \to W \text{ as } q \to 1$$

(right boundary kernels are defined analogously). Dong and Simonoff proved that the boundary kernel estimator achieves MSSE $= O(n^{-4/5}K^{-1})$ without restrictive boundary conditions. Dong and Ye (1996) advocated choosing the kernel to minimize asymptotic variance, which amounts to using a boundary-corrected uniform kernel. Rajagopalan and Lall (1995) defined a discrete kernel estimator where the coefficients of the quadratic kernel change for each cell to guarantee that the required moment conditions on the kernel are satisfied exactly (the bandwidth is also constrained so that $hK$ is an integer).

Aerts, Augustyns, and Janssen (1997a) examined the properties of local polynomial estimators for sparse multinomials. They derived the theoretical properties of the estimator and showed its improved performance over the local constant estimator in a small Monte Carlo study.

Simonoff (1983) gave the first demonstration of sparse asymptotic consistency for an estimator, using a penalized likelihood approach. The estimator is the maximizer of

$$\sum_{i=1}^{K} n_i \log p_i - \beta \sum_{i=1}^{K-1} (\log p_i - \log p_{i+1})^2, \qquad \beta > 0,$$

where $\beta$ is a smoothing parameter. Assuming appropriate smoothness of $f$, and the boundary conditions $f'(0) = f'(1)$ (these conditions were mistakenly omitted in the paper), the maximum penalized likelihood estimator $\tilde{\mathbf{p}}$ is sparse asymptotic consistent, with rate

$$\sup_{1 \le i \le K} \left| \frac{\tilde{p}_i}{p_i} - 1 \right| = O_p[K^{-2/5}(\log K)^{2/5}].$$

Monte Carlo simulations support the improved performance of $\tilde{\mathbf{p}}$ over the frequency and shrinkage estimators.

As always, penalized likelihood estimators also can be justified on Bayesian grounds, with the penalty function being proportional to the logarithm of the prior density. Bayesian justifications for penalized likelihood estimation (and related estimators) for discrete data can be found in Leonard (1973), Thorburn (1986), Lenk (1990), and Granville and Rasson (1992). The discrete maximum penalized likelihood density estimator of Scott, Tapia, and Thompson (1980) also can be viewed as a penalized likelihood estimator for categorical data.

### Section 6.3

Burman's (1987a) results on kernel estimation apply to $d$-dimensional tables, for general $d$. He showed that while the bias of the estimator remains $O(h^4 K^{-1})$, the variance is $O[(nh^d K)^{-1}]$ (taking the smoothing parameters to be converging to zero at the same rate). This result still requires boundary conditions (that all first partial derivatives are zero at the boundary). Dong and Simonoff (1995) described the construction of boundary kernels for $d$-dimensional contingency tables that achieve the same MSSE convergence rate without requiring boundary conditions.

Aerts, Augustyns, and Janssen (1997b) studied the properties of local polynomial estimation for sparse multidimensional contingency tables. They showed that the asymptotic MSSE has a form similar to that of the AMISE in regression estimation, just as it does for one-dimensional data. Using the parameterization of Sections 4.2 and 5.7 for the multivariate kernel $W$ and bandwidth matrix $H$, the asymptotic MSSE for the local linear estimator is

$$\text{AMSSE} = \frac{R(W)}{nh^d K} + \frac{h^4}{4K} \int \{\text{trace}[AA' \nabla^2 f(\mathbf{u})]\}^2 \, d\mathbf{u}.$$

Aerts *et al.* also described generalization of these results to higher order local polynomial estimators.

Simonoff (1995c) gave the MBA survey data, while Simonoff (1987) is the source of the salary data.

Dong and Simonoff (1995) proposed and analyzed the geometric combination estimator for $d$-dimensional tables. They showed that if $f$ has bounded fourth partial derivatives, then the geometric combination estimator (6.5) using $d$-dimensional boundary kernels has SSE that converges to zero in probability at the rate $n^{-8/(d+8)}K^{-1}$, while not yielding nonnegative estimates. So, for example, for a one-dimensional table, $\text{SSE}(\mathbf{p}^*) = O_p(n^{-8/9}K^{-1})$. This convergence in probability of SSE is not as strong as mean square convergence, which can be achieved under more restrictive conditions. If all second and third partial derivatives are zero in the boundary region, and $d \leq 4$, then $\mathbf{p}^*$ using non-boundary-corrected kernels has $\text{MSSE} = O(n^{-8/(d+8)}K^{-1})$, as expected.

Simonoff (1987) first proposed the marginal/conditional estimator (described in Chapter 4 for use with bivariate continuous data) for use with two-dimensional contingency tables. Let $\overline{p}_{i\cdot}$ and $\overline{p}_{\cdot j}$ be the frequency estimates of the marginal probabilities of falling in the $i$th row and $j$th column, respectively. Define $\hat{p}_{ij}^R, j = 1, \ldots, C$, to be the penalized likelihood estimates of the conditional probability of falling in the $(i, j)$th cell given being in the $i$th row (another one-dimensional multinomial smoother also could be used, but the MPLE was used in the paper). Similarly, let $\hat{p}_{ij}^C, i = 1, \ldots, R$, be the penalized likelihood estimates of the conditional probability of falling in the $(i, j)$th cell given being in the $j$th row. The marginal/conditional estimator is $\hat{p}_{ij} = (\hat{p}_{ij}^R \hat{p}_{ij}^C \overline{p}_{i\cdot} \overline{p}_{\cdot j})^{1/2}$. Assuming appropriate smoothness and boundary conditions, if $R$ and $C$ both become infinite at the rate $\sqrt{n}$, then

$$\sup_{i,j} \left| \frac{\hat{p}_{ij}}{p_{ij}} - 1 \right| = O_p[n^{-1/5}(\log n)^{2/5}].$$

Although this rate is slower than the corresponding rate for a product kernel estimator, Monte Carlo simulations indicate better performance, presumably because of the locally adaptive nature of the estimator.

Just as a smooth underlying one-dimensional probability vector can be viewed as locally uniform (showing that smoothing towards local uniformity is useful), so too a smooth underlying two-dimensional probability matrix can be viewed as exhibiting local independence. For this reason, Bayesian and penalized likelihood methods that shrink the frequency estimates towards independence will result in smoother probability estimates. Leonard (1975), Laird (1978), Simonoff (1983), and Granville and Rasson (1995) gave examples of such estimators.

A different kind of multidimensional contingency table structure is that of multivariate binary data. Here, observations fall in $2^d$ cells, a number that can become very large as $d$ increases, leading to very sparse tables. Aitchison and Aitken (1976) gave a kernel estimator for tables of this type, where the distance between two cells is a function of how many of the dimension entries (0 or 1) they have in common. That is, "closeness" of two cells is defined based on having common indices over many dimensions, rather than having close indices within a particular dimension.

Grund (1993) examined the theoretical properties of the kernel estimator for fixed dimension $d$, while Grund and Hall (1993) investigated the case of sparse tables. They defined smoothness of the underlying probabilities through a smooth function of the distance between two cells, and showed that the superiority in MSSE of the kernel estimator over the frequency estimator (as $d$ increases) increases with the smoothness of the underlying probabilities and the sparseness of the table. Grund and Hall also showed that minimizing the cross-validatory choice of smoothing parameter is asymptotically equivalent to minimizing the mean sum of squared error.

Diaconis (1983) examined projection pursuit for categorical data. He showed that the "least interesting" projection for such data is uniform, rather than Gaussian, but did not examine sparse asymptotic properties or the best way to smooth the lower-dimensional projections.

Fahrmeir and Tutz (1994, Sect. 5.2) discussed smoothed categorical regression estimators. These estimators are designed for regression data where the target variable is categorical and the predicting variables are either continuous or categorical. Kernel estimators combine local density estimation at a particular predictor value $\mathbf{x}_i$ with weighting across predictor values, using either continuous kernels or categorical kernels, as appropriate.

## Section 6.4

The background material for Section 3.3.1 described the connection between the generalized jackknife boundary kernel and the local polynomial density estimator. In particular Fan, Gijbels, Hu, and Huang (1996) and Cheng, Fan, and Marron (1997) proposed constructing local polynomial density estimates using local polynomial regression estimates on binned data. World Bank (1994) gave the water access data.

Cheng (1997) investigated plug-in bandwidth selection for the local linear density estimator. The selector is conceptually similar to the Sheather–Jones selector $\hat{h}_{\mathrm{SJ}}$ but uses a plug-in estimate of $R(f'')$ that is based on local polynomial estimates instead of kernel estimates, thereby adapting to boundary effects. The resultant estimator satisfies $\hat{h}/h_0 - 1 = O_p(n^{-\alpha})$, where $\alpha = 5/14$ if $\int f''(u)f^{(4)}(u)du < 0$ (which is true if the density has zero second and third derivatives at the boundary), the usual rate for $\hat{h}_{\mathrm{SJ}}$, or $\alpha = 2/7$ if $\int f''(u)f^{(4)}(u)du > 0$.

Wei and Chu (1994) gave a different method for adapting nonparametric regression methods to estimation of a density. The method uses a Taylor Series expansion to motivate a regression formulation based on the empirical cumulative distribution function. The method is not based on binning the data. Wei and Chu report that the resultant density estimate can have a very rough appearance in boundary regions.

Lindsey (1974a,b) proposed estimating a density using Poisson regression on binned data, although this was in the context of parametric, rather than nonparametric, modeling; see also Lindsey and Mersch (1992). Efron and Tibshirani (1996) also proposed using Poisson regression, to determine their semiparametric special exponential family estimate. Loader (1995) gave an example of density estimation using a locally adaptive Poisson regression estimate. Jones (1996) discussed the connections between the different versions of local likelihood density estimation and local polynomial density estimation.

## Computational issues

Fortran code to calculate the one-dimensional boundary kernel estimates of Dong and Simonoff (1994), and boundary kernel estimates based on generalized jackknifing, is available as the collection `dong-simonoff` in the `jcgs` directory of `statlib`.

Any of the nonparametric regression methods described in Chapter 5 can be used to smooth categorical data, taking $\overline{\mathbf{p}}$ as the response variable and $i/K, i = 1, \ldots, K$ as the equispaced predictor design. Higher dimensional tables can be smoothed the same way, with $d$ predictor variables based on an equispaced design corresponding to the cell index in that dimension divided by the number of cells in that dimension. Likelihood-based estimators can be constructed using a nonparametric regression package that also allows generalized likelihood modeling. Nonparametric regression-based and approximate local likelihood density estimators are calculated in the same way. The S–PLUS code described in Loader (1995), available at `http://cm.bell-labs.com/stat/project/locfit` using a World Wide Web browser, does this directly.

## Exercises

**Exercise 6.1.** Fit local linear, quadratic, and cubic estimators to the three versions of the univariate salary data given in Table 6.1. Do any of the higher order polynomials provide insight to these data that the kernel estimates did not?

**Exercise 6.2.** Fit local quadratic and cubic estimators to the mine explosion data. Do either of these estimates improve on the local linear estimate?

**Exercise 6.3.** A different way to guarantee nonnegative probability estimates in a contingency table is to take the square root of the counts, smooth them, and then square the resultant fitted values. Apply this strategy to local linear estimation of the MBA survey data (Table 6.4) and local quadratic estimation of the salary data (Table 6.5). How does it compare with the methods discussed in this chapter?

**Exercise 6.4.** Treat the $10 \times 10$ binning of the water access data that was used to construct the density estimates in Figs. 6.14 and 6.15 as a $10 \times 10$ contingency table. Construct different smoothing-based estimates of the probability matrix, such as local polynomials of various degrees and the geometric combination estimator. Do these estimates give the same impression of the data as the continuous density estimates do? Which methods seem to work best?

**Exercise 6.5.** Construct spline-based density estimates that correspond to the local polynomial-based ones given in Figs. 6.13, 6.16, 6.17, and 6.19. How do the two approaches compare?

**Exercise 6.6.** Construct approximate local polynomial likelihood estimates for the marathon record data of Fig. 3.12 and the earthquake depth data of Fig. 3.16. Explore the possibilities of using higher order polynomials and of locally varying the bandwidth. Is it possible to construct satisfactory density estimates using this method?

**Exercise 6.7.** Construct regression-based density estimates for the NBA data of Fig. 4.5 and the Swiss bank note data of Fig. 4.12. Explore the possibilities of using higher order polynomials and of locally varying the bandwidth. Is it possible to construct satisfactory density estimates using this method?

Chapter 7

# Further Applications of Smoothing

The focus of the previous chapters was mostly on the uses of smoothing as an exploratory tool in graphical data analysis. This chapter gives several (brief) examples of the application of smoothing methods in other, more formal applications to illustrate the general applicability of the idea of smoothing.

## 7.1 Discriminant Analysis

The discriminant analysis problem arises when one wishes to classify an object as a member of one of $M$ classes, where data have previously been sampled from those classes. The available data are a set of $M$ samples of $p$-dimensional data $\mathbf{x}$, with the $i$th sample (of size $n_i$) known to come from the $i$th class. Given these data (called the *training set*) and a new observation, the goal is to predict the actual class of the new observation.

A standard approach to this problem is by using Bayes' Theorem. Say the probability of an observation coming from the $i$th class is $\pi_i, i = 1, \ldots, M$ (these are called the prior probabilities), and the observations from the $i$th class are a random sample from a distribution with density $f_i(\mathbf{x})$. Then the posterior probability of an observation being from the $i$th class given the data $\mathbf{x}$ is

$$p_i(\mathbf{x}) = \frac{\pi_i f_i(\mathbf{x})}{f(\mathbf{x})}, \tag{7.1}$$

where $\pi_i f_i(\mathbf{x})$ is the joint density of the data $\mathbf{x}$ and being in class $i$ and $f(\mathbf{x}) = \sum_{j=1}^{M} \pi_j f_j(\mathbf{x})$ is the marginal density of $\mathbf{x}$. The Bayes rule is to classify an observation to the group with highest posterior probability (or equivalently, maximum $\pi_i f_i(\mathbf{x})$), so the discrimination problem becomes one of estimating the densities $f_i$.

The classical approach to discriminant analysis substitutes a particular parametric form for $f_i$ into (7.1). The assumption of a multivariate normal density for each class, with common covariance matrix but possibly different mean vectors, yields the *linear discriminant rule.* The common covariance matrix is estimated using a pooled estimate of the covariance,

$$\hat{\Sigma} = \frac{1}{\sum_{j=1}^{M} n_j - M} \left[ (n_1 - 1)\hat{\Sigma}_1 + \cdots + (n_M - 1)\hat{\Sigma}_M \right],$$

where $\hat{\Sigma}_j$ is the sample covariance matrix from the training sample for the $j$th class. Assuming multivariate normal densities for all classes while allowing different covariance matrices yields the *quadratic discriminant rule.* The problem with these rules, of course, is that if these parametric assumptions do not hold, the classification rule can be a complete failure.

Figure 7.1 gives a graphical representation of a univariate linear discriminant rule. The data are the previous salaries (to the nearest $500) of 91 first-year, full-time MBA students at New York University's Stern School of Business in 1989 and 1990. The data fall into the two classes of 65 male students and 26 female students, and their corresponding salaries form rugs at the bottom (male students) and top (female students) of each plot.

It is known that the gender distribution of the school's students was roughly 65% male and 35% female at the time, so the prior probabilities are taken here to be $\pi_{\text{male}} = .65$ and $\pi_{\text{female}} = .35$. The curves in Fig. 7.1 represent the joint densities of salary and gender for each gender, using a solid line for the male class and a dashed line for the female class. The mean salary for the men is $35,785, while that for the women is $35,635, and the pooled standard deviation is $15,979. The top plot of Fig. 7.1 gives the resultant linear discriminant rule; the joint density for the male class is larger than that for the female class for all salaries between 0 and $100,000, so all observations are classified as male.

This failure of the linear discriminant rule could be due to the presence of an outlier. One woman had a previous salary of $104,000, far larger than that of any other woman. If this observation is dropped from the training sample, the mean salary of women drops to $32,900, and the pooled standard deviation drops to $14,247. The bottom plot of Fig. 7.1 gives the resultant linear discriminant rule, and all observations are still classified as male.

A nonparametric discriminant analysis (the top plot of Fig. 7.2) makes clear what is going on here. The curves are again estimates of the joint density for each class but are now based on kernel density estimates (using a Gaussian kernel, with $h_{\text{male}} = 5200$ and $h_{\text{female}} = 3150$, respectively). Linear discriminant analysis fails because the variability of the two classes is different. Men's salaries cover a much wider range than women's salaries, which is a clear violation of the constant variance assumption of linear discriminant analysis.

The kernel-based discriminant analysis implies the following classification rule:

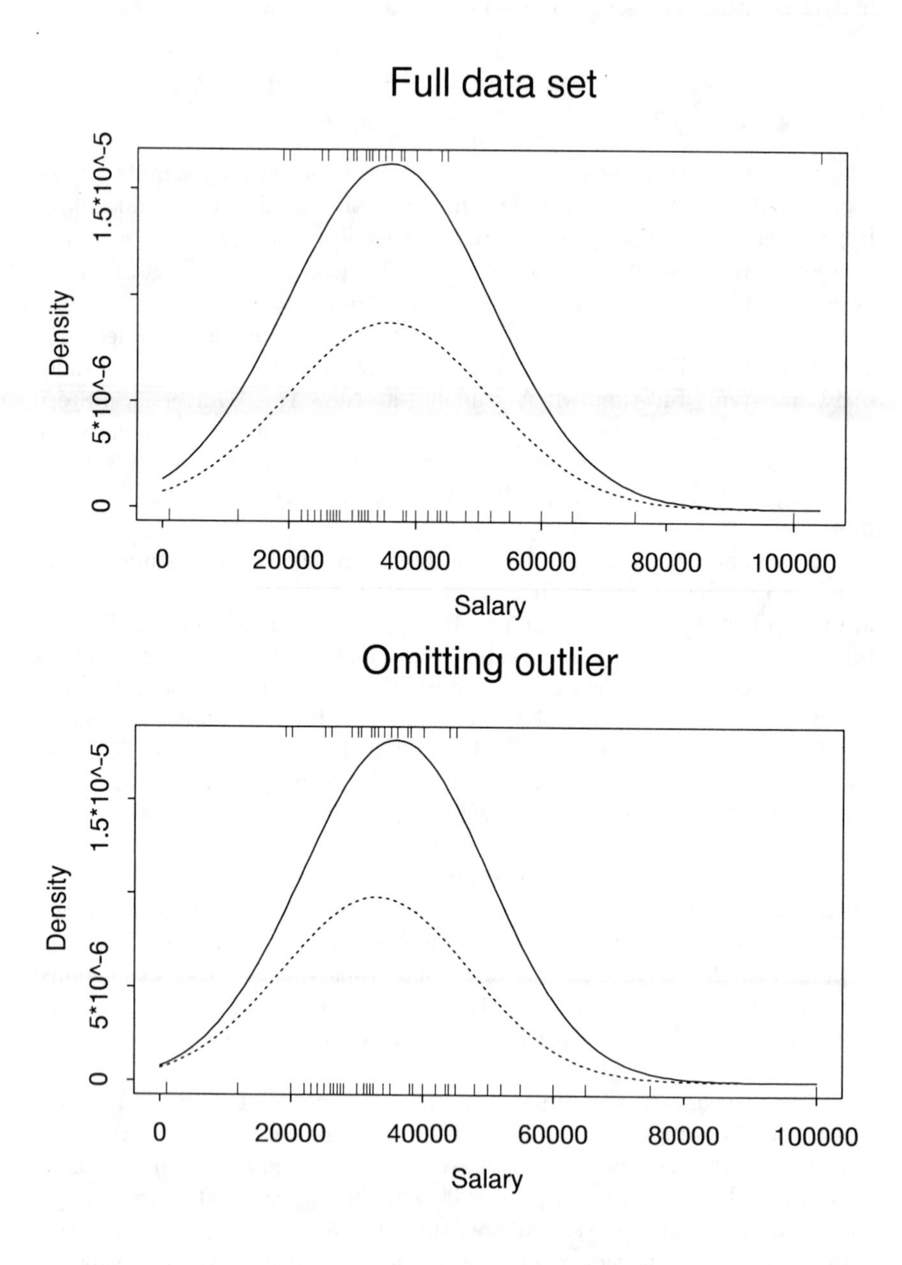


**Fig. 7.1.** Application of linear discriminant rule to MBA salary data. The solid curves are Gaussian fits to the male students, and the dashed curves are Gaussian fits to the female students. Top plot: using the full data set. Bottom plot: omitting an outlier.

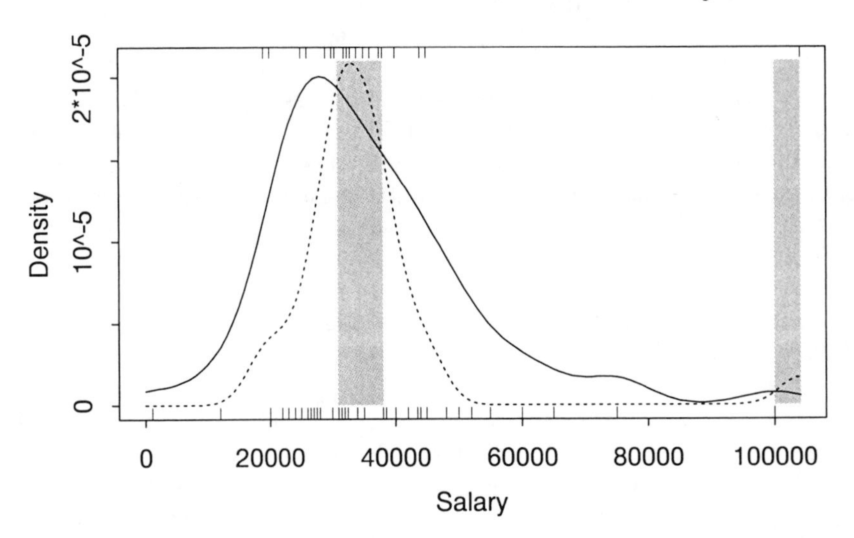


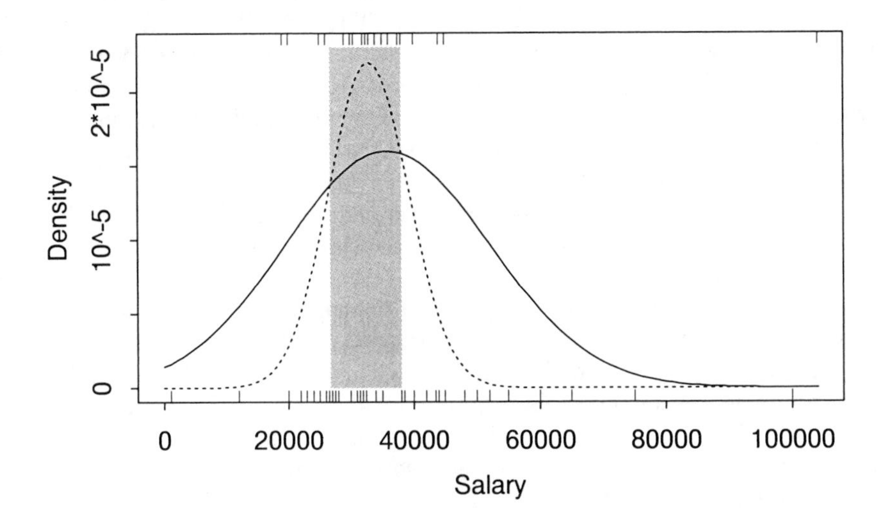


**Fig. 7.2.** Application of discriminant rules to MBA salary data. The solid curves are fits for the male students, and the dashed curves are fits for the female students. Top plot: kernel-based rule. Bottom plot: quadratic rule after omitting an outlier.

| *Salary range* | *Classification* |
|---|---|
| $< \$30,890$ | Male |
| $[\$30,890, \$38,029)$ | Female |
| $[\$38,029, \$100,017)$ | Male |
| $\geq \$100,017$ | Female |

The shaded regions in Fig. 7.2 correspond to classifications to female students. Since the densities for male and female salaries are not very non-Gaussian, the kernel-based discriminant analysis suggests trying a quadratic discriminant analysis on these data. Omitting the outlier, the standard deviation of men's salaries is $16,247, while that of women's salaries is $6,351, yielding the quadratic discriminant analysis in the bottom plot of Fig. 7.2. The results are similar to the kernel-based discriminant analysis, with an observation being classified to the female class if salary is in the range $[\$26,700, \$38,060]$.

The classification rules can be compared by examining their ability to classify the observations correctly. In order to mimic the prediction of a new observation from a training sample, a cross-validated estimate is useful, where each observation is successively omitted from the data and then classified based on the remaining observations. The two methods are very similar based on this criterion, with the nonparametric rule correctly classifying 61 of the 91 cases, and the quadratic discriminant rule correctly classifying 59 of the 90 (nonoutlying) cases. (The kernel-based rule is better for the men, with 52 of 65 correct classifications versus 44 of 65 correct classifications for quadratic rule; the quadratic rule is better for the women, with 15 of 25 correct versus 9 of 26.) Of course, the quadratic discriminant rule fails if the outlier is not omitted, while the outlier does not greatly affect the kernel-based rule.

Generalization to multivariate data proceeds by using multivariate density estimates. Figures 7.3 and 7.4 illustrate a bivariate example. The data are the score on the Graduate Management Admission Test (GMAT) and first-year grade point average (GPA) for 61 second-year MBA students at New York University's Stern School of Business in 1995, with data values for men marked by $\times$ and those for women marked by $\circ$. The GMAT is a standardized examination used by almost all American business schools as an admission criterion that is designed to predict success of students in school (with a maximum score of 800), so the relationship between these two variables is of interest to school administrators.

Figure 7.3 is a contour plot of kernel estimates of the joint densities of GMAT and GPA and gender for the 48 men (solid curves) and 13 women (dashed curves) in the training sample. Prior probabilities of gender were again set to $\pi_{\text{male}} = .65$ and $\pi_{\text{female}} = .35$, and both estimates are based on multivariate normal kernels with diagonal bandwidth matrices $H = \text{diag}(25, .15)$ for the men and $H = \text{diag}(45, .12)$ for the women, respectively.

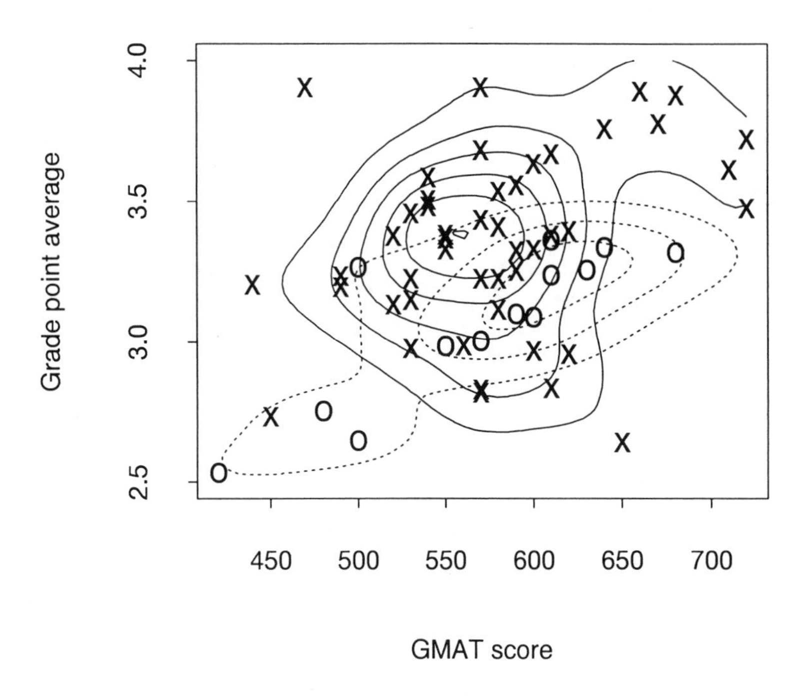


**Fig. 7.3.** Contour plots of joint density estimates of GMAT score and grade point average and gender for male (solid curves and $\times$) and female (dashed curves and $\circ$) MBA students.

The joint density for the men shows little correlation between GMAT and GPA, but the women concentrate in a narrow band showing positive correlation. These densities yield the discrimination rule given in Fig. 7.4. All regions lead to classification of a student as male, except for two regions (low GMAT and GPA, and high GMAT and moderate GPA).

The kernel-based discriminant analysis correctly classifies over 90% of the observations, although the cross-validated estimate of correct classification rate drops to roughly 79%. These are higher than the corresponding values for the linear discriminant analysis rule (74% and 72%, respectively).

The usefulness of discriminant analysis based on density estimation obviously depends on the accuracy of the underlying density estimates. Data that are roughly multivariate normally distributed with constant covariance matrix over classes are much more suitable for analysis using linear discriminant analysis, but data that violate the assumptions are potential candidates for smoothing-based discriminant analysis.

A potentially serious problem for discriminant analysis based on density estimation is the curse of dimensionality, particularly since actual discrimination data sets often include many potential predictors. Methods

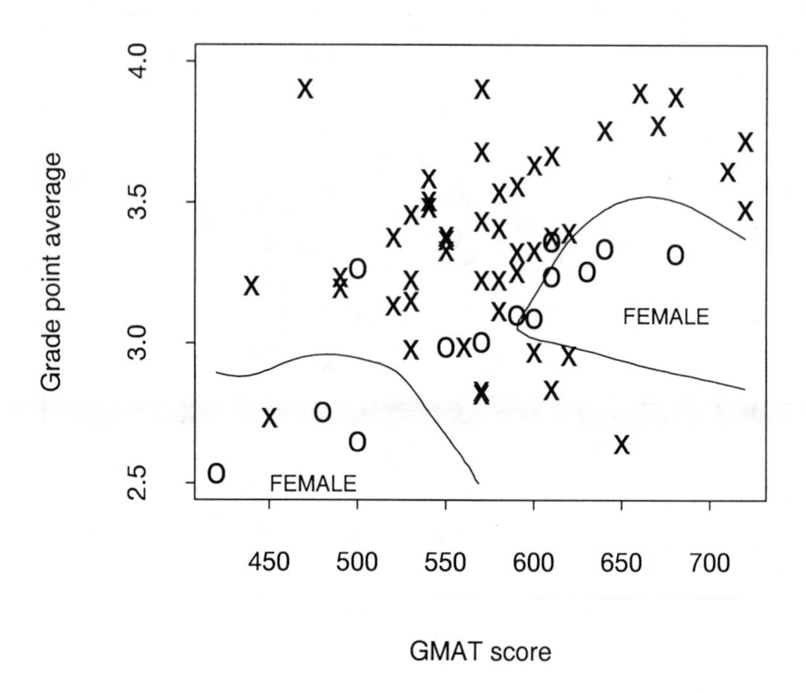


**Fig. 7.4.** Classification regions based on nonparametric discriminant analysis.

designed to address the curse of dimensionality, such as projection pursuit, can help to overcome this problem. Classification also can be viewed as a regression problem (with group membership being the response variable), and nonparametric regression methods (based on additive models, for example) also can be used for multivariate data.

## 7.2 Goodness-of-Fit Tests

In the discussion of Long Island CD rates in Chapter 1, the appearance of density estimates (a histogram or a kernel estimate) for those data argued against an underlying Gaussian density. Put another way, the density estimates showed a lack of fit of the Gaussian model to the data.

Smooth density, regression, and probability estimates can be used to assess lack of fit formally as well, using goodness-of-fit tests. A natural way to do this is by using some measure of the difference between a nonparametric estimate and a fitted parametric estimate. So, for example, the standardized distance

$$\int \frac{[\hat{f}(u) - f(u, \boldsymbol{\theta})]^2}{f(u, \boldsymbol{\theta})} \, du, \tag{7.2}$$

where $\hat{f}(u)$ is a nonparametric density estimate and $f(u, \boldsymbol{\theta})$ is a parametric fit to the data, provides a test of the hypotheses

$H_0$: The density is a member of the parametric family $f(\cdot, \boldsymbol{\theta})$

versus

$H_a$: The density is a smooth density not in the family $f(\cdot, \boldsymbol{\theta})$.

Similarly, linearity of a regression function can be tested using the test statistic

$$\frac{n}{2} \left\{ 1 - \frac{\sum_{i=1}^n [y_i - \hat{m}(x_i)]^2}{\sum_{i=1}^n [y_i - \hat{\beta}_0 - \hat{\beta}_1 x_i]^2} \right\}, \tag{7.3}$$

where $\hat{m}(\cdot)$ is a nonparametric regression estimate based on sample pairs $\{x_i, y_i\}, i = 1, \ldots, n$, and $\hat{\boldsymbol{\beta}}$ is the vector of least squares regression coefficients. The test (7.3) compares the residual sum of squares from the nonparametric fit to that of the parametric fit and takes the form of a likelihood ratio test of the linear model (5.1) versus the arbitrary smooth regression model (5.2).

Tests of this type can be more powerful than omnibus tests, which are designed to detect arbitrary alternatives to the null, since they focus on alternative functions (densities or regression curves) that are smooth. If the true function is not smooth, these tests are not appropriate, but a smooth density or regression curve under the alternative hypothesis is usually a reasonable assumption if the null function is smooth.

Goodness-of-fit based on sparse categorized data provides a good example of the benefits of testing using smoothed estimates. Consider a vector of counts $\{n_1, \ldots, n_K\}$ generated from an underlying probability vector $\mathbf{p}$. These could be inherently categorical data or a discretized version of continuous data. The classic goodness-of-fit problem tests the null hypothesis

$H_0$: $\mathbf{p} = \mathbf{p}_0$

(for some specified $\mathbf{p}_0$ that could be based on estimated parameters) versus the alternative

$H_a$: $\mathbf{p} \neq \mathbf{p}_0$.

The most frequently used statistics for this test are the Pearson $\chi^2$ statistic

$$X^2 = \sum_{i=1}^K \frac{(n_i - np_{i0})^2}{np_{i0}}$$

and the likelihood ratio $\chi^2$ statistic

$$G^2 = 2 \sum_{i=1}^K n_i \log \left( \frac{n_i}{np_{i0}} \right).$$

The usual asymptotic approximation for each of these statistics under $H_0$ is $\chi^2_{K-\nu-1}$, where $\nu$ is the number of estimated parameters under $H_0$. This is appropriate under the model $n \to \infty$ with $\inf_i np_i \to \infty$ and is thus not appropriate for a sparse table. In particular, the distribution of $G^2$ can be very far from $\chi^2$ for sparse tables, and the tests have very low power, since the frequency estimates $\overline{p}_i = n_i/n$ are poor estimates of the true $p_i$ values.

The smoothed probability estimators described in Chapter 6 can be used to construct a more powerful test. If the alternative hypothesis is restricted to be

$H_a'$: $\mathbf{p}$ is a smooth probability vector not equal to $\mathbf{p}_0$,

the local linear estimator $\hat{\mathbf{p}}$ provides an accurate estimator of $\mathbf{p}$ under both $H_0$ and $H_a'$. First, define

$$z_i = \frac{\hat{p}_i - p_{i0}}{p_{i0}}.$$

Under sparse asymptotics, under $H_0$ the asymptotic mean of $z_i$ in the interior satisfies

$$E_0(z_i) \approx \frac{h^2 p_{i0}'' \sigma_W^2}{2p_{i0}} \equiv \mu_0(z_i), \tag{7.4}$$

and the asymptotic variance of $z_i$ satisfies

$$\mathrm{Var}_0(z_i) \approx \frac{R(W)}{nhKp_{i0}},$$

where $W$ is the kernel used.

If $H_0$ does not hold, $z_i$ will no longer have asymptotic mean (7.4), which suggests that the test statistic

$$M = \sum_{i=1}^{K} \frac{|z_i - h^2 p_{i0}'' \sigma_W^2/(2p_{i0})|}{[R(W)/(nhKp_{i0})]^{1/2}} \tag{7.5}$$

is useful to identify nonnull behavior. If the true probability vector $\mathbf{p} \neq \mathbf{p}_0$,

$$E[z_i - \mu_0(z_i)] \approx \left| \frac{p_i - p_{i0}}{p_{i0}} + \frac{h^2\sigma_W^2(p_i'' - p_{i0}'')}{2p_{i0}} \right|.$$

Thus, the test $M$ effectively uses two types of measures of deviation from the null: a distance measure similar to that used by $X^2$ and a measure that compares the local smoothness of $\mathbf{p}$ and $\mathbf{p}_0$. If the true density is smooth, this additional contribution can lead to improved power.

Application of this test to the calcium carbonate data given in Table 6.3 illustrates the possible gains over standard $\chi^2$ tests. A test of the null hypothesis of uniformity

$H_0$: $p_i = 1/K, \qquad i = 1, \ldots, K,$

for these data gives $X^2 = 65.3$ with $\chi^2$ tail probability (on 49 degrees of freedom) .08. The $\chi^2$ approximation is suspect here, but the tail probability also can be estimated using Monte Carlo methods by repeatedly generating random tables with $n = 52$ and $K = 50$ from a uniform probability vector and determining the proportion of the resultant values of $X^2$ that are greater than 65.3. Here, the simulated tail probability based on 1000 replications is .046. Thus, the evidence against the uniform distribution given by $X^2$ is weak, despite the very nonuniform appearance of the probability estimates in Fig. 6.7.

This result can be contrasted with that obtained when using (7.5). The vector $\hat{\mathbf{p}}$ is the local linear estimate of Fig. 6.7, with $p''_{i0} = 0$ under the uniform null hypothesis. The statistic $M = 113.05$, with Monte Carlo tail probability less than .001. Thus, uniformity is strongly rejected here, which certainly seems to be the correct decision.

## 7.3 Smoothing-Based Parametric Estimation

The goodness-of-fit tests of the previous section are designed to identify situations where a hypothesized parametric family does not provide an adequate fit to the observed data. Smoothing methods also can be used in situations where the parametric model is reasonable (except for possibly a few unusual values) by improving the parameter estimates themselves.

Figure 7.5 illustrates the problem. The data are the normal minimum January temperatures for weather recording stations, with one station in each of the 50 states. A rug along the bottom of the plot gives the data values. The solid curve is a fitted normal density using the maximum likelihood estimates $\hat{\mu} = 23.54$ and $\hat{\sigma} = 13.38$, while the dashed curve is a kernel estimate of the density, using a Gaussian kernel with $h = 4.8$. The fitted normal density does not follow the nonparametric estimate as well as we would like, but not because of strong inherent nonnormality. Rather, the maximum likelihood estimates are apparently inflated by the two unusually high values (corresponding to January temperatures in Honolulu, Hawaii and Key West, Florida), causing the fitted Gaussian density to be centered at too high a value and to be too wide.

Smoothing-based estimators of $\mu$ and $\sigma$ can be constructed by choosing $\hat{\mu}$ and $\hat{\sigma}$ so that the fitted Gaussian density $f(\cdot, \hat{\mu}, \hat{\sigma})$ is as close as possible to the nonparametric density estimate $\tilde{f}(\cdot)$. One reasonable measure of the distance between the two curves is the Hellinger distance,

$$D^2(\hat{f}, \tilde{f}) = \int [f(u, \hat{\mu}, \hat{\sigma})^{1/2} - \tilde{f}(u)^{1/2}]^2. \tag{7.6}$$

The resultant *minimum Hellinger distance estimators* (MHDEs, the minimizers of (7.6)) are asymptotically normal and asymptotically efficient if the model family is correct (since the Hellinger distance is asymptotically equivalent to the likelihood distance in this case). In contrast to the maximum likelihood estimators however they are also robust. The local nature

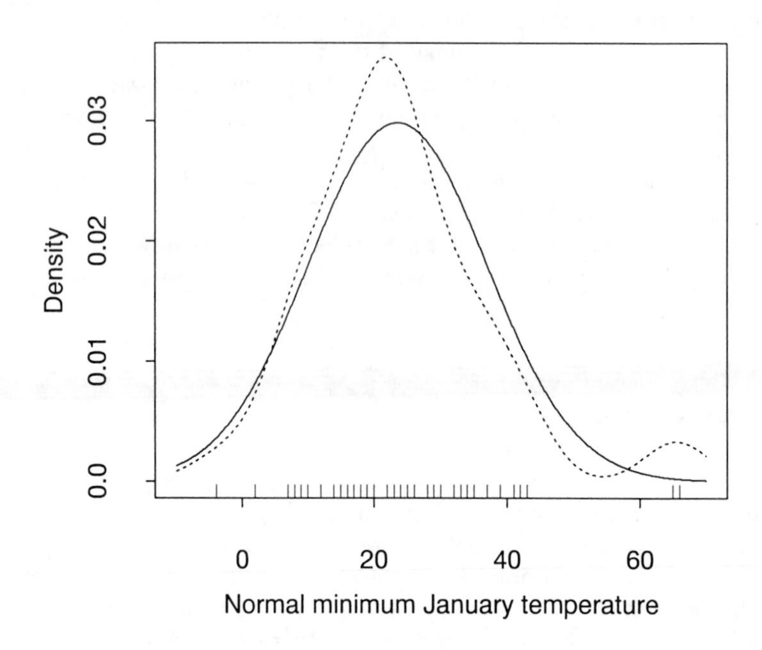


**Fig. 7.5.** Fitted Gaussian density using maximum likelihood estimates (solid curve) and kernel estimate (dashed curve) for January temperature data.

of $\tilde{f}$ means that outliers have little effect on most of the density estimate, so the outliers have a limited effect on the estimates of $\mu$ and $\sigma$. More technically, the *breakdown point* of the estimator for Gaussian data is at least .25 (the breakdown point is the smallest fraction of bad data that can cause an estimator to give an arbitrarily bad answer). This breakdown point persists for multivariate data as well. By comparison, the breakdown point for the maximum likelihood estimators is $1/n$.

The MHDEs for these data are $\hat{\mu} = 22.22$ and $\hat{\sigma} = 12.53$, with resultant fitted Gaussian density given in Fig. 7.6. The fitted parametric density is closer to the kernel estimate, and the outliers have less effect on them. The outliers do still have some effect, however, as the MHDEs with them omitted are $\hat{\mu} = 21.92$ and $\hat{\sigma} = 11.86$.

Minimum Hellinger distance estimation also can be applied to categorical data. The corresponding distance measure to (7.6) is

$$D^2(\hat{\mathbf{p}}, \tilde{\mathbf{p}}) = \sum_{i=1}^{K} [p_i(\hat{\boldsymbol{\theta}})^{1/2} - \tilde{p}_i^{1/2}]^2, \tag{7.7}$$

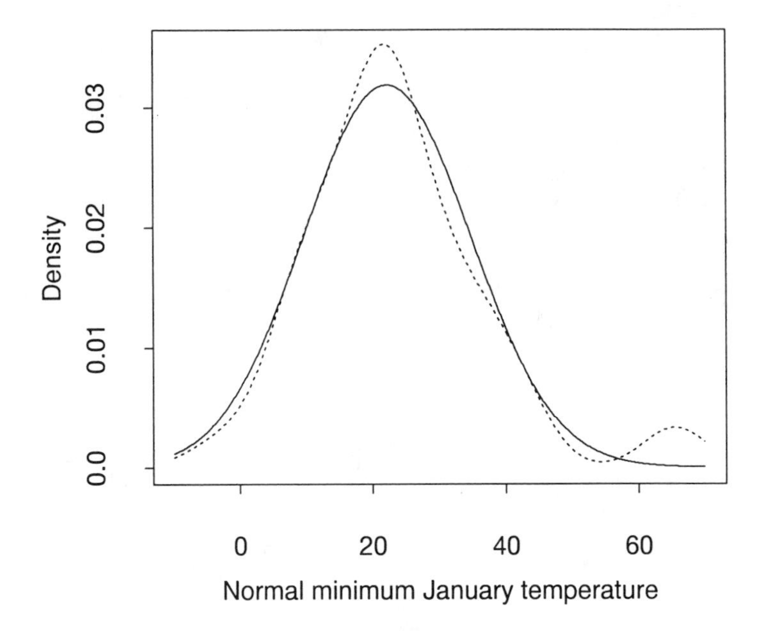


**Fig. 7.6.** Fitted Gaussian density using minimum Hellinger distance estimates (solid curve) and kernel estimate (dashed curve) for January temperature data.

where $\mathbf{p}(\hat{\boldsymbol{\theta}})$ is the estimated probability vector under the parametric model. Under standard asymptotics, the frequency estimators $\overline{\mathbf{p}}$ can be used for $\tilde{\mathbf{p}}$, and the resultant MHDEs are asymptotically efficient with breakdown .5 (the maximum value). For sparser tables with ordered categories, however, the frequency estimator can be replaced with one of the smoothed probability estimators described in Chapter 6, since they are more accurate. For nonsparse tables, the smoothed estimates are very close to the frequency estimates, so it does not matter which set is used.

Consider the following example, summarized in Figs. 7.7 and 7.8. The data are the number of accidents that 17 Asian airlines experienced over the period 1985–1994 (eight airlines had no accidents, five airlines had one accident, two airlines had three accidents, and one airline each had five and ten accidents). A Poisson model might be hypothesized for these count data,

$$P(X = x) = \frac{e^{-\lambda}\lambda^x}{x!}, \quad x = 0, 1, \ldots.$$

The maximum likelihood estimate of $\lambda$ is simply the sample mean, or $\hat{\lambda} = 1.53$. Figure 7.7 gives a plot of the frequency estimates (solid lines

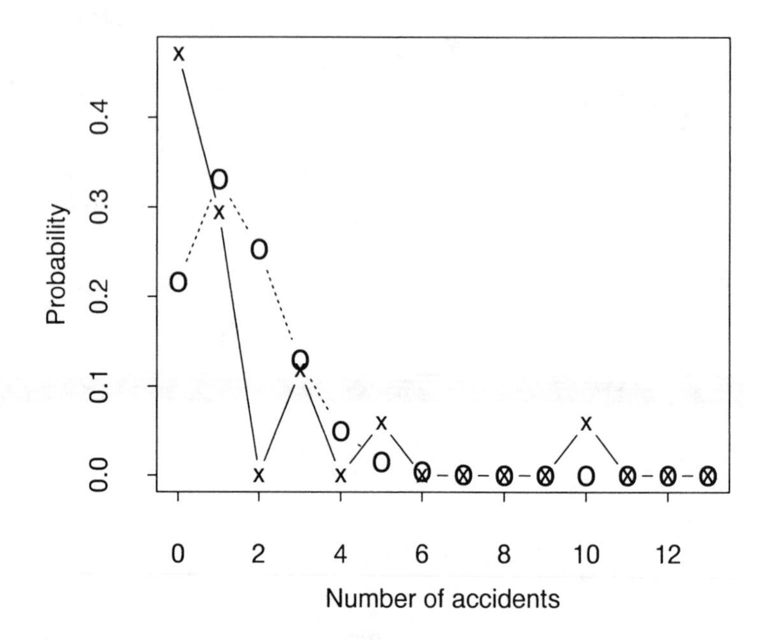


**Fig. 7.7.** Frequency estimates (solid line connecting $\times$'s) and fitted Poisson probabilities using maximum likelihood estimate (dashed line connecting $\circ$'s) for airline accident data.

connecting $\times$'s) and the resultant parametric estimates (dashed lines connecting $\circ$'s). The one airline with ten accidents (Merpati Airlines of Indonesia) has seriously inflated the estimate of $\lambda$, as the fitted probabilities are too low for zero accidents and too high for two and three accidents.

The MHDE using the frequency estimates as $\tilde{\mathbf{p}}$ is $\hat{\lambda} = .48$, and the top plot of Fig. 7.8 gives the resultant fitted probabilities. This is better than in Fig. 7.7, but now $\hat{\lambda}$ seems too small, as the fitted probabilities are too high for zero accidents and too low for three accidents. The problem is that the zero observed count at two accidents has caused $\hat{\lambda}$ to be deflated, to improve the fit there.

The MHDE based on a local linear estimate (using a Gaussian kernel and $h = .91$) is $\hat{\lambda} = .97$. The bottom plot of Fig. 7.8 gives the resultant fitted Poisson probabilities along with the local linear probability estimates. The parametric fit is similar to the nonparametric estimates and also provides a good fit to the observed data (the frequency estimates). The maximum likelihood estimate of $\lambda$ if Merpati Airlines is omitted from the data is $\hat{\lambda} = 1$, which highlights the robustness of the smoothing-based MHDE.

The Hellinger distances (7.6) and (7.7) also can be used to test para-

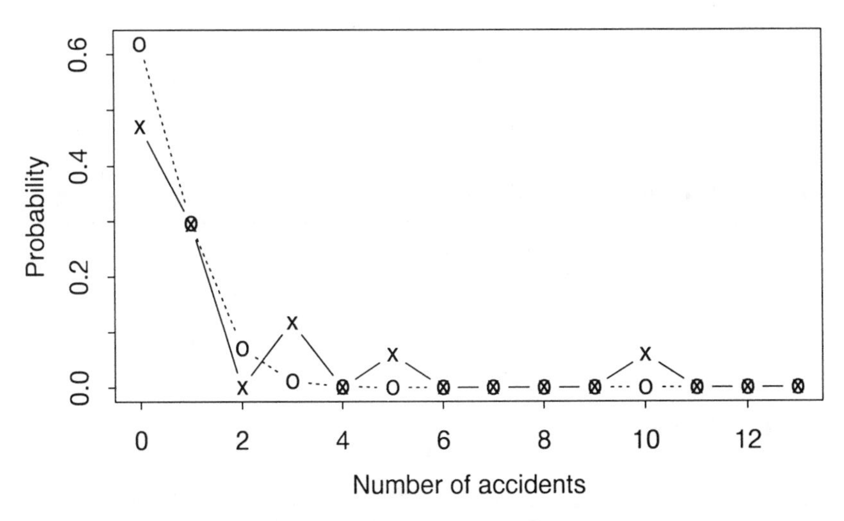


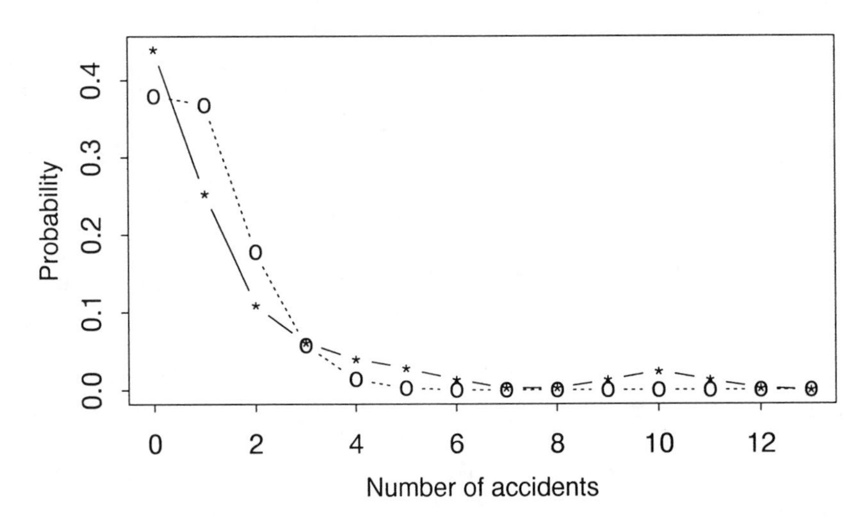


**Fig. 7.8.** Minimum Hellinger distance estimates for airline accident data. Top plot: frequency estimates (solid lines and $\times$'s) and fitted Poisson probabilities (dashed lines and $\circ$'s). Bottom plot: local linear estimates (solid lines and $\times$'s) and fitted Poisson probabilities (dashed lines and $\circ$'s).

metric hypotheses, as an alternative to likelihood ratio tests. Let $\Theta_0$ be a specified proper subset of $\Theta$, the allowable space of $\boldsymbol{\theta}$. A Hellinger-based test of the hypotheses

$H_0$: $\boldsymbol{\theta} \in \Theta_0$

versus

$H_a$: $\boldsymbol{\theta} \in \Theta \backslash \Theta_0$

can be constructed as

$$4n[D^2(\hat{f}_0, \tilde{f}) - D^2(\hat{f}, \tilde{f})],$$

where $\hat{f}_0$ is the parametric density estimate based on the MHDE over $\Theta_0$, while $\hat{f}$ is the parametric density estimate based on the MHDE over $\Theta$. This Hellinger deviance test has properties analogous to the MHDE, as it is asymptotically efficient within the parametric family while being more resistant to outliers than the likelihood ratio test.

## 7.4 The Smoothed Bootstrap

The bootstrap is a nonparametric approach to the estimation of errors in statistical estimation. The standard bootstrap technique is applied to an estimator $\hat{\boldsymbol{\theta}}(x_1, \ldots, x_n)$ by resampling $B$ times (with replacement) the observed sample $\{x_1, \ldots, x_n\}$ to form bootstrap samples $\{x_1^*, \ldots, x_n^*\}$. The distribution of the resultant values of $\hat{\boldsymbol{\theta}}^* = \hat{\boldsymbol{\theta}}(x_1^*, \ldots, x_n^*)$ is then used to assess the properties of $\hat{\boldsymbol{\theta}}$ under $F$, the distribution generating the data. For example, the standard error of $\hat{\boldsymbol{\theta}}$ can be estimated by the standard deviation of the $B$ values $\hat{\boldsymbol{\theta}}^*$. Thus, the bootstrap substitutes the empirical distribution function of the observed sample for the unknown underlying distribution function in the definition of the functional of interest.

The smoothed bootstrap changes this algorithm by using a smooth estimate of the distribution (or density) instead of the empirical estimate. Then, the bootstrap samples are generated from the smooth density estimate. Unlike the ordinary bootstrap samples, these smoothed bootstrap samples will not usually have repeat values, and they are less likely to be dominated by unusual values in the sample.

Theoretical attention to the smoothed bootstrap has focused on continuous densities. Generally speaking, smoothing has only a second order asymptotic effect on the bootstrap for statistics that depend on global properties of the underlying distribution (of course, improvements in small samples could still be substantial). All statistics that are expressible as differentiable functions of vector means, such as means, ratios of means, variances, and correlation coefficients, fall into this category.

Statistics that are functions of local properties of the underlying distribution are very different in that first order improvements are possible.

Examples include bootstrap estimation of statistical properties for the mode and for sample quantiles (such as the median).

The smoothed bootstrap also can be useful for ordered categorical data. The ordinary bootstrap cannot differentiate between a structural zero value in the $i$th cell (where the true underlying probability is zero) and a random zero in the $i$th cell (where the count is zero but the cell has positive probability), since in either case the bootstrap samples are generated from a probability vector with zero in the $i$th cell. In contrast, smoothed bootstrap samples are generated from a smoothed probability vector that is generally nonzero for cells with random zeroes.

Table 7.1 illustrates the benefits of smoothing in a simple context. The table compares the accuracy of bootstrap estimates of the standard error of the sample mean and sample median, respectively, drawn from a discrete uniform distribution on the integers from 1 to $K$ ($K$ equaling 20 or 50). The unsmoothed bootstrap samples from the frequency estimates, while the smoothed bootstrap samples from a local linear estimate of the probabilities. The table reports values of bias and standard error (s.e.) based on 500 Monte Carlo replications for each $n$ and $K$.

**Table 7.1.** Performance of unsmoothed and smoothed bootstrap estimates of standard error of sample mean and sample median, respectively, from a uniform distribution on the integers 1 to $K$, for varying sample sizes.

$K = 20$

| | **Sample mean** | | | | **Sample median** | | | |
|---|---|---|---|---|---|---|---|---|
| | *Unsmoothed* | | *Smoothed* | | *Unsmoothed* | | *Smoothed* | |
| $n$ | Bias | s.e. | Bias | s.e. | Bias | s.e. | Bias | s.e. |
| 5 | −.321 | .627 | −.266 | .545 | −.190 | 1.320 | −.342 | 1.153 |
| 10 | −.101 | .306 | −.094 | .266 | −.043 | .803 | −.116 | .696 |
| 15 | −.057 | .204 | −.052 | .188 | −.009 | .764 | −.107 | .623 |
| 20 | −.044 | .165 | −.042 | .150 | .058 | .588 | .009 | .503 |

$K = 50$

| | **Sample mean** | | | | **Sample median** | | | |
|---|---|---|---|---|---|---|---|---|
| | *Unsmoothed* | | *Smoothed* | | *Unsmoothed* | | *Smoothed* | |
| $n$ | Bias | s.e. | Bias | s.e. | Bias | s.e. | Bias | s.e. |
| 10 | −.313 | .781 | −.266 | .699 | −.308 | 2.075 | −.481 | 1.794 |
| 20 | −.113 | .409 | −.120 | .390 | .044 | 1.536 | −.060 | 1.219 |
| 30 | −.073 | .274 | −.085 | .272 | .169 | 1.167 | .007 | .913 |
| 40 | −.027 | .231 | −.052 | .225 | .208 | .975 | .110 | .758 |
| 50 | −.022 | .199 | −.010 | .188 | .157 | .853 | .077 | .637 |

For small samples, the smoothed bootstrap estimates of standard error can improve on those of the ordinary (unsmoothed) bootstrap in terms of both bias and variability, and one benefit that shows up consistently is lower standard error (that is, less variability). Interestingly, it is generally the sample size, rather than the sparseness of the discrete uniform table, that seems to determine the extent of improvement due to smoothing. The gains are larger for the estimate of standard error of the median than for that of the mean, which is consistent with the theoretical results mentioned earlier.

## Background material

### Section 7.1

Hand (1982) gave a book-length discussion of the application of kernel density estimation to discrimination, with Chapter 7 summarizing Monte Carlo comparisons with linear and quadratic discriminant analysis. Hall and Wand (1988c) described bandwidth selection using a criterion designed for classification. Posse (1992) and Polzehl (1995) discussed the application of projection pursuit to the discrimination problem. Hastie and Tibshirani (1996) proposed modeling each class using a mixture of multivariate normal densities with common covariance matrix, calling this "mixture discriminant analysis."

Breiman and Ihaka (1984) proposed additive modeling for discriminant analysis, choosing a nominal transformation of the response classes using ACE; see also Berres (1993). Hastie, Buja, and Tibshirani (1995) generalized linear discriminant analysis by proposing to choose discriminant variables to maximize the between-class covariance subject to a constraint on a penalized version of the within-class covariance, where the penalty forces smoothness of the solution. They called this "penalized discriminant analysis" and described applications to situations where there are a large number of correlated predictors or where the number of predictors is smaller but these predictors are better used via smoothing splines, tensor product splines or additive splines. The flexible discriminant analysis method of Hastie, Tibshirani, and Buja (1994) uses the more adaptive smooth regression estimators based on MARS and BRUTO.

### Section 7.2

Bickel and Rosenblatt (1973) suggested using functionals such as (7.2) based on kernel estimators to construct goodness-of-fit tests. They derived the asymptotic distributions of such tests, although they did note that the asymptotics might not be directly applicable for moderate sample sizes.

Lewis *et al.* (1977) proposed using Gamma distributions to determine critical values based on Monte Carlo simulations. Khashimov (1984), Falk (1985, 1986), Ghosh and Huang (1991), Bowman (1992), and Huang (1997) examined (variations of) tests of this type, while Rosenblatt (1975) and Bowman and Foster (1993) described generalizations to multivariate data. Tests based on other density estimators are also possible: Lii (1978) described a similar test based on the spline density estimator of Lii and Rosenblatt (1975), Mack (1982) proposed using a nearest neighbor density estimator, while Bowman and Foster (1993) proposed using variable kernel estimators. Ghorai (1980) and Eubank and LaRiccia (1992) based their tests on an orthogonal series estimator. The results of these papers suggest that smoothing-based tests can be more powerful than omnibus tests in detecting alternatives with sharp peaks (that is, high frequency alternatives).

Azzalini, Bowman, and Härdle (1989) proposed the pseudo-likelihood ratio regression test (7.3). Other related proposals include those of Yanagimoto and Yanagimoto (1987), Cox *et al.* (1988), Cox and Koh (1989), Munson and Jernigan (1989), Eubank and Spiegelman (1990), Kozek (1990, 1991), Raz (1990), Buckley (1991), Firth, Glosup, and Hinkley (1991), le Cessie and van Houwelingen (1991), Simonoff and Tsai (1991b), Staniswalis and Severini (1991), Müller (1992), Azzalini and Bowman (1993), Eubank and Hart (1993), Eubank, Hart, and LaRiccia (1993), Eubank and LaRiccia (1993), Härdle and Mammen (1993), Chen (1994a,b), and Djojosugito (1994, 1995). Eubank and Hart (1992) and Hart and Wehrly (1992) proposed tests based on the smoothing parameter, with rejection of the null occurring if the minimizer of estimated MISE leads to too much smoothing away from linearity.

Härdle and Mammen (1993) showed that the ordinary bootstrap cannot be used to determine the tail probabilities for these tests (see also Firth, Glosup, and Hinkley, 1991) and proposed using a different form of bootstrapping called the wild bootstrap. Many authors noted the slow convergence of these test statistics to their asymptotic distributions and used Monte Carlo to determine tail probabilities in practice.

Smoothing-based tests also can be constructed to compare two nonparametric regression curves. Examples of such tests include those of Hall and Hart (1990c), Härdle and Marron (1990), King, Hart, and Wehrly (1991), and Kulasekera (1995).

Bowman and Young (1996) described a graphical procedure designed to be used in conjunction with formal tests of hypotheses of equality of nonparametric regression curves, linearity of regression functions, and normality of an underlying density. The graphical method is not designed to replace formal tests of hypotheses, but rather to identify why or why not the relevant null hypotheses are rejected using the tests.

Read and Cressie (1988) gave a book-length treatment of different $\chi^2$ goodness-of-fit tests. Difficulties with the $\chi^2$ approximation for these tests for sparse tables were noted by Cochran (1954), Larntz (1978), Koehler and Larntz (1980), and Lewis, Saunders, and Westcott (1984), while Pierce and Schafer (1986) and Koehler and Gan (1990) noted the low power of $X^2$. Morris (1975), McCullagh (1985, 1986), and Koehler (1986) treated the asymptotic behavior of the tests for sparse tables. Simonoff and Tsai (1991a) developed a diagnostic to gauge how poor the $\chi^2$ approximation to the distribution of $G^2$ is for a given data set.

Simonoff (1985) proposed a statistic similar to $M$, based on penalized likelihood estimators. Monte Carlo simulations confirmed that the smoothing-based test has greater power than $X^2$ unless the alternative is very nonsmooth. A test of uniformity for the calcium carbonate data based on the penalized likelihood estimates also strongly rejects uniformity, with a tail probability less than .001.

Burman (1982, 1987b) discussed a smoothing-based test of independence in contingency tables using kernel estimators. He showed that the test is more sensitive than $X^2$ under sparse asymptotics, where the probability matrix under the alternative satisfies nonindependence that is consistent with a smooth matrix.

**Section 7.3**

The January temperature data are from Hoffman (1992, p. 186). Newsday (1995) gave the airline accident data.

Beran (1977) and Stather (1981) examined the properties of minimum Hellinger distance estimators. Beran proved consistency for compact $\Theta$, and established asymptotic normality and efficiency for compact densities and possibly random bandwidth, while Stather allowed densities with infinite support with deterministic bandwidth. Tamura and Boos (1986) studied estimation for multivariate, elliptically symmetric distributions. They provided corresponding asymptotic normality and efficiency results (including weakening the requirement of compact $\Theta$) and showed that the finite sample breakdown point (Donoho and Huber, 1983) of MHDEs of location and covariance is at least .25 for arbitrary dimension. Eslinger and Woodward (1991) studied the small sample properties of the estimator for univariate Gaussian data and found that the MHDEs do achieve robustness competitive with that of $M$-estimators while remaining efficient at the true model.

Rao (1963) established the asymptotic efficiency of the MHDE for multinomial data with a finite number of cells. Simpson (1987) discussed MHDEs for discrete data with a possibly infinite number of cells (such as a Poisson distribution). He established asymptotic normality and efficiency of the estimator and determined the asymptotic breakdown under certain conditions (for example, the asymptotic breakdown under a Poisson model is .5).

Lindsay (1994) studied the properties of minimum distance estimators for discrete data in some detail. He showed that the influence curve does not measure their robustness well and proposed the residual adjustment function as an alternative way to assess the estimator's tradeoff of efficiency and robustness. (See also Basu and Lindsay, 1994, for a discussion of Gaussian models and Basu, Markatou, and Lindsay, 1993, for a discussion of regression models.)

These results for discrete data are based on using the frequency estimates for $\tilde{\mathbf{p}}$ in (7.7). Harris and Basu (1994) showed that in this case the MHDE minimizes a penalized version of the Kullback–Leibler distance, where the penalty equals

$$h \sum_{n_i=0} p_i(\boldsymbol{\theta}),$$

with $h = 1$. For nonsparse data, there are few cells with $n_i = 0$, so the Hellinger distance is close to the Kullback–Leibler distance (and hence the MHDE is similar to the MLE). For sparse data, however, there can be many such cells, resulting in behavior different from that of the MLE at the true model. They proposed taking $h = .5$, which forces the so-called penalized MHDE to treat the empty cells the same way that the MLE does.

Park, Basu, and Basu (1995) generalized this penalized approach, suggesting defining estimators to minimize a blended weight Hellinger distance (Lindsay, 1994)

$$\frac{1}{2} \sum_i \frac{[\overline{p}_i - p_i(\boldsymbol{\theta})]^2}{[\alpha \overline{p}_i^{1/2} + (1-\alpha) p_i(\boldsymbol{\theta})^{1/2}]^2}, \tag{7.8}$$

for $\alpha \in [0,1]$ (taking $\alpha = .5$ gives twice the squared Hellinger distance). They also described a generalization of this distance that includes the distance of Harris and Basu (1994).

Simpson (1989) examined the asymptotic properties of Hellinger deviance tests, establishing asymptotic equivalence to likelihood ratio tests. Let $\epsilon_1$ be the smallest fraction of contamination that can make the test inconsistent. For tests of the Gaussian mean and Poisson mean, $\epsilon_1 = 0$ for the likelihood ratio test, while $\epsilon_1 > 0$ for the Hellinger deviance test. The exact value depends on the specific null and alternative hypotheses, but generally speaking, the stronger the evidence against the null, the higher the breakdown point. See also Beran (1977) and Eslinger and Woodward (1991).

Basu and Sarkar (1994) discussed the use of (7.8) for testing goodness-of-fit and recommended taking $\alpha = 1/9$ in that context. The penalized Hellinger distance of Harris and Basu (1994) also can be used to construct goodness-of-fit tests, and Basu, Harris, and Basu (1996) described corresponding deviance, score, and Wald tests.

A different way to estimate an unknown parameter $\boldsymbol{\theta}$ is by choosing the parameters to minimize an estimate of the integrated squared error between the parametric and true densities. Since

$$\begin{aligned}\text{ISE} &= \int [f(u,\boldsymbol{\theta}) - f(u)]^2\,du \\ &= \int f(u,\boldsymbol{\theta})^2\,du - 2\int f(u,\boldsymbol{\theta})f(u)du + \int f(u)^2\,du,\end{aligned}$$

estimating $2\int f(u,\boldsymbol{\theta})f(u)du$ by $2n^{-1}\sum_{i=1}^{n} f(x_i,\boldsymbol{\theta})$ gives as the estimate the minimizer $\hat{\boldsymbol{\theta}}$ of

$$\int f(u,\boldsymbol{\theta})^2\,du - \frac{2}{n}\sum_{i=1}^{n} f(x_i,\boldsymbol{\theta}).$$

The connection with smoothing of this approach is that it is equivalent for a Gaussian model to choosing $\hat{\mu}$ and $\hat{\sigma}$ to minimize the integrated squared distance between the Gaussian density and a histogram of the data with bin width $h \to 0$ (Brown and Hwang, 1993; Jones and Hjort, 1994). The resultant estimators of $\mu$ and $\sigma$ are inefficient, with asymptotic variances 54% and 85% higher, respectively, than those of the sample mean and sample standard deviation.

## Section 7.4

In his original formulation of the bootstrap, Efron (1979) also described the smoothed bootstrap as an alternative to the ordinary (unsmoothed) version. See also Efron (1982, Sect. 5.3), De Angelis and Young (1992), and Efron and Tibshirani (1993, Sect. 16.5). Application of the method requires the ability to simulate from density estimates, which is described in Efron (1979) and Silverman (1986, Sect. 6.4).

Silverman and Young (1987) described conditions under which the smoothed bootstrap is more accurate than the unsmoothed bootstrap in estimating standard errors. Hall, DiCiccio, and Romano (1989) showed that the smoothed bootstrap can provide only second order improvement over the unsmoothed bootstrap if the statistic being examined is a differentiable function of vector means. Despite this, marked improvement can still be seen in small samples, as in Efron (1982, Table 5.2).

Statistics whose properties depend on local properties of the underlying distribution are much more likely to benefit from smoothed bootstrapping. Such examples include sample quantiles (Hall, DiCiccio, and Romano, 1989; De Angelis, Hall, and Young, 1993a), the mode (Romano, 1988), and least absolute values regression (De Angelis, Hall, and Young, 1993b).

In situations where smoothing is helpful, higher order smoothing also can be theoretically advantageous. Lee and Young (1994) discussed how to construct smoothed bootstrap samples from higher order kernel estimates (with possibly negative values).

Simonoff, Hochberg, and Reiser (1986) studied the problem of estimating $P(X < Y)$ for $X$ and $Y$ variables that are discretized into categorical variables. They used the bootstrap to construct confidence intervals for different estimators of $\lambda = P(X < Y) - P(Y < X)$, including ones based on smooth (penalized likelihood) probability estimates for $X$ and $Y$. They also noted that the smoothed bootstrap in this context refers to generating multinomial samples from a smoothed probability vector based on the original data.

## Computational issues

S–PLUS and Fortran code for discriminant analysis using ACE is available as the collection `gdiscr` in the `S` directory of `statlib`. The file `mda.shar.Z`, available via anonymous `ftp` at the address `playfair.stanford.edu` in the directory `pub/hastie`, contains Ratfor and S–PLUS code to carry out mixture discriminant analysis, flexible discriminant analysis and penalized discriminant analysis.

## Exercises

**Exercise 7.1.** The default setting for most discriminant analysis programs is to take the prior probabilities to be equal for all classes. How do the results of the linear, quadratic, and kernel discriminant rules for the MBA salary data change if the prior probabilities are taken to be .5 for each gender class?

**Exercise 7.2.** Nonparametric discriminant analysis need not be based on a kernel density estimator but can be based on any estimation method. Do the results for the MBA salary data change if a variable kernel estimator is used in the algorithm? What about if a transformation-based estimator is used, based on a logarithmic transformation?

**Exercise 7.3.** The actual distribution of genders at the Stern School of Business in 1995 was 74% male and 26% female. Does taking these values as the prior probabilities change the results of the linear and kernel discriminant analyses for the GMAT and GPA data?

**Exercise 7.4.** The apparent difference in covariances between GMAT and GPA for men and women in Fig. 7.3 suggests that quadratic discriminant analysis might be useful for these data. Is that the case? How do the results of the quadratic discriminant analysis compare with those of the kernel discriminant analysis?

**Exercise 7.5.** Test the fit of a Poisson model to the airline accident data of Figs. 7.7 and 7.8 using the goodness-of-fit statistics $X^2$, $G^2$, and $M$. Do the tests give different impressions of whether the Poisson distribution fits these data? Which test(s) is (are) most trustworthy in this context?

**Exercise 7.6.** Test the fit of an exponential distribution to the mine explosion data given in Table 6.2 using $X^2$, $G^2$, and $M$. Does the exponential model fit the data? Which test do you trust most for these data?

**Exercise 7.7.** Fit separate Gaussian distributions to the men's and women's salaries from the MBA salary data of Figs. 7.1 and 7.2 using minimum Hellinger distance estimation. How do the MHDEs compare with the sample mean and standard deviation for these variables? Would you expect there to be very much difference between the estimators for either of these data sets?

**Exercise 7.8.** Construct estimates of the standard error of $\hat{\lambda}$ using the unsmoothed and smoothed bootstrap for the mine explosion data of Table 6.2, where $\hat{\lambda}$ is either the maximum likelihood or minimum Hellinger distance estimator of the time between explosions based on an exponential fit to the data. Which version of the bootstrap do you prefer for these data? Which estimator of $\lambda$ has a smaller estimated standard error for these data?

# Appendices

## A. Descriptions of the Data Sets

This appendix gives descriptions of the data sets used in the book. All of the data sets can be obtained electronically over the Internet from the `statlib` server. One way to obtain them is to send the message

**send smoothmeth from datasets**

to the Internet address `statlib@lib.stat.cmu.edu`.

The files also can be accessed from the World Wide Web, by connecting to the URL address

`http://www.stern.nyu.edu/SOR/SmoothMeth`

using a Web browser, such as `Mosaic` or `Netscape`. The latter method also provides access to updated information about this book. The data files are written in plain ASCII (character) text, so it should be possible to import them into virtually any statistical, database management or spreadsheet package. Note, however, that for many of the data sets the last variable is a character variable that labels the observations, which should not be input as a numerical variable. It is also likely that such files would have to be input using fixed, rather than free, format, if the case labels were to be included.

---

**File name**

`adptvisa.dat`

**Description of data**

Changes in visas issued in 37 countries or regions by the Immigration and Naturalization Service for the purpose of adoption by U.S. residents

**Variables in file**

(1) Log[(1991 visas + 1)/(1988 visas + 1)]
(2) Log[(1992 visas + 1)/(1991 visas + 1)]
(3) Country or region

**Data source**

Chatterjee, Handcock, and Simonoff (1995)

---

**File name**

`airaccid.dat`

**Description of data**

Airline accidents for 17 Asian airlines for 1985–1994

**Variables in file**

(1) Number of accidents
(2) Airline

**Data source**

Newsday (1995)

---

**File name**

`basesal.dat`

**Description of data**

Salaries of the 118 Major League baseball players who were eligible for salary arbitration in 1993

**Variables in file**

(1) Salary (thousands of dollars)
(2) Name

**Data source**

Newsday (1993a)

---

**File name**

`baskball.dat`

**Description of data**

Performance of the 96 National Basketball Association players who played the guard position during the 1992–1993 season and played at least 10 minutes per game

**Variables in file**

(1) Points scored per minute played
(2) Assists credited per minute played
(3) Height (cm)
(4) Minutes played per game
(5) Age (years)
(6) Name

**Data source**

Chatterjee, Handcock, and Simonoff (1995)

---

**File name**

`birthrt.dat`

**Description of data**

Monthly birth rates in the U.S. for January 1940 through December 1947 (96 months)

**Variables in file**

(1) Monthly time index starting at January 1940
(2) Number of births

**Data source**

Cook and Weisberg (1994)

---

**File name**

`caco2.dat`

**Description of data**

Discretization of 52 ranges of percentage concentrations of calcium carbonate in five samples, taken from a mixing plant of raw metal

**Variables in file**

(1) Cell number
(2) Count

**Data source**

Simonoff (1985)

---

**File name**

`calibrat.dat`

**Description of data**

Relationship of radioactivity counts to hormone level for 14 immunoassay calibration values

**Variables in file**

(1) Concentration of dosage of TSH (micro units/ml of incubator mixture)
(2) Radioactivity counts

**Data source**

Tiede and Pagano (1979)

---

**File name**

`cars93.dat`

**Description of data**

Properties of 93 1993 new-model automobiles

**Variables in file**

(1) Price of basic version of automobile (thousands of dollars)
(2) EPA city mileage (miles per gallon)
(3) EPA highway mileage (miles per gallon)
(4) Engine size (liters)
(5) Maximum horsepower
(6) Fuel tank capacity (gallons)
(7) Weight (pounds)
(8) Make and model

**Data source**

Lock (1993)

---

**File name**

`cdrate.dat`

**Description of data**

Three-month CD rates for 69 Long Island banks and thrifts in August 1989

**Variables in file**

(1) Return on CD
(2) Type of institution: commercial bank (0) or thrift (1)

**Data source**

Newsday (1989)

---

**File name**

`chondrit.dat`

**Description of data**

Percentage of silica in 22 chondrite meteors

**Variable in file**

(1) Percentage of silica

**Data sources**

Ahrens (1965), Good and Gaskins (1980)

---

**File name**

`diabetes.dat`

**Description of data**

Factors affecting patterns of insulin-dependent diabetes mellitus in 43 children

**Variables in file**

(1) Age (years)
(2) Logarithm of C-peptide concentration (pmol/ml) at diagnosis

**Data sources**

Sockett *et al.* (1987), Hastie and Tibshirani (1990)

---

**File name**

`elusage.dat`

**Description of data**

Electricity usage in an all-electric home for 55 months

**Variables in file**

(1) Average daily temperature (degrees Fahrenheit)
(2) Average daily electricity usage (kilowatt hours)
(3) Month

**Data source**

Chatterjee, Handcock, and Simonoff (1995)

---

**File name**

`ethanol.dat`

**Description of data**

Engine exhaust for 88 burnings of ethanol in an automobile test engine

**Variables in file**

(1) Concentration of nitric oxides in engine exhaust, normalized by engine work
(2) Equivalence ratio

**Data source**

Brinkman (1981)

---

**File name**

`galaxy.dat`

**Description of data**

Velocities relative to the Milky Way of 82 galaxies

**Variable in file**

(1) Velocity (km/sec)

**Data source**

Roeder (1990)

---

**File name**

`gascons.dat`

**Description of data**

Price indices and gasoline consumption in the U.S. over the 27 years 1960–1986

**Variables in file**

(1) Year
(2) Consumption of gasoline (tens of millions of 1967 dollars)
(3) Price index for gasoline (1967 dollars)
(4) Per capita disposable income (1967 dollars)
(5) Price index for used cars (1967 dollars)

**Data source**

Chatterjee, Handcock, and Simonoff (1995)

---

**File name**

`geyser.dat`

**Description of data**

Characteristics of 222 eruptions of the "Old Faithful Geyser" during August 1978 and August 1979

**Variables in file**

(1) Duration of eruption (minutes)
(2) Time until following eruption (minutes)

**Data source**

Weisberg (1985)

---

**File name**

`hckshoot.dat`

**Description of data**

Shooting percentage of the 292 National Hockey League players who played at least 60 games in the 1991–1992 season and scored at least one goal

**Variables in file**

(1) Shooting percentage
(2) Position played: defenseman (0) or forward (1)
(3) Name

**Data source**

National Hockey League (1992)

---

**File name**

`jantemp.dat`

**Description of data**

Normal minimum January temperature for weather stations in each of the 50 U.S. states

**Variables in file**

(1) Normal minimum January temperature (degrees Fahrenheit)
(2) Weather station

**Data source**

Hoffman (1992)

---

**File name**

`marathon.dat`

**Description of data**

National men's record times for the marathon for 55 countries

**Variables in file**

(1) Marathon record time (minutes)
(2) Country

**Data source**

Dawkins (1989)

---

**File name**

`mbagrade.dat`

**Description of data**

Gender, GMAT score, and first-year grade point average of 61 second-year MBA students

**Variables in file**

(1) Gender: male (0) or female (1)
(2) GMAT score
(3) Grade point average

---

**File name**

`mbasalry.dat`

**Description of data**

Previous salaries of 91 MBA students by gender

**Variables in file**

(1) Gender: male (0) or female (1)
(2) Previous salary (dollars, to nearest $500)

---

**File name**

`mbasurv.dat`

**Description of data**

Regression form of contingency table of survey on opinions of 55 MBA students on the importance of statistics and economics in business education (Table 6.4)

**Variables in file**

(1) Row (Statistics) cell
(2) Column (Economics) cell
(3) Count

**Data source**

Simonoff (1995c)

---

**File name**

`mineacci.dat`

**Description of data**

Twenty-eight time intervals between accidents causing fatalities in Division 5 of the Great Britain National Coal Board in 1950

**Variable in file**

(1) Time interval (days)

**Data source**

Maguire, Pearson, and Wynn (1951)

---

**File name**

`mineexpl.dat`

**Description of data**

Discretization of 109 time intervals between explosions in mines involving at least 10 men killed in Great Britain from December 8, 1875 to May 29, 1951

**Variables in file**

(1) Cell number
(2) Count

**Data source**

Simonoff (1983)

---

**File name**

`newscirc.dat`

**Description of data**

Circulation figures for 19 newspapers

**Variables in file**

(1) Sunday circulation (thousands)
(2) Daily circulation (thousands)
(3) Newspaper

**Data source**

Berenson and Levine (1992)

---

**File name**

`quake.dat`

**Description of data**

Characteristics of 2178 earthquakes with magnitude at least 5.8 on the Richter scale occurring between January 1964 and February 1986

**Variables in file**

(1) Focal depth (km)
(2) Latitude (degrees)
(3) Longitude (degrees)
(4) Body wave magnitude on the Richter scale

**Data sources**

Bulletin of the International Seismological Centre, as discussed in Frohlich and Davis (1990)

---

**File name**

`racial.dat`

**Description of data**

Racial distribution of students in the 56 public school districts in Nassau County in the 1992–1993 school year

**Variables in file**

(1) Proportion of white students
(2) District

**Data source**

Chatterjee, Handcock, and Simonoff (1995)

---

**File name**

`safewatr.dat`

**Description of data**

Population access to safe water in rural areas of 70 countries in 1985 and 1990

**Variables in file**

(1) Percentage of rural population with access to safe water in 1985
(2) Percentage of rural population with access to safe water in 1990
(3) Country

**Data source**

World Bank (1994)

---

**File name**

`salary.dat`

**Description of data**

Discretizations of monthly salaries of 147 nonsupervisory female employees holding the Bachelor's degree (but no higher) who were practicing mathematics or statistics in 1981

**Variables in file**

(1) Cell number
(2) Count for 6-cell discretization (cells 7–28 coded as `M`)
(3) Count for 12-cell discretization (cells 13–28 coded as `M`)
(4) Count for 28-cell discretization

**Data sources**

Department of Energy (1982), Simonoff (1987), Simonoff and Tsai (1991a)

---

**File name**

`salmon.dat`

**Description of data**

Annual recruits and spawners of the Skeena River sockeye salmon stock for the 28 years 1940–1967

**Variables in file**

(1) Year
(2) Spawners (thousands)
(3) Recruits (thousands)

**Data source**

Carroll and Ruppert (1988)

---

**File name**

`salyear.dat`

**Description of data**

Regression form of contingency table of 147 female employees of salaries and years since degree (Table 6.5)

**Variables in file**

(1) Row (Salary) cell number
(2) Column (Years since degree) cell number
(3) Count

**Data source**

Simonoff (1987)

---

**File name**

`schlvote.dat`

**Description of data**

Budget votes for 38 Long Island school districts in 1993

**Variables in file**

(1) Voting result: failure to pass (0) or pass (1)
(2) Proposed average equalized property tax rate (dollars per $100 assessed valuation)
(3) Total proposed budget (dollars)
(4) Percentage change in budget
(5) Percentage change in tax rate
(6) Average full property wealth per student (dollars) (missing value coded as `M`)
(7) District

**Data source**

Newsday (1993b)

---

**File name**

`sulfate.dat`

**Description of data**

Relationship of distance between 3321 pairs of measuring stations and correlation of adjusted sulfate wet deposition levels

**Variables in file**

(1) Distance between stations (km)
(2) Correlation of adjusted sulfate wet deposition levels

**Data source**

Gary W. Oehlert, similar to data in Oehlert (1993)

---

**File name**

`swissmon.dat`

**Description of data**

Characteristics of 100 real and 100 forged Swiss bank notes

**Variables in file**

(1) Width of bottom margin (mm), forged bills
(2) Image diagonal length (mm), forged bills
(3) Width of bottom margin (mm), real bills
(4) Image diagonal length (mm), real bills

**Data source**

Flury and Riedwyl (1988)

---

**File name**

`vineyard.dat`

**Description of data**

Grape yields for the 52 rows of a vineyard for different harvest years

**Variables in file**

(1) Row number
(2) Number of lugs for 1989 harvest
(3) Number of lugs for 1990 harvest
(4) Number of lugs for 1991 harvest

**Data source**

Chatterjee, Handcock, and Simonoff (1995)

---

**File name**

`votfraud.dat`

**Description of data**

Democratic over Republican pluralities of voting machine and absentee votes for 22 Philadelphia County elections

**Variables in file**

(1) Democratic plurality in machine votes
(2) Democratic plurality in absentee votes

**Data source**

Chatterjee, Handcock, and Simonoff (1995)

---

**File name**

`whale.dat`

**Description of data**

Nursing patterns for two beluga whale calves born in captivity at the New York Aquarium for 228 and 223 six-hour time periods postpartum, respectively

**Variables in file**

(1) Six-hour time period postpartum index
(2) Nursing time for Hudson (seconds)
(3) Nursing time for Casey (seconds) (extra time periods 224–228 coded as `M`)

**Data source**

Chatterjee, Handcock, and Simonoff (1995)

---

## B. More on Computational Issues

The **Computational issues** sections include sources for computer code, since it is only using smoothing methods that reveals their real power (and their strengths and weaknesses). This information comes from promotional literature, software reviews, news groups, Internet searches, and personal knowledge. If a paper includes an invitation for readers to obtain code from the author, that fact is also noted in the section. Naturally, one's own biases come in here; S–PLUS was used to construct all the figures in this book, so it should not be a surprise to find a fairly thorough discussion of S–PLUS-related software. I apologize for any omissions or errors in descriptions of the software, or of locations where code can be obtained electronically. I would be happy to have readers let me know of such errors or omissions; I will keep a list of updates and corrections on the WWW server described in Appendix A.

Discussion of any software does not imply any endorsement of any kind about that software, and I provide no warranty of any kind on the correctness or usefulness of any software mentioned, or of the accuracy of my descriptions of the software. Users of any software should consider the software as being used at their own risk.

The following list provides trademark information about software mentioned in the book. All other trademarks are the property of their respective owners.

# References


Abdous, B. (1993) Note on the minimum mean integrated squared error of kernel estimates of a distribution function and its derivatives. *Communications in Statistics — Theory and Methods*, **22**, 603–609.

Abdous, B. (1995) Computationally efficient classes of higher-order kernel functions. *Canadian Journal of Statistics*, **23**, 21–27.

Abou-Jaoude, S. (1976a) Sur une condition nécessaire et suffisante de $L_1$-convergence presque complète de l'estimateur de la partition fixe pour une densité. *Comptes Rendus de l'Académie des Sciences de Paris Série A*, **283**, 1107–1110.

Abou-Jaoude, S. (1976b) Conditions nécessaires et suffisantes de convergence $L_1$ en probabilité de l'histogramme pour une densité. *Annales de l'Institut Henri Poincaré*, **12**, 213–231.

Abramson, I.S. (1982) On bandwidth variation in kernel estimates — a square root law. *Annals of Statistics*, **10**, 1217–1223.

Aerts, M., Augustyns, I., and Janssen, P. (1997a) Smoothing sparse multinomial data using local polynomial fitting. *Journal of Nonparametric Statistics*, **8**, 127–147.

Aerts, M., Augustyns, I., and Janssen, P. (1997b) Local polynomial estimation of contingency table cell probabilities. *Statistics*, **30**, 127–148.

Ahrens, L.H. (1965) Observations on the Fe-Si-Mg relationship in chondrites. *Geochimica et Cosmochimica Acta*, **29**, 801–806.

Aitchison, J. and Aitken, C.G.G. (1976) Multivariate binary discrimination by the kernel method. *Biometrika*, **63**, 413–420.

Aitken, C.G.G. (1983) Kernel methods for the estimation of discrete distributions. *Journal of Statistical Computation and Simulation*, **16**, 189–200.

Akaike, H. (1954) An approximation to the density function. *Annals of the Institute of Statistical Mathematics*, **6**, 127–132.

Akaike, H. (1970) Statistical predictor information. *Annals of the Institute of Statistical Mathematics*, **22**, 203–217.

Akaike, H. (1973) Information theory and an extension of the maximum likelihood principle. In *Second International Symposium on Information Theory*, eds. B.N. Petrov and P. Cźaki, Akademiai Kiadó, Budapest, 267–281.

Akaike, H. (1974) A new look at the statistical model identification. *IEEE Transactions of Automatic Control AC*, **19**, 716–723.

Allen, D.M. (1974) The relationship between variable selection and data augmentation and a method for prediction. *Technometrics*, **16**, 125–127.

Altman, N.S. (1990) Kernel smoothing of data with correlated errors. *Journal of the American Statistical Association*, **85**, 749–759.

Altman, N.S. (1992) An introduction to kernel and nearest-neighbor nonparametric regression. *American Statistician*, **46**, 175–185.

Altman, N.S. (1993) Estimating error correlation in nonparametric regression. *Statistics and Probability Letters*, **18**, 213–218.

Ammann, L.P. (1993) Robust singular value decompositions: a new approach to projection pursuit. *Journal of the American Statistical Association*, **88**, 505–514.

Anderssen, R.S. and Bloomfield, P. (1974) A time series approach to numerical differentiation. *Technometrics*, **16**, 69–75.

Anderssen, R.S., Bloomfield, P., and McNeil, D.R. (1974) Spline functions in data analysis. Technical Report No. 69, Ser. 2, Department of Statistics, Princeton University.

Antoch, J. and Janssen, P. (1989) Nonparametric regression $M$-quantiles. *Statistics and Probability Letters*, **8**, 355–362.

Antoniadis, A. (1994) Smoothing noisy data with coiflets. *Statistica Sinica*, **4**, 651–678.

Antoniadis, A., Gregoire, G., and McKeague, I.W. (1994) Wavelet methods for curve estimation. *Journal of the American Statistical Association*, **89**, 1340–1353.

Aoki, M. (1990) *State Space Modeling of Time Series*, 2nd ed., Springer–Verlag, New York.

Atilgan, T. (1990) On derivation and application of AIC as a data-based criterion for histograms. *Communications in Statistics — Theory and Methods*, **19**, 885–903.

Azzalini, A. and Bowman, A.W. (1993) On the use of nonparametric regression for checking linear relationships. *Journal of the Royal Statistical Society, Ser. B*, **55**, 549–557.

Azzalini, A., Bowman, A.W., and Härdle, W. (1989) On the use of nonparametric regression for model checking. *Biometrika*, **76**, 1–11.

Banks, D.L., Maxion, R.A., and Olszewski, R. (1995) Comparing methods for multivariate nonparametric regression. Unpublished manuscript.

Barnett, V. and Lewis, T. (1994) *Outliers in Statistical Data*, 3rd ed., John Wiley, Chichester.

Barron, A.R. and Sheu, C.-H. (1991) Approximation of density functions by sequences of exponential families. *Annals of Statistics*, **19**, 1347–1369.

Barry, D. (1983) Nonparametric Bayesian regression. Unpublished Ph.D. thesis, Yale University.

Barry, D. (1993) Testing for additivity of a regression function. *Annals of Statistics*, **21**, 235–254.

Barry, D. (1995) A Bayesian analysis for a class of penalized likelihood estimates. *Communications in Statistics — Theory and Methods*, **24**, 1057–1071.

Bartlett, M.S. (1963) Statistical estimation of density functions. *Sankhyā, Ser. A*, **25**, 245–254.

Basu, A., Harris, I.R., and Basu, S. (1996) Tests of hypotheses in discrete models based on the penalized Hellinger distance. *Statistics and Probability Letters*, **27**, 367–373.

Basu, A. and Lindsay, B.G. (1994) Minimum disparity estimation for continuous model: efficiency, distributions and robustness. *Annals of the Institute of Statistical Mathematics*, **46**, 683–705.

Basu, A., Markatou, M., and Lindsay, B.G. (1993) Robustness via weighted likelihood estimating equations. Technical report, Columbia University Department of Statistics.

Basu, A. and Sarkar, S. (1994) On disparity based goodness-of-fit tests for multinomial models. *Statistics and Probability Letters*, **19**, 307–312.

Bates, D.M., Lindstrom, M.J., Wahba, G., and Yandell, B.S. (1987) GCVPACK — Routines for generalized cross validation. *Communications in Statistics — Simulation and Computation*, **16**, 263–297.

Becker, R.A., Chambers, J.M., and Wilks, A.R. (1988) *The New S Language: A Programming Environment for Data Analysis and Graphics*, Wadsworth and Brooks/Cole, Pacific Grove, CA.

Bellman, R.E. (1961) *Adaptive Control Processes*, Princeton University Press, Princeton, NJ.

Benedetti, J.K. (1977) On the nonparametric estimation of regression functions. *Journal of the Royal Statistical Society, Ser. B*, **39**, 248–253.

Bennett, J. (1974) Estimation of multivariate probability density functions using B-splines. Unpublished Ph.D. thesis, Rice University.

Beran, J. (1994) *Statistics for Long-Memory Processes*, Chapman and Hall, New York.

Beran, R. (1977) Minimum Hellinger distance estimates for parametric models. *Annals of Statistics*, **5**, 445–463.

Berenson, M.L. and Levine, D.M. (1992) *Basic Business Statistics: Concepts and Applications*, Prentice–Hall, Englewood Cliffs, N.J.

Berlinet, A. (1990) Reproducing kernels and finite order kernels. In *Nonparametric Functional Estimation and Related Topics*, ed. G.G. Roussas, Klüwer Academic Publishers, Dordrecht, 3–18.

Berres, M. (1993) A comparison of some linear and nonlinear discrimination methods. *Computational Statistics*, **8**, 223–239.

Bertrand-Retali, R. (1978) Convergence uniforme d'une estimateur de la densité par la method du noyau. *Revue Roumaine de Mathematiques Pures et Appliquées*, **23**, 361–385.

Bickel, P.J. and Rosenblatt, M. (1973) On some global measures of the deviations of density function estimates. *Annals of Statistics*, **1**, 1071–1095.

Bickel, P.J. and Zhang, P. (1992) Variable selection in nonparametric regression with categorical covariates. *Journal of the American Statistical Association*, **87**, 90–97.

Bishop, Y.M.M., Fienberg, S.E., and Holland, P.W. (1975) *Discrete Multivariate Analysis*, MIT Press, Cambridge, MA.

Boente, G. and Fraiman, R. (1989) Robust nonparametric regression estimation. *Journal of Multivariate Analysis*, **29**, 180–198.

Boneva, L.I., Kendall, D., and Stefanov, I. (1971) Spline transformations: three new diagnostic aids for the statistical data-analyst (with discussion). *Journal of the Royal Statistical Society, Ser. B*, **33**, 1–70.

Bosq, D. (1995) Optimal asymptotic quadratic error of density estimators for strong mixing or chaotic data. *Statistics and Probability Letters*, **22**, 339–347.

Bowman, A.W. (1980) A note on consistency of the kernel method for the analysis of categorical data. *Biometrika*, **67**, 682–684.

Bowman, A.W. (1984) An alternative method of cross-validation for the smoothing of density estimates. *Biometrika*, **71**, 353–360.

Bowman, A.W. (1985) A comparative study of some kernel-based nonparametric density estimators. *Journal of Statistical Computation and Simulation*, **21**, 313–327.

Bowman, A.W. (1992) Density-based tests for goodness of fit. *Journal of Statistical Computation and Simulation*, **40**, 1–13.

Bowman, A.W. and Foster, P.J. (1993) Adaptive smoothing and density-based tests of multivariate normality. *Journal of the American Statistical Association*, **88**, 529–537.

Bowman, A.W., Hall, P., and Titterington, D.M. (1984) Cross-validation in nonparametric estimation of probabilities and probability densities. *Biometrika*, **71**, 341–351.

Bowman, A.W. and Young, S.G. (1996) Graphical comparison of nonparametric curves. *Applied Statistics*, **45**, 83–98.

Box, G.E.P. (1980) Sampling and Bayes' inference in scientific modelling and robustness. *Journal of the Royal Statistical Society, Ser. A*, **143**, 383–430.

Boyd, D.W. and Steele, J.M. (1978) Lower bounds for nonparametric density estimation rates. *Annals of Statistics*, **6**, 932–934.

Breiman, L. and Friedman, J.H. (1985) Estimating optimal transformations for multiple regression and correlation (with discussion). *Journal of the American Statistical Association*, **80**, 580–619.

Breiman, L. and Ihaka, R. (1984) Nonlinear discriminant analysis via scaling and ACE. Technical Report No. 40, University of California at Berkeley Department of Statistics, Berkeley, CA.

Breiman, L., Meisel, W., and Purcell, E. (1977) Variable kernel estimates of multivariate densities. *Technometrics*, **19**, 135–144.

Brillinger, D.R. (1977) Discussion of "Consistent nonparametric regression." *Annals of Statistics*, **5**, 622–623.

Brillinger, D.R. and Krishnaiah, P.R. (eds.) (1984) *Handbook of Statistics (Volume 3): Time Series in the Frequency Domain*, Elsevier Science, Amsterdam.

Brinkman, N.D. (1981) Ethanol fuel — a single cylinder engine study of efficiency and exhaust emissions. *SAE Transactions*, **90**, No. 810345, 1410–1427.

Brockmann, M., Gasser, T., and Herrmann, E. (1993) Locally adaptive bandwidth choice for kernel regression estimators. *Journal of the American Statistical Association*, **88**, 1302–1309.

Brockwell, P. and Davis, R.A. (1991) *Time Series: Theory and Methods*, 2nd ed., Springer-Verlag, New York.

Broniatowski, M., Deheuvels, P., and Devroye, L. (1989) On the relationship between stability of extreme order statistics and convergence of the maximum likelihood kernel density estimate. *Annals of Statistics*, **17**, 1070–1086.

Brown, L.D. and Hwang, J.T.G. (1993) How to approximate a histogram by a normal density. *American Statistician*, **47**, 251–255.

Brown, P.J. and Rundell, P.W.K. (1985) Kernel estimates for categorical data. *Technometrics*, **27**, 293–299.

Bruce, A.G. and Gao, H.-Y. (1995) S + WAVELETS Toolkit Statistics, a division of MathSoft Inc., Seattle, WA.

Buckland, S.T. (1992a) Fitting density functions with polynomials. *Applied Statistics*, **41**, 63–76.

Buckland, S.T. (1992b) Algorithm AS 270: maximum likelihood fitting of Hermite and simple polynomial densities. *Applied Statistics*, **41**, 241–266.

Buckley, M.J. (1991) Detecting a smooth signal: optimality of cusum based procedures. *Biometrika*, **78**, 253–262.

Buckley, M.J., Eagleson, G.K., and Silverman, B.W. (1988) The estimation of residual variance in nonparametric regression. *Biometrika*, **75**, 189–199.

Buja, A., Hastie, T., and Tibshirani, R. (1989) Linear smoothers and additive models (with discussion). *Annals of Statistics*, **17**, 453–555.

Buja, A. and Stuetzle, W. (1985) Discussion of "Projection pursuit." *Annals of Statistics*, **13**, 484–490.

Burman, P. (1982). Smoothing in discrete multivariate analysis. Unpublished Ph.D. thesis, University of California at Berkeley, Berkeley, CA.

Burman, P. (1985) A data dependent approach to density estimation. *Zeitschrift für Wahrscheinlichkeitstheorie und verwandte Gebiete*, **69**, 609–628.

Burman, P. (1987a) Smoothing sparse contingency tables. *Sankhyā, Ser. A*, **49**, 24–36.

Burman, P. (1987b). Central limit theorem for quadratic forms for sparse tables. *Journal of Multivariate Analysis*, **22**, 258–277.

Cacoullos, T. (1966) Estimation of a multivariate density. *Annals of the Institute of Statistical Mathematics*, **18**, 179–189.

Cao, R., Cuevas, A., and González-Manteiga, W. (1994) A comparative study of several smoothing methods in density estimation. *Computational Statistics and Data Analysis*, **17**, 153–176.

Cao, R., Quintela del Río, A., and Vilar Fernández, J.M. (1993) Bandwidth selection in nonparametric density estimation under dependence: a simulation study. *Computational Statistics*, **8**, 313–332.

Carmody, T.J. (1988) Diagnostics for multivariate smoothing splines. *Journal of Statistical Planning and Inference*, **19**, 171–186.

Carr, D.B., Littlefield, R.J., Nicholson, W.L., and Littlefield, J.S. (1987) Scatterplot matrix techniques for large *n*. *Journal of the American Statistical Association*, **82**, 424–436.

Carroll, R.J., Fan, J., Gijbels, I., and Wand, M.P. (1997) Generalized partially linear single-index models. *Journal of the American Statistical Association*, **92**, 477–489.

Carroll, R.J. and Ruppert, D. (1988) *Transformations and Weighting in Regression*, Chapman and Hall, New York.

Carter, C.K. and Eagleson, G.K. (1992) A comparison of variance estimators in nonparametric regression. *Journal of the Royal Statistical Society, Ser. B*, **54**, 773–780.

Chatterjee, S. and Hadi, A.S. (1988) *Sensitivity Analysis in Linear Regression*, John Wiley, New York.

Chatterjee, S., Handcock, M.S., and Simonoff, J.S. (1995) *A Casebook for a First Course in Statistics and Data Analysis*, John Wiley, New York.

Chen, H. (1988) Convergence rates for parametric components in a partly linear model. *Annals of Statistics*, **16**, 136–146.

Chen, H. (1991) Estimation of a projection-pursuit type regression model. *Annals of Statistics*, **19**, 142–157.

Chen, H. and Shiau, J.-J.H. (1991) A two-stage spline smoothing method for partially linear models. *Journal of Statistical Planning and Inference*, **27**, 187–201.

Chen, J.-C. (1994a) Testing for no effect in nonparametric regression via spline smoothing techniques. *Annals of the Institute of Statistical Mathematics*, **46**, 251–265.

Chen, J.-C. (1994b) Testing goodness of fit of polynomial models via spline smoothing techniques. *Statistics and Probability Letters*, **19**, 65–76.

Chen, Z. (1987) A stepwise approach for purely periodic interaction spline models. *Communications in Statistics — Theory and Methods*, **16**, 877–895.

Chen, Z. (1993) Fitting multivariate regression functions by interaction spline models. *Annals of Statistics*, **55**, 473–491.

Cheng, K.F. and Lin, P.E. (1981) Nonparametric estimation of a regression function. *Zeitschrift für Wahrscheinlichkeitstheorie und verwandte Gebiete*, **57**, 223–233.

Cheng, M.-Y. (1997) A bandwidth selector for local linear density estimators. *Annals of Statistics*, **25**, 1001–1013.

Cheng, M.-Y., Fan, J., and Marron, J.S. (1997) On automatic boundary corrections. *Annals of Statistics*, **25**, 1691–1708.

Chiu, S.-T. (1989) Bandwidth selection for kernel estimate with correlated noise. *Statistics and Probability Letters*, **8**, 347–354.

Chiu, S.-T. (1990) Why do bandwidth selectors tend to choose smaller bandwidths and a remedy. *Biometrika*, **77**, 222–226.

Chiu, S.-T. (1991a) Bandwidth selection for kernel density estimation. *Annals of Statistics*, **19**, 1883–1905.

Chiu, S.-T. (1991b) The effect of discretization error on bandwidth selection for kernel density estimation. *Biometrika*, **78**, 436–441.

Chiu, S.-T. (1991c) Some stabilized bandwidth selectors for nonparametric regression. *Annals of Statistics*, **19**, 1528–1546.

Chiu, S.-T. (1992) An automatic bandwidth selector for kernel density estimate. *Biometrika*, **79**, 771–782.

Chow, Y.-S., Geman, S., and Wu, L.-D. (1983) Consistent cross-validated density estimation. *Annals of Statistics*, **11**, 25–38.

Chu, C.-K. and Marron, J.S. (1991a) Choosing a kernel regression estimator (with discussion). *Statistical Science*, **6**, 404–436.

Chu, C.-K. and Marron, J.S. (1991b) Comparison of two bandwidth selectors with dependent errors. *Annals of Statistics*, **19**, 1906–1918.

Clark, R.M. (1975) A calibration curve for radiocarbon dates. *Antiquity*, **49**, 251–266.

Cleveland, W.S. (1979) Robust locally weighted regression and smoothing scatterplots. *Journal of the American Statistical Association*, **74**, 829–836.

Cleveland, W.S. (1993) *Visualizing Data.* Hobart Press, Summit, NJ.

Cleveland, W.S. and Devlin, S.J. (1988) Locally weighted regression: an approach to regression analysis by local fitting. *Journal of the American Statistical Association*, **83**, 596–610.

Cleveland, W.S. and Grosse, E.H. (1991) Computational methods for local regression. *Statistics and Computing*, **1**, 47–62.

Cleveland, W.S., Grosse, E.H., and Shyu, W.M. (1992) Local regression models. In *Statistical Models in S*, eds. J.M. Chambers and T. Hastie, Wadsworth and Brooks/Cole, Pacific Grove, CA, 309–376.

Cleveland, W.S. and Loader, C. (1996) Smoothing by local regression: principles and methods (with discussion). In *Statistical Theory and Computational Aspects of Smoothing*, eds. W. Hardle and M.G. Schimek, Physica-Verlag, Heidelberg, 10–49; 80–102; 113–120.

Cleveland, W.S. and McGill, R. (1984) The many faces of a scatterplot. *Journal of the American Statistical Association*, **79**, 807–822.

Cline, D.B.H. (1988) Admissible kernel estimators of a multivariate density. *Annals of Statistics*, **16**, 1421–1427.

Cochran, W.G. (1954) Some methods for strengthening the common $\chi^2$ tests. *Biometrics*, **10**, 417–451.

Cook, D., Buja, A., and Cabrera, J. (1993) Projection pursuit indexes based on orthonormal function expansions. *Journal of Computational and Graphical Statistics*, **2**, 225–250.

Cook, D., Buja, A., Cabrera, J., and Hurley, C. (1995) Grand tour and projection pursuit. *Journal of Computational and Graphical Statistics*, **4**, 155–172.

Cook, R.D. (1986) Assessment of local influence (with discussion). *Journal of the Royal Statistical Society, Ser. B*, **48**, 133–169.

Cook, R.D. and Weisberg, S. (1982) *Residuals and Influence in Regression*, Chapman and Hall, New York.

Cook, R.D. and Weisberg, S. (1994) *An Introduction to Regression Graphics*, John Wiley, New York.

Copas, J.B. (1995) Local likelihood based on kernel censoring. *Journal of the Royal Statistical Society, Ser. B*, **57**, 221–235.

Cox, D.D. (1983) Asymptotics for $M$-type smoothing splines. *Annals of Statistics*, **11**, 530–551.

Cox, D.D. (1988) Approximation of least squares regression on nested subspaces. *Annals of Statistics*, **16**, 713–732.

Cox, D.D. and Koh, E. (1989) A smoothing spline based test of model adequacy in polynomial regression. *Annals of the Institute of Statistical Mathematics*, **41**, 383–400.

Cox, D.D., Koh, E., Wahba, G., and Yandell, B.S. (1988) Testing the (parametric) null model hypothesis in (semiparametric) partial and generalized spline models. *Annals of Statistics*, **16**, 113–119.

Cox, D.D. and O'Sullivan, F. (1990) Asymptotic analysis of penalized likelihood and related estimators. *Annals of Statistics*, **18**, 1676–1695.

Craven, P. and Wahba, G. (1979) Smoothing noisy data with spline functions. *Numerische Mathematik*, **31**, 377–403.

Csörgő, S. and Mielniczuk, J. (1995a) Density estimation under long-range dependence. *Annals of Statistics*, **23**, 990–999.

Csörgő, S. and Mielniczuk, J. (1995b) Nonparametric regression under long-range dependent normal errors. *Annals of Statistics*, **23**, 1000–1014.

Cuevas, A. and González-Manteiga, W. (1991) Data driven smoothing based on convexity properties. In *Nonparametric Functional Estimation and Related Topics*, ed. G.G. Roussas, Klüwer Academic Publishers, Dordrecht, 225–240.

Daly, J.E. (1988) The construction of optimal histograms. *Communications in Statistics — Theory and Methods*, **17**, 2921–2931.

Davies, O.L. and Goldsmith, P.L. (1980) *Statistical Methods in Research and Production*, Longman Group Ltd., New York.

Davis, K.B. (1975) Mean square error properties of density estimates. *Annals of Statistics*, **3**, 1025–1030.

Davis, K.B. (1977) Mean integrated square error properties of density estimates. *Annals of Statistics*, **5**, 530–535.

Davis, K.B. (1981) Estimation of the scaling parameter for a kernel-type density estimate. *Journal of the American Statistical Association*, **76**, 632–636.

Dawkins, B. (1989) Multivariate analysis of national track records. *American Statistician*, **43**, 110–115.

De Angelis, D., Hall, P., and Young, G.A. (1993a) A note on coverage error of bootstrap confidence intervals for quantiles. *Mathematics Proceedings of the Cambridge Philosophical Society*, **114**, 517–531.

De Angelis, D., Hall, P., and Young, G.A. (1993b), Analytical and bootstrap approximations to estimator distributions in $L^1$ regression. *Journal of the American Statistical Association*, **88**, 1310–1322.

De Angelis, D. and Young, G.A. (1992) Smoothing the bootstrap. *International Statistical Review*, **60**, 45–56.

Deheuvels, P. (1977a) Estimation nonparamétrique de la densité par histogrammes généralisés. *Revue de Statistique Appliquée*, **25/3**, 5–42.

Deheuvels, P. (1977b) Estimation nonparamétrique de la densité par histogrammes généralisés (II). *Publications de l'Institut Statistique de l'Université de Paris*, **22**, 1–23.

de Montricher, G.M., Tapia, R.A., and Thompson, J.R. (1975) Nonparametric maximum likelihood estimation of probability densities by penalty function methods. *Annals of Statistics*, **3**, 1329–1348.

Department of Energy (1982) *1981 National Survey of Compensation Paid Scientists and Engineers Engaged in Research and Development Activities* (DOE/TIC-11501 (US-70)), U.S. Government Printing Office, Washington, DC.

Devroye, L. (1985) A note on the $L_1$ consistency of variable kernel estimates. *Annals of Statistics*, **13**, 1041–1049.

Devroye, L. (1987) *A Course in Density Estimation*, Birkhäuser, Boston.

Devroye, L. (1989) The double kernel method in density estimation. *Annales de l'Institut Henri Poincaré*, **25**, 533–580.

Devroye, L. (1992) A note on the usefulness of superkernels in density estimation. *Annals of Statistics*, **20**, 2037–2056.

Devroye, L. and Györfi, L. (1985) *Nonparametric Density Estimation: The $L_1$ View*, John Wiley, New York.

Devroye, L. and Penrod, C.S. (1986) The strong uniform convergence of multivariate variable kernel estimates. *Canadian Journal of Statistics*, **14**, 211–219.

Devroye, L. and Wagner, T.J. (1977) The strong uniform consistency of nearest neighbor density estimates. *Annals of Statistics*, **5**, 536–540.

Diaconis, P. (1983) Projection pursuit for discrete data. Technical Report 193, Department of Statistics, Stanford University.

Diaconis, P. and Freedman, D. (1984) Asymptotics of graphical projection pursuit. *Annals of Statistics*, **12**, 793–815.

Diaconis, P. and Shahshahani, M. (1984) On nonlinear functions of linear combinations. *SIAM Journal on Scientific and Statistical Computing*, **5**, 175–190.

Diggle, P. (1985a) A kernel method for smoothing point process data. *Applied Statistics*, **34**, 138–147.

Diggle, P. (1985b) Discussion of "Some aspects of the spline smoothing approach to non-parametric regression curve fitting." *Journal of the Royal Statistical Society, Ser. B*, **47**, 28–29.

Diggle, P. and Hutchinson, M.F. (1989) On spline smoothing with autocorrelated errors. *Australian Journal of Statistics*, **31**, 166–182.

Diggle, P. and Marron, J.S. (1988) Equivalence of smoothing parameter selectors in density and intensity estimation. *Journal of the American Statistical Association*, **83**, 793–800.

Dixon, W.J., ed. (1988) *BMDP Statistical Software Manual*, University of California Press, Berkeley, CA.

Djojosugito, R.A. (1994) On the use of cubic spline smoothing for testing parametric linear regression models. *Computational Statistics*, **9**, 213–231.

Djojosugito, R.A. (1995) A cubic smoothing spline based lack of fit test for nonlinear regression models. *Communications in Statistics — Theory and Methods*, **24**, 2183–2197.

Doane, D.P. (1976) Aesthetic frequency classifications. *American Statistician*, **30**, 181–183.

Dodge, Y. (1986) Some difficulties involving nonparametric estimation of a density function. *Journal of Official Statistics*, **2**, 193–202.

Dong, J. and Simonoff, J.S. (1994) The construction and properties of boundary kernels for sparse multinomials. *Journal of Computational and Graphical Statistics*, **3**, 57–66.
Dong, J. and Simonoff, J.S. (1995) A geometric combination estimator for $d$-dimensional ordinal sparse contingency tables. *Annals of Statistics*, **23**, 1143–1159.
Dong, J. and Ye, Q. (1996) A minimum variance kernel estimator and a discrete frequency polygon estimator for ordinal sparse contingency tables. *Communications in Statistics — Theory and Methods*, **25**, 3217–3245.
Donoho, D.L. (1982) Breakdown properties of multivariate location estimators. Unpublished Ph.D. qualifying paper, Harvard University.
Donoho, D.L. and Gasko, M. (1992) Breakdown properties of location estimates based on halfspace depth and projected outlyingness. *Annals of Statistics*, **20**, 1803–1827.
Donoho, D.L. and Huber, P.J. (1983) The notion of breakdown point. In *A Festschrift for Erich L. Lehmann*, eds. P. Bickel, K. Doksum, and J.L. Hodges, Jr., Wadsworth, Belmont, CA, 157—184.
Donoho, D.L. and Johnstone, I.M. (1986) Regression approximation using projections and isotropic kernels. *Contemporary Mathematics*, **59**, 153–167.
Donoho, D.L. and Johnstone, I.M. (1989) Projection-based approximation and a duality with kernel methods. *Annals of Statistics*, **17**, 58–106.
Donoho, D.L. and Johnstone, I.M. (1994) Ideal spatial adaptation via wavelet shrinkage. *Biometrika*, **81**, 425–455.
Donoho, D.L. and Johnstone, I.M. (1995) Adapting to unknown smoothness via wavelet shrinkage. *Journal of the American Statistical Association*, **90**, 1200–1224.
Donoho, D.L., Johnstone, I.M., Kerkyacharian, G., and Picard, D. (1995) Wavelet shrinkage: asymptopia? (with discussion). *Journal of the Royal Statistical Society, Ser. B*, **57**, 301–369.
Donoho, D.L., Johnstone, I.M., Kerkyacharian, G., and Picard, D. (1996) Density estimation by wavelet thresholding. *Annals of Statistics*, **24**, 508–539.
Duda, R.O. and Hart, P.E. (1973) *Pattern Classification and Scene Analysis*, John Wiley, New York.
Duin, R.P.W. (1976) On the choice of smoothing parameters for Parzen estimators of density functions. *IEEE Transactions on Computing*, **C25**, 1175–1179.
Efron, B. (1979) Bootstrap methods: another look at the jackknife. *Annals of Statistics*, **7**, 1–26.
Efron, B. (1982) *The Jackknife, the Bootstrap and Other Resampling Plans*, SIAM, Philadelphia.
Efron, B. and Tibshirani, R. (1993) *An Introduction to the Bootstrap*, Chapman and Hall, New York.
Efron, B. and Tibshirani, R. (1996) Using specially designed exponential families for density estimation. *Annals of Statistics*, **24**, 2431–2461.
Eilers, P.H.C. and Marx, B.D. (1992) Generalized linear models with P-splines. In *Advances in GLIM and Statistical Modelling. Proceedings of the GLIM92 Conference and the 7th International Workshop on Statistical Modelling, Munich*, eds. L. Fahrmeir, B. Francis, R. Gilchrist, and G. Tutz, Lecture Notes in Statistics 78, Springer-Verlag, New York, 72–77.
Eilers, P.H.C. and Marx, B.D. (1996) Flexible smoothing with B-splines and penalties (with discussion). *Statistical Science*, **11**, 89–121.
Elphinstone, C.D. (1983) A target distribution model for nonparametric density estimation. *Communications in Statistics — Theory and Methods*, **12**, 161–198.
Engel, J., Herrmann, E., and Gasser, T. (1994) An iterative bandwidth selector for kernel estimation of densities and their derivatives. *Journal of Nonparametric Statistics*, **4**, 21–34.
Epanechnikov, V.A. (1969) Non-parametric estimation of a multivariate probability density. *Theory of Probability and Its Applications*, **14**, 153–158.
Escobar, M.D. and West, M. (1995) Bayesian density estimation and inference using mixtures. *Journal of the American Statistical Association*, **90**, 577–588.

Eslinger, P.W. and Woodward, W.A. (1991) Minimum Hellinger distance estimation for normal models. *Journal of Computation and Simulation*, **39**, 95–114.

Eubank, R.L. (1984) The hat matrix for smoothing splines. *Statistics and Probability Letters*, **2**, 9–14.

Eubank, R.L. (1985) Diagnostics for smoothing splines. *Journal of the Royal Statistical Society, Ser. B*, **47**, 332–341.

Eubank, R.L. (1988) *Spline Smoothing and Nonparametric Regression*, Marcel Dekker, New York.

Eubank, R.L. and Hart, J.D. (1992) Testing goodness-of-fit in regression via order selection criteria. *Annals of Statistics*, **20**, 1412–1425.

Eubank, R.L. and Hart, J.D. (1993) Commonality of cusum, von Neumann and smoothing-based goodness-of-fit tests. *Biometrika*, **80**, 89–98.

Eubank, R.L., Hart, J.D., and LaRiccia, V.N. (1993) Testing goodness of fit via nonparametric function estimation techniques. *Communications in Statistics — Theory and Methods*, **22**, 3327–3354.

Eubank, R.L. and LaRiccia, V.N. (1992) Asymptotic comparison of Cramér–von Mises and nonparametric function estimation techniques for testing goodness-of-fit. *Annals of Statistics*, **20**, 2071–2086.

Eubank, R.L. and LaRiccia, V.N. (1993) Testing for no effect in nonparametric regression. *Journal of Statistical Planning and Inference*, **36**, 1–14.

Eubank, R.L. and Speckman, P.L. (1990) Curve fitting by polynomial-trigonometric regression. *Biometrika*, **77**, 1–9.

Eubank, R.L. and Speckman, P.L. (1993) Confidence bands in nonparametric regression. *Journal of the American Statistical Association*, **88**, 1287–1301.

Eubank, R.L. and Spiegelman, C.H. (1990) Testing the goodness-of-fit of a linear model via nonparametric regression techniques. *Journal of the American Statistical Association*, **85**, 387–392.

Eubank, R.L. and Thomas, W. (1993) Detecting heteroscedasticity in nonparametric regression. *Journal of the Royal Statistical Society, Ser. B*, **55**, 145–155.

Fahrmeir, L. and Tutz, G. (1994) *Multivariate Statistical Modelling Based on Generalized Linear Models*, Springer-Verlag, New York.

Falk, M. (1985) Rates of convergence for a global measure of performance of kernel density estimators. *South African Statistical Journal*, **19**, 1–19.

Falk, M. (1986) A quick global measure of performance of kernel density estimators. *South African Statistical Journal*, **20**, 165–172.

Fan, J. (1992) Design-adaptive nonparametric regression. *Journal of the American Statistical Association*, **87**, 998–1004.

Fan, J. (1993) Local linear regression smoothers and their minimax efficiencies. *Annals of Statistics*, **21**, 196–216.

Fan, J., Gasser, T., Gijbels, I., Brockmann, M., and Engel, J. (1997) Local polynomial fitting: optimal kernels and minimax efficiency. *Annals of the Institute of Statistical Mathematics*, **49**, 79–99.

Fan, J. and Gijbels, I. (1992) Variable bandwidth and local linear regression smoothers. *Annals of Statistics*, **20**, 2008–2036.

Fan, J. and Gijbels, I. (1995a) Data-driven bandwidth selection in local polynomial fitting: variable bandwidth and spatial adaptation. *Journal of the Royal Statistical Society, Ser. B*, **57**, 371–394.

Fan, J. and Gijbels, I. (1995b) Adaptive order polynomial fitting: bandwidth robustification and bias reduction. *Journal of Computational and Graphical Statistics*, **4**, 213–227.

Fan, J. and Gijbels, I. (1996) *Local Polynomial Modelling and Its Applications*, Chapman and Hall, London.

Fan, J., Gijbels, I., Hu, T.-C., and Huang, L.-S. (1996) A study of variable bandwidth selection for local polynomial regression. *Statistica Sinica*, **6**, 113–127.

Fan, J. and Hall, P. (1994) On curve estimation by minimizing mean absolute deviation and its implications. *Annals of Statistics*, **22**, 867–885.

Fan, J., Hall, P., Martin, M., and Patil, P. (1996) On local smoothing of nonparametric curve estimators. *Journal of the American Statistical Association*, **91**, 258–266.

Fan, J., Heckman, N.E., and Wand, M.P. (1995) Local polynomial kernel regression for generalized linear models and quasi-likelihood functions. *Journal of the American Statistical Association*, **90**, 141–150.

Fan, J. and Hu, T.-C. (1992) Bias correction and higher order kernel functions. *Statistics and Probability Letters*, **13**, 235–243.

Fan, J., Hu, T.-C., and Truong, Y.K. (1994) Robust non-parametric function estimation. *Scandinavian Journal of Statistics*, **21**, 433–446.

Fan, J. and Marron, J.S. (1992) Best possible constant for bandwidth selection. *Annals of Statistics*, **20**, 2057–2070.

Fan, J. and Marron, J.S. (1994) Fast implementations of nonparametric curve estimators. *Journal of Computational and Graphical Statistics*, **3**, 35–56.

Fan, J. and Müller, M. (1995) Density and regression smoothing. In *XploRe: An Interactive Statistical Computing Environment*, eds W. Härdle, S. Klinke and B.A. Turlach, Springer-Verlag, New York, 77–99.

Faraway, J. (1990a) Implementing semiparametric density estimation. *Statistics and Probability Letters*, **10**, 141–143.

Faraway, J. (1990b) Bootstrap selection of bandwidth and confidence bands for nonparametric regression. *Journal of Statistical Computation and Simulation*, **37**, 37–44.

Fenstad, G.U. and Hjort, N.L. (1995) Comparison of two Hermite expansion density estimators with the kernel method. Unpublished manuscript.

Ferguson, T.S. (1973) A Bayesian analysis of some nonparametric problems. *Annals of Statistics*, **1**, 209–230.

Ferguson, T.S. (1983) Bayesian density estimation by mixtures of normal distributions. In *Recent Advances in Statistics, Chernoff Festschrift Volume*, eds. H. Rizvi, J.S. Rustagi, and D.O. Siegmund, Academic Press, New York, 287–302.

Fienberg, S.E. and Holland, P.W. (1973) Simultaneous estimation of multinomial cell probabilities. *Journal of the American Statistical Association*, **68**, 683–691.

Firth, D., Glosup, J., and Hinkley, D.V. (1991) Model checking with nonparametric curves. *Biometrika*, **78**, 245–252.

Fix, E. and Hodges, J.L., Jr. (1951) Nonparametric discrimination: consistency properties. Report Number 4, USAF School of Aviation Medicine, Randolph Field, Texas.

Flury, B. and Riedwyl, H. (1988) *Multivariate Statistics: A Practical Approach*, Chapman and Hall, London.

Földes, A. and Révész, P. (1974) A general method for density estimation. *Studia Scientiarum Mathematicarum Hungarica*, **9**, 81–92.

Foster, P.J. (1995) A comparative study of some bias correction techniques for kernel-based density estimators. *Journal of Statistical Computation and Simulation*, **51**, 137–152.

Freedman, D. and Diaconis, P. (1981) On the histogram as a density estimator: $L_2$ theory. *Zeitschrift für Wahrscheinlichkeitstheorie und verwandte Gebiete*, **57**, 453–476.

Friedman, J.H. (1987) Exploratory projection pursuit. *Journal of the American Statistical Association*, **82**, 249–266.

Friedman, J.H. (1991) Multivariate adaptive regression splines (with discussion). *Annals of Statistics*, **19**, 1–141.

Friedman, J.H. and Silverman, B.W. (1989) Flexible parsimonious smoothing and additive modeling (with discussion). *Technometrics*, **31**, 3–39.

Friedman, J.H. and Stuetzle, W. (1981) Projection pursuit regression. *Journal of the American Statistical Association*, **76**, 817–823.

Friedman, J.H., Stuetzle, W., and Schroeder, A. (1984) Projection pursuit density estimation. *Journal of the American Statistical Association*, **79**, 599–608.

Friedman, J.H. and Tukey, J.W. (1974) A projection pursuit algorithm for exploratory data analysis. *IEEE Transactions on Computers*, **C-23**, 881–889.

Friendly, M. and Fox, J. (1994) Using APL2 to create an object-oriented environment for statistical computation. *Journal of Computational and Graphical Statistics*, **3**, 387–407.

Frohlich, C. (1995) Personal communication, April 20, 1995.

Frohlich, C. and Davis, S.D. (1990) Single-link cluster analysis as a method to evaluate spatial and temporal properties of earthquake catalogues. *Geophysical Journal International*, **100**, 19–32.
Fukunaga, K. (1972) *Introduction to Statistical Pattern Recognition*, Academic Press, New York.
Fukunaga, K. and Hostetler, L.D. (1975) The estimation of the gradient of a density function, with applications in pattern recognition. *IEEE Transactions on Information Theory*, **IT-21**, 32–40.
Gajek, L. (1986) On improving density estimators which are not bona fide functions. *Annals of Statistics*, **14**, 1612–1618.
Gasser, T., Kneip, A., and Köhler, W. (1991) A flexible and fast method for automatic smoothing. *Journal of the American Statistical Association*, **86**, 643–652.
Gasser, T. and Müller, H.-G. (1979) Kernel estimation of regression functions. In *Smoothing Techniques for Curve Estimation*, eds. T. Gasser and M. Rosenblatt, Springer-Verlag, Berlin, 23–68.
Gasser, T. and Müller, H.-G. (1984) Estimating regression functions and their derivatives by the kernel method. *Scandinavian Journal of Statistics*, **11**, 171–185.
Gasser, T., Müller, H.-G., and Mammitzsch, V. (1985) Kernels for nonparametric curve estimation. *Journal of the Royal Statistical Society, Ser. B*, **47**, 238–252.
Gasser, T., Sroka, L., and Jennen-Steinmetz, C. (1986) Residual variance and residual pattern in nonlinear regression. *Biometrika*, **73**, 625–633.
Gawronski, W. and Stadtmüller, U. (1980) On density estimation by means of Poisson's distribution. *Scandinavian Journal of Statistics*, **7**, 90–94.
Gawronski, W. and Stadtmüller, U. (1981) Smoothing histograms by means of lattice- and continuous distributions. *Metrika*, **28**, 155–164.
Geffroy, J. (1974) Sur l'estimation d'une densité dans un espace métrique. *Comptes Rendus de l'Académie des Sciences de Paris Série A*, **278**, 1449–1452.
Geman, S. and Hwang, C.-R. (1982) Nonparametric and maximum likelihood estimation by the method of sieves. *Annals of Statistics*, **10**, 401–414.
Georgiev, A.A. (1984a) Nonparametric system identification by kernel methods. *IEEE Transactions on Automatic Control*, **29**, 356–358.
Georgiev, A.A. (1984b) Speed of convergence in nonparametric kernel estimation of a regression function and its derivatives. *Annals of the Institute of Statistical Mathematics*, **36**, 455–462.
Ghorai, J.K. (1980) Asymptotic normality of a quadratic measure of orthogonal series type density estimate. *Annals of the Institute of Statistical Mathematics*, **32**, 341–350.
Ghorai, J.K. and Rubin, H. (1979) Computational procedure for maximum penalized likelihood estimate. *Journal of Statistical Computation and Simulation*, **10**, 65–78.
Ghorai, J.K. and Rubin, H. (1982) Bayes risk consistency of nonparametric Bayes density estimates. *Australian Journal of Statistics*, **24**, 51–66.
Ghosh, B.K. and Huang, W.-M. (1991) The power and optimal kernel of the Bickel–Rosenblatt test for goodness-of-fit. *Annals of Statistics*, **19**, 999–1009.
González-Manteiga, W., Sánchez-Sellero, C., and Wand, M.P. (1996) Accuracy of binned kernel functional approximations. *Computational Statistics and Data Analysis*, **22**, 1–16.
Good, I.J. and Gaskins, R.A. (1971) Nonparametric roughness penalties for probability densities. *Biometrika*, **58**, 255–277.
Good, I.J. and Gaskins, R.A. (1972) Global nonparametric estimation of probability densities. *Virginia Journal of Science*, **23/4**, 171–193.
Good, I.J. and Gaskins, R.A. (1980) Density estimation and bump-hunting by the penalized likelihood method exemplified by scattering and meteorite data (with discussion). *Journal of the American Statistical Association*, **75**, 42–73.
Good, I.J., Holtzman, G.I., Deaton, M.L., and Bernstein, L.H. (1989) Diagnosis of heart attack from two enzyme measurements by means of bivariate probability density estimation: statistical details. *Journal of Statistical Computation and Simulation*, **32**, 68–76.

Granville, V. and Rasson, J.P. (1992) Density estimation on a finite regular lattice. *Computational Statistics*, **7**, 129–136.
Granville, V. and Rasson, J.P. (1995) Multivariate discriminant analysis and maximum penalized likelihood density estimation. *Journal of the Royal Statistical Society, Ser. B.* **57**, 501–518.
Greblicki, W. (1974) Asymptotically optimal probabilistic algorithms for pattern recognition and identification (in Polish). *Prace Naukowe Instytutu Cybernetyki Technicznej No. 18, Seria: Monografi No. 3*, Wroclaw.
Greblicki, W. and Krzyzak, A. (1980) Asymptotic properties of kernel estimates of a regression function. *Journal of Statistical Planning and Inference*, **4**, 81–90.
Green, P.J. (1987) Penalized likelihood for general semi-parametric regression models. *International Statistical Review*, **55**, 245–259.
Green, P.J. and Silverman, B.W. (1994) *Nonparametric Regression and Generalized Linear Models: A Roughness Penalty Approach*, Chapman and Hall, London.
Gregory, G.G. and Schuster, E.F. (1979) Contributions to non-parametric maximum likelihood methods of density estimation. In *Computing Science and Statistics: Proceedings of the 12th Symposium on the Interface*, ed. J. Gentleman, University of Waterloo, Ontario, 427–431.
Grund, B. (1993) Kernel estimators for cell probabilities. *Journal of Multivariate Analysis*, **46**, 283–308.
Grund, B. and Hall, P. (1993) On the performance of kernel estimators for high-dimensional, sparse binary data. *Journal of Multivariate Analysis*, **44**, 321–344.
Grund, B., Hall, P., and Marron, J.S. (1994) Loss and risk in smoothing parameter selection. *Journal of Nonparametric Statistics*, **4**, 107–132.
Gu, C. (1992) Diagnostics for nonparametric regression models with additive terms. *Journal of the American Statistical Association*, **87**, 1051–1058.
Gu, C. (1993) Smoothing spline density estimation: a dimensionless automatic algorithm. *Journal of the American Statistical Association*, **88**, 495–504.
Gu, C. (1995a) Model indexing and smoothing parameter selection in nonparametric function estimation. Unpublished manuscript.
Gu, C. (1995b) Smoothing spline density estimation: conditional distribution. *Statistica Sinica*, **5**, 709–726.
Gu, C. and Qiu, C. (1993) Smoothing spline density estimation: theory. *Annals of Statistics*, **21**, 217–234.
Gu, C. and Qiu, C. (1994) Penalized likelihood regression: a simple asymptotic analysis. *Statistica Sinica*, **4**, 297–304.
Gu, C. and Wahba, G. (1991) Minimizing GCV/GML scores with multiple smoothing parameters via the Newton method. *SIAM Journal on Scientific and Statistical Computing*, **12**, 383–398.
Gu, C. and Wahba, G. (1993a) Semiparametric ANOVA with tensor product thin plate splines. *Journal of the Royal Statistical Society, Ser. B*, **55**, 353–368.
Gu, C. and Wahba, G. (1993b) Smoothing spline ANOVA with componentwise Bayesian "confidence intervals." *Journal of Computational and Graphical Statistics*, **2**, 97–117.
Györfi, L., Härdle, W., Sarda, P., and Vieu, P. (1990) *Nonparametric Curve Estimation from Time Series, Lecture Notes in Statistics No. 60*, Springer-Verlag, New York.
Habbema, J.D.F., Hermans, J., and Remme, J. (1978) Variable kernel density estimation in discriminant analysis. In *Compstat 1978*, eds. L.C.A. Corsten and J. Hermans, Physica-Verlag, Vienna, 178–185.
Habbema, J.D.F., Hermans, J., and van den Broek, K. (1974) A stepwise discriminant analysis program using density estimation. In *Compstat 1974*, ed. G. Bruckmann, Physica-Verlag, Vienna, 101–110.
Hadi, A.S. and Simonoff, J.S. (1993) Procedures for the identification of multiple outliers in linear models. *Journal of the American Statistical Association*, **88**, 1264–1272.
Hadi, A.S. and Simonoff, J.S. (1994) Improving the estimation and outlier identification properties of the least median of squares and minimum volume ellipsoid estimators. *Parisankhyan Samikkha*, **1**, 61–70.

Hall, P. (1981) On nonparametric multivariate binary discrimination. *Biometrika*, **68**, 287–294.
Hall, P. (1982) Cross-validation in density estimation. *Biometrika*, **69**, 383–390.
Hall, P. (1983a) Large sample optimality of least squares cross-validation in density estimation. *Annals of Statistics*, **11**, 1156–1174.
Hall, P. (1983b) On near neighbor estimates of a multivariate density. *Journal of Multivariate Analysis*, **13**, 24–39.
Hall, P. (1985) Asymptotic theory of minimum integrated square error for multivariate density estimation. In *Multivariate Analysis — VI*, ed. P.R. Krishnaiah, Elsevier Science, Amsterdam, 289–309.
Hall, P. (1987) On Kullback–Leibler loss and density estimation. *Annals of Statistics*, **15**, 1491–1519.
Hall, P. (1989a) On polynomial-based projection indices for exploratory projection pursuit. *Annals of Statistics*, **17**, 589–605.
Hall, P. (1989b) On projection pursuit regression. *Annals of Statistics*, **17**, 573–588.
Hall, P. (1990a) Akaike's information criterion and Kullback–Leibler loss for histogram density estimation. *Probability Theory and Related Fields*, **85**, 449–467.
Hall, P. (1990b) On the bias of variable bandwidth curve estimators. *Biometrika*, **77**, 529–535.
Hall, P. (1992a) On global properties of variable bandwidth density estimators. *Annals of Statistics*, **20**, 762–778.
Hall, P. (1992b) On bootstrap confidence intervals in nonparametric regression. *Annals of Statistics*, **20**, 695–711.
Hall, P. (1993) On plug-in rules for local smoothing of density estimators. *Annals of Statistics*, **21**, 694–710.
Hall, P., DiCiccio, T.J., and Romano, J.P. (1989) On smoothing and the bootstrap. *Annals of Statistics*, **17**, 692–704.
Hall, P. and Hannan, E.J. (1988) On stochastic complexity and nonparametric density estimation. *Biometrika*, **75**, 705–714.
Hall, P. and Hart, J.D. (1990a) Convergence rates in density estimation for data from infinite-order moving average processes. *Probability Theory and Related Fields*, **87**, 253–274.
Hall, P. and Hart, J.D. (1990b) Nonparametric regression with long-range dependence. *Stochastic Processes and Their Applications*, **36**, 339–351.
Hall, P. and Hart, J.D. (1990c) Bootstrap test for difference between means in nonparametric regression. *Journal of the American Statistical Association*, **85**, 1039–1049.
Hall, P., Hu, T.-C., and Marron, J.S. (1995) Improved variable window kernel estimates of probability densities. *Annals of Statistics*, **23**, 1–10.
Hall, P. and Johnstone, I.M. (1992) Empirical functionals and efficient smoothing parameter selection (with discussion). *Journal of the Royal Statistical Society, Ser. B*, **54**, 475–530.
Hall, P. and Jones, M.C. (1990) Adaptive $M$-estimation in nonparametric regression. *Annals of Statistics*, **18**, 1712–1728.
Hall, P., Kay, J.W., and Titterington, D.M. (1990) Asymptotically optimal difference-based estimation of variance in nonparametric regression. *Biometrika*, **77**, 521–528.
Hall, P. and Marron, J.S. (1987a) On the amount of noise inherent in bandwidth selection for a kernel density estimator. *Annals of Statistics*, **15**, 163–181.
Hall, P. and Marron, J.S. (1987b) Extent to which least-squares cross-validation minimises integrated square error in nonparametric density estimation. *Probability Theory and Related Fields*, **74**, 567–581.
Hall, P. and Marron, J.S. (1988a) Variable window width kernel estimates of probability densities. *Probability Theory and Related Fields*, **80**, 37–49.
Hall, P. and Marron, J.S. (1988b) Choice of kernel order in density estimation. *Annals of Statistics*, **16**, 161–173.
Hall, P. and Marron, J.S. (1990) On variance estimation in nonparametric regression. *Biometrika*, **77**, 415–419.

Hall, P. and Marron, J.S. (1991a) Lower bounds for bandwidth selection in density estimation. *Probability Theory and Related Fields*, **90**, 149–173.

Hall, P. and Marron, J.S. (1991b) Local minima in cross-validation functions. *Journal of the Royal Statistical Society, Ser. B*, **53**, 245–252.

Hall, P., Marron, J.S., and Park, B.U. (1992) Smoothed cross-validation. *Probability Theory and Related Fields*, **92**, 1–20.

Hall, P., Marron, J.S., and Titterington, D.M. (1995) On partial local smoothing rules for curve estimation. *Biometrika*, **82**, 575–587.

Hall, P. and Murison, R.D. (1993) Correcting the negativity of high-order kernel density estimates. *Journal of Multivariate Analysis*, **47**, 103–122.

Hall, P. and Patil, P. (1995) Formulae for mean integrated squared error of nonlinear wavelet-based density estimators. *Annals of Statistics*, **23**, 905–928.

Hall, P. and Schucany, W.R. (1989) A local cross-validation algorithm. *Statistics and Probability Letters*, **8**, 109–117.

Hall, P., Sheather, S.J., Jones, M.C., and Marron, J.S. (1991) On optimal data-based bandwidth selection in kernel density estimation. *Biometrika*, **78**, 263–269.

Hall, P. and Titterington, D.M. (1987a) Common structure of techniques for choosing smoothing parameters in regression problems. *Journal of the Royal Statistical Society, Ser. B*, **49**, 184–198.

Hall, P. and Titterington, D.M. (1987b) On smoothing sparse multinomial data. *Australian Journal of Statistics*, **29**, 19–37.

Hall, P. and Titterington, D.M. (1988) On confidence bands in nonparametric density estimation and regression. *Journal of Multivariate Analysis*, **27**, 228–254.

Hall, P. and Turlach, B.A. (1996) Contribution to discussion of papers by Seifert and Gasser, Marron, and Cleveland and Loader. In *Statistical Theory and Computational Aspects of Smoothing*, eds. W. Hardle and M.G. Schimek, Physica-Verlag, Heidelberg, 80–84.

Hall, P. and Wand, M.P. (1988a) Minimizing $L_1$ distance in nonparametric density estimation. *Journal of Multivariate Analysis*, **26**, 59–88.

Hall, P. and Wand, M.P. (1988b) On the minimization of absolute distance in kernel density estimation. *Statistics and Probability Letters*, **6**, 311–314.

Hall, P. and Wand, M.P. (1988c) On nonparametric discrimination using density differences. *Biometrika*, **75**, 541–547.

Hall, P. and Wand, M.P. (1996) On the accuracy of binned kernel density estimators. *Journal of Multivariate Analysis*, **56**, 165–184.

Hampel, F.R., Ronchetti, E.M., Rousseeuw, P.J., and Stahel, W.A. (1986) *Robust Statistics: The Approach Based on Influence Functions*, John Wiley, New York.

Hand, D.J. (1982) *Kernel Discriminant Analysis*, Research Studies Press, Chichester.

Härdle, W. (1984) Robust regression function estimation. *Journal of Multivariate Analysis*, **14**, 169–180.

Härdle, W. (1986) A note on jackknifing kernel regression function estimators. *IEEE Transactions on Information Theory*, **32**, 298–300.

Härdle, W. (1987) Resistant smoothing using the Fast Fourier Transform. *Applied Statistics*, **36**, 104–111.

Härdle, W. (1990) *Applied Nonparametric Regression*, Cambridge University Press, Cambridge.

Härdle, W. (1991) *Smoothing Techniques with Implementation in S*, Springer-Verlag, New York.

Härdle, W. and Bowman, A.W. (1988) Bootstrapping in nonparametric regression: local adaptive smoothing and confidence bands. *Journal of the American Statistical Association*, **83**, 102–110.

Härdle, W. and Gasser, T. (1984) Robust non-parametric function fitting. *Journal of the Royal Statistical Society, Ser. B*, **46**, 42–51.

Härdle, W., Hall, P., and Ichimura, H. (1993) Optimal smoothing in single-index models. *Annals of Statistics*, **21**, 157–178.

Härdle, W., Hall, P., and Marron, J.S. (1988) How far are automatically chosen regression smoothing parameters from their optimum? (with discussion). *Journal of the American Statistical Association*, **83**, 86–101.

Härdle, W., Hall, P., and Marron, J.S. (1992) Regression smoothing parameters that are not far from their optimum. *Journal of the American Statistical Association*, **87**, 227–233.

Härdle, W., Klinke, S., and Turlach, B.A., eds. (1995) *XploRe: An Interactive Statistical Computing Environment*, Springer-Verlag, New York.

Härdle, W. and Mammen, E. (1993) Comparing nonparametric versus parametric regression fits. *Annals of Statistics*, **21**, 1926–1947.

Härdle, W. and Marron, J.S. (1985) Optimal bandwidth selection in nonparametric regression function estimation. *Annals of Statistics*, **13**, 1465–1481.

Härdle, W. and Marron, J.S. (1986) Random approximations to an error criterion of nonparametric statistics. *Journal of Multivariate Analysis*, **20**, 91–113.

Härdle, W. and Marron, J.S. (1990) Semiparametric comparison of regression curves. *Annals of Statistics*, **18**, 63–89.

Härdle, W. and Marron, J.S. (1991) Bootstrap simultaneous error bars for nonparametric regression. *Annals of Statistics*, **19**, 778–796.

Härdle, W. and Marron, J.S. (1995) Fast and simple scatterplot smoothing. *Computational Statistics and Data Analysis*, **20**, 1–17.

Härdle, W. and Scott, D.W. (1992) Smoothing by weighted average of rounded points. *Computational Statistics*, **7**, 97–128.

Härdle, W. and Tsybakov, A.B. (1988) Robust nonparametric regression with simultaneous scale curve estimation. *Annals of Statistics*, **16**, 120–135.

Harris, I.R. and Basu, A. (1994) Hellinger distance as a penalized log likelihood. *Communications in Statistics — Simulation and Computation*, **23**, 1097–1113.

Hart, J.D. (1984) Efficiency of a kernel density estimator under an autoregressive dependence model. *Journal of the American Statistical Association*, **79**, 110–117.

Hart, J.D. (1985) Data-based choice of the smoothing parameter for a kernel density estimator. *Australian Journal of Statistics*, **27**, 44–52.

Hart, J.D. (1991) Kernel regression estimation with time series errors. *Journal of the Royal Statistical Society, Ser. B*, **53**, 173–187.

Hart, J.D. (1994) Automated kernel smoothing of dependent data by using time series cross-validation. *Journal of the Royal Statistical Society, Ser. B*, **56**, 529–542.

Hart, J.D. and Vieu, P. (1990) Data-driven bandwidth choice for density estimation based on dependent data. *Annals of Statistics*, **18**, 873–890.

Hart, J.D. and Wehrly, T.E. (1986) Kernel regression estimation using repeated measurement data. *Journal of the American Statistical Association*, **81**, 1080–1088.

Hart, J.D. and Wehrly, T.E. (1992) Kernel regression estimation when the boundary is large, with an application to testing the adequacy of polynomial models. *Journal of the American Statistical Association*, **87**, 1018–1024.

Hastie, T. (1989) Discussion of "Flexible parsimonious smoothing and additive modeling." *Technometrics*, **31**, 23–29.

Hastie, T. (1992) Generalized additive models. In *Statistical Models in S*, eds. J.M. Chambers and T. Hastie, Wadsworth and Brooks/Cole, Pacific Grove, CA, 249–307.

Hastie, T., Buja, A., and Tibshirani, R. (1995) Penalized discriminant analysis. *Annals of Statistics*, **23**, 73–102.

Hastie, T. and Loader, C. (1993) Local regression: automatic kernel carpentry (with discussion). *Statistical Science*, **8**, 120–143.

Hastie, T. and Tibshirani, R. (1986) Generalized additive models (with discussion). *Statistical Science*, **1**, 297–318.

Hastie, T. and Tibshirani, R. (1987) Generalized additive models: some applications. *Journal of the American Statistical Association*, **82**, 371–386.

Hastie, T. and Tibshirani, R. (1990) *Generalized Additive Models*, Chapman and Hall, London.

Hastie, T. and Tibshirani, R. (1996) Discriminant analysis by Gaussian mixtures. *Journal of the Royal Statistical Society, Ser. B*, **58**, 155–176.

Hastie, T., Tibshirani, R., and Buja, A. (1994) Flexible discriminant analysis by optimal scoring. *Journal of the American Statistical Association*, **89**, 1255–1270.

Hawkins, D.M. (1980) *Identification of Outliers*, Chapman and Hall, New York.

Heckman, N.E. (1986) Spline smoothing in a partly linear model. *Journal of the Royal Statistical Society, Ser. B*, **48**, 244–248.

Heckman, N.E. (1988) Minimax estimates in a semiparametric model. *Journal of the American Statistical Association*, **83**, 1090–1096.

Herrmann, E. (1997) Local bandwidth choice in kernel regression estimation. *Journal of Computational and Graphical Statistics*, **6**, 35–54.

Herrmann, E., Gasser, T., and Kneip, A. (1992) Choice of bandwidth for kernel regression when residuals are correlated. *Biometrika*, **79**, 783–795.

Herrmann, E., Wand, M.P., Engel, J., and Gasser, T. (1995) A bandwidth selector for bivariate kernel regression. *Journal of the Royal Statistical Society, Ser. B*, **57**, 171–180.

Hjort, N.L. (1986) On frequency polygons and average shifted histograms in higher dimensions. Unpublished manuscript.

Hjort, N.L. (1996) Bayesian approaches to non- and semiparametric density estimation. In *Bayesian Statistics 5. Proceedings of the Fifth International Valencia Meeting on Bayesian Statistics*, eds. J.M. Bernardo, J.O. Berger, A.P. Dawid, and A.F.M. Smith, Oxford University Press, Oxford, 223–254.

Hjort, N.L. and Glad, I.K. (1995) Nonparametric density estimation with a parametric start. *Annals of Statistics*, **23**, 882–904.

Hjort, N.L. and Jones, M.C. (1996) Locally parametric density estimation. *Annals of Statistics*, **24**, 1619–1647.

Hodges, J.L. and Lehmann, E.L. (1956) The efficiency of some nonparametric competitors of the *t*-test. *Annals of Mathematical Statistics*, **27**, 324–335.

Hoffman, M.S., ed. (1992) *The World Almanac and Book of Facts 1993*, Pharos Books, New York.

Hössjer, O. and Ruppert, D. (1995) Asymptotics for the transformation kernel density estimator. *Annals of Statistics*, **23**, 1198–1222.

Hu[illegible] -S. (1997) Testing goodness-of-fit based on a roughness measure. *Journal of the American Statistical Association*, **92**, 1399–1402.

Huber, P.J. (1979) Robust smoothing. In *Robustness in Statistics*, eds. R.L. Launer and G.N. Wilkinson, Academic Press, New York, 33–48.

Huber, P.J. (1981) *Robust Statistics*, John Wiley, New York.

Huber, P.J. (1985) Projection pursuit (with discussion). *Annals of Statistics*, **13**, 435–525.

Hurvich, C.M. and Tsai, C.-L. (1995) Relative rates of convergence for efficient model selection criteria in linear regression. *Biometrika*, **82**, 418–425.

Hurvich, C.M. and Zeger, S.L. (1990) A frequency domain selection criterion for regression with autocorrelated errors. *Journal of the American Statistical Association*, **85**, 705–714.

Hüsemann, J.A. and Stevens, R.R. (1991) Computation of optimal parameters for bivariate histograms. *Journal of Statistical Computation and Simulation*, **38**, 109–125.

Hüsemann, J.A. and Terrell, G.R. (1991) Optimal parameter choice for error minimization in bivariate histograms. *Journal of Multivariate Analysis*, **37**, 85–103.

Ibragimov, I.A. and Khasminskii, R.Z. (1982) Estimation of distribution density belonging to a class of entire functions. *Theory of Probability and Its Applications*, **27**, 551–562.

Ishiguro, M. and Sakamoto, Y. (1984) A Bayesian approach to the probability density estimation. *Annals of the Institute of Statistical Mathematics*, **36**, 523–538.

Ishikawa, K. (1986) *Guide to Quality Control*, Unipub, Kraus International Publications, White Plains, NY.

Izenman, A.J. (1991) Recent developments in nonparametric density estimation. *Journal of the American Statistical Association*, **86**, 205–224.

Janssen, P., Marron, J.S., Veraverbeke, N., and Sarle, W. (1995) Scale measures for bandwidth selection. *Journal of Nonparametric Statistics*, **5**, 359–380.

Jee, J.R. (1987) Exploratory projection pursuit using nonparametric density estimation. *Proceedings of the Statistical Computing Section of the American Statistical Association*, 335–339.

Jennen-Steinmetz, C. and Gasser, T. (1988) A unifying approach to nonparametric regression estimation. *Journal of the American Statistical Association*, **83**, 1084–1089.

Jhun, M. (1988) $L^1$ bandwidth selection in kernel regression function estimation. *Journal of the Korean Statistical Society*, **17**, 1–8.

Johansen, S. and Johnstone, I.M. (1990) Hotelling's theorem on the volume of tubes: some illustrations in simultaneous inference and data analysis. *Annals of Statistics*, **18**, 652–684.

Johnston, G.J. (1982) Probabilities of maximal deviations for nonparametric regression function estimates. *Journal of Multivariate Analysis*, **12**, 402–414.

Jones, M.C. (1989) Discretized and interpolated density estimates. *Journal of the American Statistical Association*, **84**, 733–741.

Jones, M.C. (1990) Variable kernel density estimates and variable kernel density estimates. *Australian Journal of Statistics*, **32**, 361–371.

Jones, M.C. (1991) The roles of ISE and MISE in density estimation. *Statistics and Probability Letters*, **12**, 51–56.

Jones, M.C. (1992) Potential for automatic bandwidth choice in variations on kernel density estimation. *Statistics and Probability Letters*, **13**, 351–356.

Jones, M.C. (1993a) Simple boundary correction for kernel density estimation. *Statistics and Computing*, **3**, 135–146.

Jones, M.C. (1993b) Kernel density estimation when the bandwidth is large. *Australian Journal of Statistics*, **35**, 319–326.

Jones, M.C. (1993c) Do not weight for heteroscedasticity in nonparametric regression. *Australian Journal of Statistics*, **35**, 89–92.

Jones, M.C. (1995a) On two recent papers of Y. Kanazawa. *Statistics and Probability Letters*, **24**, 269–271.

Jones, M.C. (1995b) On higher order kernels. *Journal of Nonparametric Statistics*, **5**, 215–221.

Jones, M.C. (1996) On close relations of local likelihood density estimation. *Test*, **5**, 345–356.

Jones, M.C., Davies, S.J., and Park, B.U. (1994) Versions of kernel-type regression estimators. *Journal of the American Statistical Association*, **89**, 825–832.

Jones, M.C. and Foster, P.J. (1993) Generalized jackknifing and higher order kernels. *Journal of Nonparametric Statistics*, **3**, 81–94.

Jones, M.C. and Foster, P.J. (1996) A simple nonnegative boundary correction method for kernel density estimation. *Statistica Sinica*, **6**, 1005–1013.

Jones, M.C. and Hjort, N.L. (1994) Comment on "How to approximate a histogram by a normal density." *American Statistician*, **48**, 353–354.

Jones, M.C. and Hjort, N.L. (1995) Local fitting of regression models by likelihood: what's important? Unpublished manuscript.

Jones, M.C. and Kappenman, R.F. (1991) On a class of kernel density estimate bandwidth selectors. *Scandinavian Journal of Statistics*, **19**, 337–349.

Jones, M.C., Linton, O., and Nielsen, J.P. (1995) A simple bias reduction method for density estimation. *Biometrika*, **82**, 327–338.

Jones, M.C. and Lotwick, H. (1983) On the errors involved in computing the empirical characteristic function. *Journal of Statistical Computation and Simulation*, **17**, 133–149.

Jones, M.C., Marron, J.S., and Park, B.U. (1991) A simple root $n$ bandwidth selector. *Annals of Statistics*, **19**, 1919–1932.

Jones, M.C., Marron, J.S., and Sheather, S.J. (1996) A brief survey of bandwidth selection for density estimation. *Journal of the American Statistical Association*, **91**, 401–407.

Jones, M.C., McKay, I.J., and Hu, T.-C. (1994) Variable location and scale kernel density estimation. *Annals of the Institute of Statistical Mathematics*, **46**, 521–535.

Jones, M.C. and Rice, J.A. (1992) Displaying the important features of large collections of similar curves. *American Statistician*, **46**, 140–145.

Jones, M.C., Samiuddin, M., Al-Harbey, A.H., and Maatouk, T.A.H. (1998). Piecewise linear smoothed histograms. *Biometrika*, **85**, to appear.

Jones, M.C. and Sibson, R. (1987) What is projection pursuit? (with discussion). *Journal of the Royal Statistical Society, Ser. B*, **15**, 1–36.

Jones, M.C. and Signorini, D.F. (1997) A comparison of higher-order bias kernel density estimators. *Journal of the American Statistical Association*, **92**, 1063–1073.

Kanazawa, Y. (1988) An optimal variable cell histogram. *Communications in Statistics — Theory and Methods*, **17**, 1401–1422.

Kanazawa, Y. (1992) An optimal variable cell histogram based on the sample spacings. *Annals of Statistics*, **20**, 291–304.

Kanazawa, Y. (1993) Hellinger distance and Akaike's information criterion for the histogram. *Statistics and Probability Letters*, **17**, 293–298.

Katkovnik, V.Y. (1979) Linear and nonlinear methods of nonparametric regression analysis. *Soviet Automatic Control*, **5**, 25–34.

Kerkyacharian, G. and Picard, D. (1992) Density estimation in Besov spaces. *Statistics and Probability Letters*, **13**, 15–24.

Kerkyacharian, G. and Picard, D. (1993) Density estimation by kernel and wavelet methods: optimality in Besov spaces. *Statistics and Probability Letters*, **18**, 327–336.

Khashimov, S.A. (1984) On the convergence rate of the distribution of a quadratic measure of deviation of nonparametric density estimates. *Theory of Probability and Its Applications*, **29**, 163–169.

Kim, W.C., Park, B.U., and Marron, J.S. (1994) Asymptotically best bandwidth selectors in kernel density estimation. *Statistics and Probability Letters*, **19**, 119–127.

Kimeldorf, G.S. and Wahba, G. (1970a) A correspondence between Bayesian estimation on stochastic processes and smoothing by splines. *Annals of Mathematical Statistics*, **41**, 495–502.

Kimeldorf, G.S. and Wahba, G. (1970b) Spline functions and stochastic processes. *Sankhyā, Ser. A*, **32**, 173–180.

King, E.C., Hart, J.D., and Wehrly, T.E. (1991) Testing the equality of two regression curves using linear smoothers. *Statistics and Probability Letters*, **12**, 239–247.

Klonias, V.K. (1982) Consistency of two nonparametric penalized likelihood estimators of the probability density function. *Annals of Statistics*, **10**, 811–824.

Klonias, V.K. (1984) On a class of nonparametric density and regression estimators. *Annals of Statistics*, **12**, 1263–1284.

Klonias, V.K. and Nash, S.G. (1987) Numerical techniques in nonparametric estimation. *Journal of Statistical Computation and Simulation*, **28**, 97–126.

Knafl, G., Sacks, J., and Ylvisaker, D. (1985) Confidence bands for regression functions. *Journal of the American Statistical Association*, **80**, 683–691.

Kneip, A. (1994) Ordered linear smoothers. *Annals of Statistics*, **22**, 835–866.

Koehler, K.J. (1986) Goodness-of-fit tests for log-linear models in sparse contingency tables. *Journal of the American Statistical Association*, **81**, 483–493.

Koehler, K.J. and Gan, F.F. (1990) Chi-squared goodness-of-fit tests: cell selection and power. *Communications in Statistics — Simulation and Computation*, **19**, 1265–1278.

Koehler, K.J. and Larntz, K. (1980) An empirical investigation of goodness-of-fit statistics for sparse multinomials. *Journal of the American Statistical Association*, **75**, 336–344.

Kogure, A. (1987) Asymptotically optimal cells for a histogram. *Annals of Statistics*, **15**, 1023–1030.

Kohn, R., Ansley, C.F., and Tharm, D. (1991) The performance of cross-validation and maximum likelihood estimators of spline smoothing parameters. *Journal of the American Statistical Association*, **86**, 1042–1050.

Kohn, R., Ansley, C.F., and Wong, C.-M. (1992) Nonparametric spline regression with autoregressive moving average errors. *Biometrika*, **79**, 335–346.

Konakov, V.D. (1973) Nonparametric estimation of density functions. *Theory of Probability and Its Applications*, **17**, 361–362.

Kooperberg, C. and Stone, C.J. (1991) A study of logspline density estimation. *Computational Statistics and Data Analysis*, **12**, 327–347.

Koshkin, G.M. (1988) Improved non-negative kernel estimate of a density. *Theory of Probability and Its Applications*, **33**, 759–764.

Kozek, A.S. (1990) A nonparametric test of fit of a linear model. *Communications in Statistics — Theory and Methods*, **19**, 169–179.

Kozek, A.S. (1991) A nonparametric test of fit of a parametric model. *Journal of Multivariate Analysis*, **37**, 66–75.

Kozek, A.S. (1992) A new nonparametric estimation method: local and nonlinear. In *Computing Science and Statistics: Proceedings of the 24th Symposium on the Interface*, ed. H.J. Newton, Interface Foundation of North America, Fairfax, VA, 388–393.

Kraft, C.H., Lepage, Y., and van Eeden, C. (1983) Some finite-sample-size properties of Rosenblatt density estimates. *Canadian Journal of Statistics*, **11**, 95–104.

Kruskal, J.B. (1969) Toward a practical method which helps uncover the structure of a set of multivariate observations by finding the linear transformation which optimizes a new "index of condensation." In *Statistical Computation*, eds. R.C. Milton and J.A. Nelder, Academic Press, New York, 427–440.

Kruskal, J.B. (1972) Linear transformation of multivariate data to reveal clustering. In *Multidimensional Scaling: Theory and Applications in the Behavioral Sciences, Vol. 1*, eds. R.N. Shepard, A.K. Romney, and S.B. Nerlove, Seminar Press, London, 179–191.

Kulasekera, K.B. (1995) Comparison of regression curves using quasi-residuals. *Journal of the American Statistical Association*, **90**, 1085–1093.

Laird, N.M. (1978) Empirical Bayes methods for two-way contingency tables. *Biometrika*, **65**, 581–590.

Larntz, K. (1978) Small-sample comparisons of exact levels for chi-squared goodness-of-fit statistics. *Journal of the American Statistical Association*, **73**, 253–263.

Larson, H.J. (1975) *Statistics: An Introduction*, John Wiley, New York.

le Cessie, S. and van Houwelingen, J.C. (1991) A goodness-of-fit test for binary regression models, based on smoothing methods. *Biometrics*, **47**, 1267–1282.

Lecoutre, J.-P. (1985) The $L_2$-optimal cell width for the histogram. *Statistics and Probability Letters*, **3**, 303–306.

Lee, S. and Young, G.A. (1994) Practical higher-order smoothing of the bootstrap. *Statistica Sinica*, **4**, 445–459.

Lejeune, M. (1985) Estimation non-paramétrique par noyaux: régression polynomiale mobile. *Revue de Statistique Appliquée*, **33**, 43–67.

Lejeune, M. and Sarda, P. (1992) Smooth estimators of distribution and density functions. *Computational Statistics and Data Analysis*, **14**, 457–471.

Lenk, P.J. (1988) The logistic normal distribution for Bayesian, nonparametric, predictive densities. *Journal of the American Statistical Association*, **83**, 509–516.

Lenk, P.J. (1990) Bayesian predictive distributions under multinomial sampling. In *Bayesian and Likelihood Methods in Statistics and Econometrics*, eds. S. Geisser, J.S. Hodges, S.J. Press, and A. Zellner, Elsevier Science, Amsterdam, 357–370.

Lenk, P.J. (1991) Towards a practicable Bayesian density estimator. *Biometrika*, **78**, 531–543.

Lenk, P.J. (1993) A Bayesian nonparametric density estimator. *Journal of Nonparametric Statistics*, **3**, 53–69.

Leonard, T. (1973) A Bayesian method for histograms. *Biometrika*, **60**, 297–308.

Leonard, T. (1975) Bayesian estimation methods for two-way contingency tables. *Journal of the Royal Statistical Society, Ser. B*, **37**, 23–37.
Leonard, T. (1978) Density estimation, stochastic processes and prior information (with discussion). *Journal of the Royal Statistical Society, Ser. B*, **40**, 113–146.
Lewis, P.A.W., Liu, L.H., Robinson, D.W., and Rosenblatt, M. (1977) Empirical sampling study of a goodness of fit statistic for density function estimation. In *Multivariate Analysis — IV*, ed. P.R. Krishnaiah, North-Holland, Amsterdam, 159–174.
Lewis, T., Saunders, I.W., and Westcott, M. (1984) The moments of the Pearson chi-square statistic and the minimum expected value in two-way tables. *Biometrika*, **71**, 515–522.
Li, G.-Y. and Chen, Z. (1985) Projection pursuit approach to robust dispersion matrices and principal components: primary theory and Monte Carlo. *Journal of the American Statistical Association*, **80**, 759–766.
Li, G.-Y. and Cheng, P. (1993) Some recent developments in projection pursuit in China. *Statistica Sinica*, **3**, 35–51.
Lii, K.-S. (1978) A global measure of a spline density estimate. *Annals of Statistics*, **6**, 1138–1148.
Lii, K.-S. and Rosenblatt, M. (1975) Asymptotic behavior of a spline estimate of a density function. *Computers and Mathematics with Applications*, **1**, 223–235.
Lindsay, B.G. (1994) Efficiency versus robustness: the case for minimum Hellinger distance and related methods. *Annals of Statistics*, **22**, 1081–1114.
Lindsey, J. (1974a) Comparison of probability distributions. *Journal of the Royal Statistical Society, Ser. B*, **36**, 38–47.
Lindsey, J. (1974b) Construction and comparison of statistical models. *Journal of the Royal Statistical Society, Ser. B*, **36**, 418–425.
Lindsey, J. and Mersch, G. (1992) Fitting and comparing probability distributions with log linear models. *Computational Statistics and Data Analysis*, **13**, 373–384.
Lo, A.Y. (1984) On a class of Bayesian nonparametric estimates: I. Density estimates. *Annals of Statistics*, **12**, 351–357.
Loader, C.R. (1993) Nonparametric regression, confidence bands and bias correction. *Computing Science and Statistics: Proceedings of the 25th Symposium on the Interface*, Interface Foundation of North America, Fairfax, VA, 131–136.
Loader, C.R. (1994) Computing nonparametric function estimates. *Computing Science and Statistics: Proceedings of the 26th Symposium on the Interface*, Interface Foundation of North America, Fairfax, VA, 356–361.
Loader, C.R. (1995) Old Faithful erupts: bandwidth selection reviewed. Unpublished manuscript.
Loader, C.R. (1996) Local likelihood density estimation. *Annals of Statistics*, **24**, 1602–1618.
Lock, R.H. (1993) 1993 new cars data. *Journal of Statistics Education*, **1**, No. 1.
Loftsgaarden, D.O. and Quesenberry, C.P. (1965) A nonparametric estimate of a multivariate density function. *Annals of Mathematical Statistics*. **36**, 1049–1051.
Lugosi, G. and Nobel, A. (1996) Consistency of data-driven histogram methods for density estimation and classification. *Annals of Statistics*, **24**, 687–706.
Macaulay, F.R. (1931) *The Smoothing of Time Series*, National Bureau of Economic Research, New York.
Mächler, M.B. (1989) "Parametric" smoothing quality in nonparametric regression: shape control by penalizing inflection points. Unpublished Ph.D. thesis, Swiss Federal Institute of Technology (ETH), Zürich.
Mächler, M.B. (1995a) Estimating distributions with a fixed number of modes. In *Robust Statistics, Data Analysis, and Computer Intensive Methods — Workshop in honor of Peter J. Huber, on his 60th birthday*, ed. H. Rieder, Springer-Verlag, Berlin, 263–272.
Mächler, M.B. (1995b) Variational solution of penalized likelihood problems and smooth curve estimation. *Annals of Statistics*, **23**, 1496–1517.
Mack, Y.P. (1982) On a goodness-of-fit problem of some nonparametric density estimates. *Scandinavian Journal of Statistics*, **9**, 179–182.

Mack, Y.P. and Rosenblatt, M. (1979) Multivariate $k$-nearest neighbor density estimates. *Journal of Multivariate Analysis*, **9**, 1–15.
Maguire, B.A., Pearson, E.S., and Wynn, A.H.A. (1952) The time intervals between industrial accidents. *Biometrika*, **39**, 168–180.
Mallows, C. (1973) Some comments on $C_p$. *Technometrics*, **15**, 661–675.
Mammen, E. (1990) A short note on optimal bandwidth selection for kernel estimators. *Statistics and Probability Letters*, **9**, 23–25.
Mammen, E. (1991) Nonparametric regression under qualitative smoothness assumptions. *Annals of Statistics*, **19**, 741–759.
Mammen, E. (1995) On qualitative smoothness of kernel density estimates. *Statistics*, **26**, 253–267.
Manchester, L. (1996) Empirical influence for robust smoothing. *Australian Journal of Statistics*, **38**, 275–290.
Manchester, L. and Trueman, D. (1993) Resistance of nonlinear scatterplot smoothers. Unpublished manuscript.
Marchette, D.J., Priebe, C.E., Rogers, G.W., and Solka, J.L. (1996) Filtered kernel density estimation. *Computational Statistics*, **11**, 95–112.
Marron, J.S. (1985a) An asymptotically efficient solution to the bandwidth problem of kernel density estimation. *Annals of Statistics*, **13**, 1011–1023.
Marron, J.S. (1985b) Discussion of "Some aspects of the spline smoothing approach to non-parametric regression curve fitting." *Journal of the Royal Statistical Society, Ser. B*, **47**, 38–39.
Marron, J.S. (1986) Will the art of smoothing ever become a science? *Contemporary Mathematics*, **59**, 169–178.
Marron, J.S. (1987) A comparison of cross-validation techniques in density estimation. *Annals of Statistics*, **15**, 152–162.
Marron, J.S. (1991) Root $n$ bandwidth selection. In *Nonparametric Functional Estimation and Related Topics*, ed. G.G. Roussas, Klüwer Academic Publishers, Dordrecht, 251–260.
Marron, J.S. (1992a) Discussion of "The performance of six popular bandwidth selection methods on some real data sets." *Computational Statistics*, **7**, 271–273.
Marron, J.S. (1992b) Bootstrap bandwidth selection. In *Exploring the Limits of Bootstrap*, eds. R. LePage and L. Billard, John Wiley, New York, 249–262.
Marron, J.S. (1993) Discussion of "Practical performance of several data driven bandwidth selectors." *Computational Statistics*, **8**, 17–19.
Marron, J.S. (1994) Visual understanding of higher-order kernels. *Journal of Computational and Graphical Statistics*, **3**, 447–458.
Marron, J.S. (1995) Presentation of smoothers: the family approach. Unpublished manuscript.
Marron, J.S. (1996) A personal view of smoothing and statistics (with discussion). In *Statistical Theory and Computational Aspects of Smoothing*, eds. W. Hardle and M.G. Schimek, Physica-Verlag, Heidelberg, 1–9; 80–112.
Marron, J.S. (1998) Assessing bandwidth selectors with visual error criteria. *Computational Statistics*, **13**, to appear.
Marron, J.S. and Ruppert, D. (1994) Transformations to reduce boundary bias in kernel density estimation. *Journal of the Royal Statistical Society, Ser. B*, **56**, 653–671.
Marron, J.S. and Tsybakov, A.B. (1995) Visual error criteria for qualitative smoothing. *Journal of the American Statistical Association*, **90**, 499–507.
Marron, J.S. and Udina, F. (1995) Interactive local bandwidth choice. Economics Working Paper Series No. 109, Universitat Pompeu Fabra, Barcelona, Spain.
Marron, J.S. and Wand, M.P. (1992) Exact mean integrated squared error. *Annals of Statistics*, **20**, 712–736.
Marx, B.D. and Eilers, P.H.C. (1994) Direct generalized additive modelling with penalized likelihood. Unpublished manuscript.
Masry, E. (1983) Probability density estimation from sampled data. *IEEE Transactions on Information Theory*, **29**, 696–709.
McCullagh, P. (1985) On the asymptotic distribution of Pearson's statistic in linear exponential-family models. *International Statistical Review*, **53**, 61–67.

McCullagh, P. (1986) The conditional distribution of goodness-of-fit statistics for discrete data. *Journal of the American Statistical Association*, **81**, 104–107.

McCullagh, P. and Nelder, J.A. (1989) *Generalized Linear Models*, 2nd ed., Chapman and Hall, London.

McDonald, J.A. and Owen, A.B. (1986) Smoothing with split linear fits. *Technometrics*, **28**, 195–208.

McKay, I.J. (1993) A note on bias reduction in variable-kernel density estimates. *Canadian Journal of Statistics*, **21**, 367–375.

McLachlan, G.J. and Basford, K.E. (1988) *Mixture Models: Inference and Applications to Clustering*, Marcel Dekker, New York.

Messer, K. (1991) A comparison of a spline estimate to its equivalent kernel estimate. *Annals of Statistics*, **19**, 817–829.

Messer, K. and Goldstein, L. (1993) A new class of kernels for nonparametric curve estimation. *Annals of Statistics*, **21**, 179–195.

Mielniczuk, J., Sarda, P., and Vieu, P. (1989) Local data-driven bandwidth choice for density estimation. *Journal of Statistical Planning and Inference*, **23**, 53–69.

Minnotte, M.C. (1996) The bias-optimized frequency polygon. *Computational Statistics*, **11**, 35–48.

Minnotte, M.C. and Scott, D.W. (1993) The mode tree: a tool for visualization of nonparametric density features. *Journal of Computational and Graphical Statistics*, **2**, 51–68.

Moore, D.S. and Yackel, J.W. (1977) Consistency properties of nearest neighbor density function estimators. *Annals of Statistics*, **5**, 143–154.

Morris, C.L. (1975) Central limit theorems for multinomial sums. *Annals of Statistics*, **3**, 165–188.

Morton, S. (1989) Interpretable projection pursuit. Technical Report 106, Stanford University, Laboratory for Computational Statistics.

Mosteller, F. and Tukey, J.W. (1977) *Data Analysis and Regression*, Addison–Wesley, Reading, MA.

Müller, H.-G. (1984) Smooth optimum kernel estimators of densities, regression curves and modes. *Annals of Statistics*, **12**, 766–774.

Müller, H.-G. (1987) Weighted local regression and kernel methods for nonparametric curve fitting. *Journal of the American Statistical Association*, **82**, 231–238.

Müller, H.-G. (1992) Goodness-of-fit diagnostics for regression models. *Scandinavian Journal of Statistics*, **19**, 157–172.

Müller, H.-G. and Stadtmüller, U. (1987) Variable bandwidth kernel estimators of regression curves. *Annals of Statistics*, **15**, 182–201.

Munson, P.J. and Jernigan, R.W. (1989) A cubic spline extension of the Durbin–Watson test. *Biometrika*, **76**, 39–47.

Nadaraya, E.A. (1964) On estimating regression. *Theory of Probability and Its Applications*, **9**, 141–142.

Nadaraya, E.A. (1974) On the integral mean square error of some nonparametric estimates for the density function. *Theory of Probability and Its Applications*, **19**, 133–141.

Nason, G. (1995) Three-dimensional projection pursuit. *Applied Statistics*, **44**, 411–430.

National Hockey League (1992) *The National Hockey League Official Guide and Record Book 1992–93*, Triumph Books, Chicago.

Newsday (1989) *Three Month CD Rates for Long Island Banks and Thrifts*, August 23, 1989, page 49.

Newsday (1993a) *Players Cash In Big-Time*, February 21, 1993, page 8 (Sports Section).

Newsday (1993b) *First Wave of School Votes*, May 6, 1993, page 37.

Newsday (1995) *Safety in the Air*, May 7, 1995, page A7.

Nguyen, H.T. (1979) Density estimation in a continuous-time stationary Markov process. *Annals of Statistics*, **7**, 341–348.

Numerical Algorithms Group (1986) *The GLIM System Release 3.77 Manual*, Royal Statistical Society, London.

Nychka, D. (1988) Bayesian confidence intervals for a smoothing spline. *Journal of the American Statistical Association*, **83**, 1134–1143.
Nychka, D. (1990) The average posterior variance of a smoothing spline and a consistent estimate of the average squared error. *Annals of Statistics*, **18**, 415–428.
Nychka, D. (1995) Splines as local smoothers. *Annals of Statistics*, **23**, 1175–1197.
Oehlert, G.W. (1992) Relaxed boundary smoothing splines. *Annals of Statistics*, **20**, 146–160.
Oehlert, G.W. (1993) Regional trends in sulfate wet deposition. *Journal of the American Statistical Association*, **88**, 390–399.
Olkin, I. and Spiegelman, C.H. (1987) A semiparametric approach to density estimation. *Journal of the American Statistical Association*, **82**, 858–865.
O'Sullivan, F. (1985) Discussion of "Some aspects of the spline smoothing approach to non-parametric regression curve fitting." *Journal of the Royal Statistical Society, Ser. B*, **47**, 39–40.
O'Sullivan, F. (1986) A statistical perspective on ill-posed inverse problems (with discussion). *Statistical Science*, **1**, 505–527.
O'Sullivan, F. (1988) Fast computation of fully automated log-density and log-hazard estimators. *SIAM Journal on Scientific and Statistical Computation*, **9**, 363–379.
O'Sullivan, F. and Pawitan, Y. (1993) Multidimensional density estimation by tomography. *Journal of the Royal Statistical Society, Ser. B*, **55**, 509–521.
O'Sullivan, F., Yandell, B.S., and Raynor, W.J., Jr. (1986) Automatic smoothing of regression functions in generalized linear models. *Journal of the American Statistical Association*, **81**, 96–103.
Park, B.U. and Marron, J.S. (1990) Comparison of data-driven bandwidth selectors. *Journal of the American Statistical Association*, **85**, 66–72.
Park, B.U. and Marron, J.S. (1992) On the use of pilot estimators in bandwidth selection. *Journal of Nonparametric Statistics*, **1**, 231–240.
Park, B.U. and Turlach, B.A. (1992) Practical performance of several data driven bandwidth selectors (with discussion). *Computational Statistics*, **7**, 251–270; 275–277; 283–285.
Park, C., Basu, A., and Basu, S. (1995) Robust minimum distance inference based on combined distances. *Communications in Statistics — Simulation and Computation*, **24**, 653–673.
Parzen, E. (1962) On estimation of a probability density function and mode. *Annals of Mathematical Statistics*, **33**, 1065–1076.
Pierce, D.A. and Schafer, D.W. (1986) Residuals in generalized linear models. *Journal of the American Statistical Association*, **81**, 977–986.
Polzehl, J. (1995) Projection pursuit discriminant analysis. *Computational Statistics and Data Analysis*, **20**, 141–157.
Posse, C. (1990) An effective two-dimensional projection pursuit algorithm. *Communications in Statistics — Simulation and Computation*, **19**, 1143–1164.
Posse, C. (1992) Projection pursuit discriminant analysis for two groups. *Communications in Statistics — Theory and Methods*, **21**, 1–19.
Posse, C. (1995a) Projection pursuit exploratory data analysis. *Computational Statistics and Data Analysis*, **20**, 669–687.
Posse, C. (1995b) Tools for two-dimensional exploratory projection pursuit. *Journal of Computational and Graphical Statistics*, **4**, 83–100.
Priebe, C.E. (1994) Adaptive mixtures. *Journal of the American Statistical Association*, **89**, 796–806.
Priestley, M.B. and Chao, M.T. (1972) Nonparametric function fitting. *Journal of the Royal Statistical Society, Ser. B*, **34**, 385–392.
Quintela del Río, A. (1994) A plug-in technique in nonparametric regression with dependence. *Communications in Statistics — Theory and Methods*, **23**, 2581–2603.
Quintela del Río, A. and Vilar Fernández, J.M. (1992) A local cross-validation algorithm for dependent data. *Test*, **1**, 123–153.
Rajagopalan, B. and Lall, U. (1995) A kernel estimator for discrete distributions. *Journal of Nonparametric Statistics*, **4**, 409–426.

Rao, C.R. (1963) Criteria of estimation in large samples. *Sankhyā*, **25**, 189–206.
Ray, B.K. and Tsay, R.S. (1996) Iterative bandwidth selection for nonparametric regression with long-range dependent errors. In *Proceedings of the International Conference on Applied Probability and Time Series, Volume II (1995)*, ed. M. Rosenblatt, Springer-Verlag, New York, 339–351.
Raz, J. (1990) Testing for no effect when estimating a smooth function by nonparametric regression: a randomization approach. *Journal of the American Statistical Association*, **85**, 132–138.
Read, T.R.C. and Cressie, N.A. (1988) *Goodness-of-Fit Statistics for Discrete Multivariate Data*, Springer-Verlag, New York.
Reinsch, C. (1967) Smoothing by spline functions. *Numerische Mathematik*, **10**, 177–183.
Rice, J. (1984a) Boundary modification for kernel regression. *Communications in Statistics — Theory and Methods*, **13**, 893–900.
Rice, J. (1984b) Bandwidth choice for nonparametric regression. *Annals of Statistics*, **12**, 1215–1230.
Rice, J. and Rosenblatt, M. (1983) Smoothing splines: regression, derivatives and deconvolution. *Annals of Statistics*, **11**, 141–156.
Riedel, K.S. (1995) Piecewise convex function estimation and model selection. In *Approximation Theory VIII*, eds. C.K. Chui and L.L. Schumaker, World Scientific Publishing, River Edge, NJ, 467–475.
Rodriguez, C.C. and Van Ryzin, J. (1985) Maximum entropy histograms. *Statistics and Probability Letters*, **3**, 117–120.
Roeder, K. (1990) Density estimation with confidence sets exemplified by superclusters and voids in the galaxies. *Journal of the American Statistical Association*, **85**, 617–624.
Roeder, K. (1992) Semiparametric estimation of normal mixture densities. *Annals of Statistics*, **20**, 929–943.
Romano, J.P. (1988) On weak convergence and optimality of kernel density estimates of the mode. *Annals of Statistics*, **16**, 629–647.
Roosen, C.B. and Hastie, T. (1994) Automatic smoothing spline projection pursuit. *Journal of Computational and Graphical Statistics*, **3**, 235–248.
Rosenblatt, M. (1956) Remarks on some nonparametric estimates of a density function. *Annals of Mathematical Statistics*, **27**, 832–837.
Rosenblatt, M. (1970) Density estimates and Markov sequences. In *Nonparametric Techniques in Statistical Inference*, ed. M. Puri, Cambridge University Press, Cambridge, 199–210.
Rosenblatt, M. (1971) Curve estimates. *Annals of Mathematical Statistics*, **42**, 1815–1842.
Rosenblatt, M. (1975) A quadratic measure of deviation of two-dimensional density estimates and a test of independence. *Annals of Statistics*, **3**, 1–14.
Rosenblatt, M. (1979) Global measures of deviation for kernel and nearest-neighbor density estimates. In *Smoothing Techniques for Curve Estimation*, eds. T. Gasser and M. Rosenblatt, Springer-Verlag, Berlin, 181–190.
Roussas, G. (1969) Nonparametric estimation in Markov processes. *Annals of the Institute of Statistical Mathematics*, **21**, 73–87.
Roussas, G. (1988) Nonparametric estimation in mixing sequences of random variables. *Journal of Statistical Planning and Inference*, **18**, 135–149.
Rudemo, M. (1982) Empirical choice of histograms and kernel density estimators. *Scandinavian Journal of Statistics*, **9**, 65–78.
Ruppert, D. and Cline, D.B.H. (1994) Bias reduction in kernel density estimation by smoothed empirical transformations. *Annals of Statistics*, **22**, 185–210.
Ruppert, D., Sheather, S.J., and Wand, M.P. (1995) An effective bandwidth selector for local least squares regression. *Journal of the American Statistical Association*, **90**, 1257–1270.
Ruppert, D. and Wand, M.P. (1992) Correcting for kurtosis in density estimation. *Australian Journal of Statistics*, **34**, 19–29.
Ruppert, D. and Wand, M.P. (1994) Multivariate locally weighted least squares regression. *Annals of Statistics*, **22**, 1346–1370.

Russell, J.M., Simonoff, J.S., and Nightingale, J. (1997) Nursing behaviors of beluga calves (delphinapterus leucas) born in captivity. *Zoo Biology*, **16**, 247–262.

Sain, S.R., Baggerly, K.A., and Scott, D.W. (1994) Cross-validation of multivariate densities. *Journal of the American Statistical Association*, **89**, 807–817.

Samiuddin, M. and El-Sayyad, G.M. (1990) On nonparametric kernel density estimates. *Biometrika*, **77**, 865–874.

Samiuddin, M., Jones, M.C., and El-Sayyad, G.M. (1993) On bin-based density estimation. *Journal of Statistical Computation and Simulation*, **47**, 241–252.

Sarda, P. (1991) Estimating smooth distribution functions. In *Nonparametric Functional Estimation and Related Topics*, ed. G.G. Roussas, Klüwer Academic Publishers, Dordrecht, 261–270.

Schäfer, H. and Trampisch, H.J. (1982) A note on the variable kernel estimate. *Biometric Journal*, **24**, 607–612.

Schilling, A. and Stute, W. (1987) The consistency of variable-knot density estimates. In *New Perspectives in Theoretical and Applied Statistics*, eds. M.L. Puri, J. Perez Vilaplana, and W. Wertz, John Wiley, New York, 277–286.

Schilling, M.F. and Watkins, A.E. (1994) A suggestion for sunflower plots. *American Statistician*, **48**, 303–305.

Schimek, M.G. (1988) A roughness penalty regression approach for statistical graphics. In *Compstat 1988. Proceedings in Computational Statistics*, eds. D. Edwards and N.E. Raun, Physica-Verlag, Heidelberg, 37–43.

Schimek, M.G. (1992) Serial correlation in spline smoothing: a simulation study. *Computational Statistics*, **7**, 309–327.

Schimek, M.G. and Schmaranz, K.G. (1994) Dependent error regression smoothing: a new method and PC program. *Computational Statistics and Data Analysis*, **17**, 457–464.

Schoenberg, I.J. (1964) Spline functions and the problem of graduation. *Proceedings of the National Academy of Sciences of the United States*, **52**, 947–950.

Schucany, W.R. (1989) Locally optimal window widths for kernel density estimation with large samples. *Statistics and Probability Letters*, **7**, 401–405.

Schucany, W.R. (1995) Adaptive bandwidth choice for kernel regression. *Journal of the American Statistical Association*, **90**, 535–540.

Schucany, W.R., Gray, H.L., and Owen, D.B. (1971) On bias reduction in estimation. *Journal of the American Statistical Association*, **66**, 524–533.

Schucany, W.R. and Sommers, J.P. (1977) Improvement of kernel type density estimators. *Journal of the American Statistical Association*, **72**, 420–423.

Schuster, E.F. and Gregory, G.G. (1981) On the nonconsistency of maximum likelihood density estimators. In *Computer Science and Statistics: Proceedings of the 13th Symposium on the Interface*, ed. W.F. Eddy, Springer-Verlag, New York, 295–298.

Schuster, E. and Yakowitz, S. (1985) Parametric/nonparametric mixture density estimation with application to flood-frequency analysis. *Water Resources Bulletin*, **21**, 797–804.

Scott, D.W. (1979) On optimal and data-based histograms. *Biometrika*, **66**, 605–610.

Scott, D.W. (1982) Optimal meshes for histograms using variable-width bins. Paper presented at annual American Statistical Association meeting, Cincinnati, OH.

Scott, D.W. (1985a) Frequency polygons: theory and application. *Journal of the American Statistical Association*, **80**, 348–354.

Scott, D.W. (1985b) Average shifted histograms: effective nonparametric density estimators in several dimensions. *Annals of Statistics*, **13**, 1024–1040.

Scott, D.W. (1988) A note on choice of bivariate histogram bin shape. *Journal of Official Statistics*, **4**, 47–51.

Scott, D.W. (1991) Discussion of "Transformations in density estimation." *Journal of the American Statistical Association*, **86**, 359.

Scott, D.W. (1992) *Multivariate Density Estimation: Theory, Practice, and Visualization*, John Wiley, New York.

Scott, D.W. and Sheather, S.J. (1985) Kernel density estimation with binned data. *Communications in Statistics — Theory and Methods*, **14**, 1353–1359.

Scott, D.W., Tapia, R.A., and Thompson, J.R. (1978) Multivariate density estimation by discrete penalized likelihood methods. In *Graphical Representation of Multivariate Data*, ed. P.C.C. Wang, Academic Press, New York, 169–181.

Scott, D.W., Tapia, R.A., and Thompson, J.R. (1980) Nonparametric probability density estimation by discrete maximum penalized likelihood criteria. *Annals of Statistics*, **8**, 820–832.

Scott, D.W. and Terrell, G.R. (1987) Biased and unbiased cross-validation in density estimation, *Journal of the American Statistical Association*, **82**, 1131–1146.

Scott, D.W. and Wand, M.P. (1991) Feasibility of multivariate density estimates. *Biometrika*, **78**, 197–205.

Seifert, B., Brockmann, M., Engel, J., and Gasser, T. (1994) Fast algorithms for nonparametric curve estimation. *Journal of Computational and Graphical Statistics*, **3**, 192–213.

Seifert, B. and Gasser, T. (1996a) Variance properties of local polynomials and ensuing modifications (with discussion). In *Statistical Theory and Computational Aspects of Smoothing*, eds. W. Hardle and M.G. Schimek, Physica-Verlag, Heidelberg, 50–102; 121–127.

Seifert, B. and Gasser, T. (1996b) Finite sample variance of local polynomials: analysis and solutions. *Journal of the American Statistical Association*, **91**, 267–275.

Seifert, B., Gasser, T., and Wolf, A. (1993) Nonparametric estimation of residual variance revisited. *Biometrika*, **80**, 373–383.

Sheather, S.J. (1986) An improved data-based algorithm for choosing the window width when estimating the density at a point. *Computational Statistics and Data Analysis*, **4**, 61–65.

Sheather, S.J. (1992) The performance of six popular bandwidth selection methods on some real data sets (with discussion). *Computational Statistics*, **7**, 225–250; 271–281.

Sheather, S.J. and Jones, M.C. (1991) A reliable data-based bandwidth selection method for kernel density estimation. *Journal of the Royal Statistical Society, Ser. B*, **53**, 683–690.

Shibata, R. (1981) An optimal selection of regression variables. *Biometrika*, **68**, 45–54.

Sidorenko, A. and Riedel, K.S. (1994) Optimal boundary kernels and weightings for local polynomial regression. Unpublished manuscript.

Silverman, B.W. (1981) Using kernel density estimates to investigate multimodality. *Journal of the Royal Statistical Society, Ser. B*, **43**, 97–99.

Silverman, B.W. (1982a) Algorithm AS176: kernel density estimation using the Fast Fourier Transform. *Applied Statistics*, **31**, 93–99.

Silverman, B.W. (1982b) On the estimation of a probability density function by the maximum penalized likelihood method. *Annals of Statistics*, **10**, 795–810.

Silverman, B.W. (1984) Spline smoothing: the equivalent variable kernel method. *Annals of Statistics*, **12**, 898–916.

Silverman, B.W. (1985) Some aspects of the spline smoothing approach to nonparametric regression curve fitting (with discussion). *Journal of the Royal Statistical Society, Ser. B*, **47** 1–52.

Silverman, B.W. (1986) *Density Estimation for Statistics and Data Analysis*, Chapman and Hall, London.

Silverman, B.W. and Jones, M.C. (1989) E. Fix and J.L. Hodges (1951): an important contribution to nonparametric discriminant analysis and density estimation. Commentary on Fix and Hodges (1951). *International Statistical Review*, **57**, 233–247.

Silverman, B.W. and Young, G.A. (1987) The bootstrap: to smooth or not to smooth? *Biometrika*, **74**, 469–479.

Simonoff, J.S. (1983) A penalty function approach to smoothing large sparse contingency tables. *Annals of Statistics*, **11**, 208–218.

Simonoff, J.S. (1985) An improved goodness-of-fit statistic for sparse multinomials. *Journal of the American Statistical Association*, **80**, 671–677.

Simonoff, J.S. (1987) Probability estimation via smoothing in sparse contingency tables with ordered categories. *Statistics and Probability Letters*, **5**, 55-63.

Simonoff, J.S. (1995a) The anchor position of histograms and frequency polygons: quantitative and qualitative smoothing. *Communications in Statistics — Simulation and Computation*, **24**, 691–710.

Simonoff, J.S. (1995b) A simple, automatic and adaptive bivariate density estimator based on conditional densities. *Statistics and Computing*, **5**, 245–252.

Simonoff, J.S. (1995c) Smoothing categorical data. *Journal of Statistical Planning and Inference*, **47**, 41–69.

Simonoff, J.S., Hochberg, Y., and Reiser, B. (1986) Alternative estimation procedures for $\Pr(X < Y)$ in categorized data. *Biometrics*, **42**, 895–907.

Simonoff, J.S. and Hurvich, C.M. (1993) A study of the effectiveness of simple density estimation methods. *Computational Statistics*, **8**, 259–278.

Simonoff, J.S. and Tsai, C.-L. (1991a) Higher order effects in log-linear and log-non-linear models for contingency tables with ordered categories. *Applied Statistics*, **40**, 449–458.

Simonoff, J.S. and Tsai, C.-L. (1991b) Assessing the influence of individual observations on a goodness-of-fit test based on nonparametric regression. *Statistics and Probability Letters*, **12**, 9–17.

Simonoff, J.S. and Udina, F. (1997) Measuring the stability of histogram appearance when the anchor position is changed. *Computational Statistics and Data Analysis*, **23**, 335–353.

Simpson, D.G. (1987) Mininum Hellinger distance estimation for the analysis of count data. *Journal of the American Statistical Association*, **82**, 802–807.

Simpson, D.G. (1989) Hellinger deviance tests: efficiency, breakdown points, and examples. *Journal of the American Statistical Association*, **84**, 107–113.

Singh, R.S. (1987) MISE of kernel estimates of a density and its derivatives. *Statistics and Probability Letters*, **5**, 153–159.

Smith, M. and Kohn, R. (1996) Nonparametric regression using Bayesian variable selection. *Journal of Econometrics*, **75**, 317–343.

Sockett, E.B., Daneman, D., Clarson, C., and Ehrich, R.M. (1987) Factors affecting and patterns of residual insulin secretion during the first year of type I (insulin dependent) diabetes mellitus in children. *Diabetes*, **30**, 453–459.

Solow, A.R. (1995) An exploratory analysis of a record of El Niño events, 1800–1987. *Journal of the American Statistical Association*, **90**, 72–77.

Speckman, P.L. (1988) Kernel smoothing in partial linear models. *Journal of the Royal Statistical Society, Ser. B*, **50**, 413–436.

Spencer, J. (1904) On the graduation of the rates of sickness and mortality. *Journal of the Institute of Actuaries*, **38**, 334–347.

Stadtmüller, U. (1983) Asymptotic distributions of smoothed histograms. *Metrika*, **30**, 145–158.

Stahel, W.A. (1981) Robuste schätzungen: infinitesimale optimalität und schätzungen von kovarianzmatrizen. Unpublished Ph.D. thesis, Swiss Federal Institute of Technology (ETH), Zürich.

Staniswalis, J.G. (1989a) Local bandwidth selection for kernel estimates. *Journal of the American Statistical Association*, **84**, 284–288.

Staniswalis, J.G. (1989b) The kernel estimate of a regression function in likelihood-based models. *Journal of the American Statistical Association*, **84**, 276–283.

Staniswalis, J.G. and Severini, T.A. (1991) Diagnostics for assessing regression models. *Journal of the American Statistical Association*, **86**, 684–692.

Stather, C.R. (1981) Robust statistical inference using Hellinger distance methods. Unpublished Ph.D. thesis, LaTrobe University, Australia.

Stoker, T.M. (1993) Smoothing bias in density derivative estimation. *Journal of the American Statistical Association*, **88**, 855–863.

Stone, C.J. (1977) Consistent nonparametric regression (with discussion). *Annals of Statistics*, **5**, 595–645.

Stone, C.J. (1984) An asymptotically optimal window selection rule for kernel density estimates. *Annals of Statistics*, **12**, 1285–1297.

Stone, C.J. (1985a) An asymptotically optimal histogram selection rule. In *Proceedings of the Berkeley Conference in Honor of Jerzy Neyman and Jack Kiefer, Vol. II*, eds. L.M. Le Cam and R.A. Olshen, Wadsworth, Pacific Grove, CA, 513–520.

Stone, C.J. (1985b) Additive regression and other nonparametric models. *Annals of Statistics*, **13**, 689–705.

Stone, C.J. (1986) The dimensionality reduction principle for generalized additive models. *Annals of Statistics*, **14**, 590–606.

Stone, C.J. (1990) Large-sample inference for log-spline models. *Annals of Statistics*, **18**, 717–741.

Stone, C.J., Hansen, M., Kooperberg, C., and Truong, Y.K. (1997) Polynomial splines and their tensor products in extended linear modeling (with discussion). *Annals of Statistics*, **25**, 1371–1470.

Stone, C.J. and Koo, C.-Y. (1986) Logspline density estimation. *Contemporary Mathematics*, **59**, 1–15.

Stone, M. (1974) Cross-validatory choice and assessment of statistical predictions (with discussion). *Journal of the Royal Statistical Society, Ser. B*, **36**, 111–147.

Stuetzle, W. and Mittal, Y. (1979) Some comments on the asymptotic behavior of robust smoothers. In *Techniques for Curve Estimation*, eds. T. Gasser and M. Rosenblatt (Lecture Notes in Mathematics 757), Springer-Verlag, Berlin, 191–195.

Sturges, H.A. (1926) The choice of a class interval. *Journal of the American Statistical Association*, **21**, 65–66.

Sun, J. (1993) Some practical aspects of exploratory projection pursuit. *SIAM Journal on Scientific Computing*, **14**, 68–80.

Sun, J. and Loader, C.R. (1994) Simultaneous confidence bands in linear regression and smoothing. *Annals of Statistics*, **22**, 1328–1345.

Tamura, R.N. and Boos, D.D. (1986) Minimum Hellinger distance estimation for multivariate location and covariance. *Journal of the American Statistical Association*, **81**, 223–229.

Tapia, R.A. and Thompson, J.R. (1978) *Nonparametric Probability Density Estimation*, Johns Hopkins University Press, Baltimore, MD.

Tarter, M.E. and Lock, M.D. (1993) *Model-Free Curve Estimation*, Chapman and Hall, New York.

Taylor, C.C. (1987) Akaike's information criterion and the histogram. *Biometrika*, **74**, 636–639.

Terrell, G.R. (1983) The multilinear frequency spline. Unpublished manuscript.

Terrell, G.R. (1990) The maximal smoothing principle in density estimation. *Journal of the American Statistical Association*, **85**, 470–477.

Terrell, G.R. (1992) Discussion of "The performance of six popular bandwidth selection methods on some real data sets" and "Practical performance of several data driven bandwidth selectors." *Computational Statistics*, **7**, 275–277.

Terrell, G.R. and Scott, D.W. (1980) On improving convergence rates for nonnegative kernel density estimators. *Annals of Statistics*, **8**, 1160–1163.

Terrell, G.R. and Scott, D.W. (1985) Oversmoothed nonparametric density estimates. *Journal of the American Statistical Association*, **80**, 209–214.

Terrell, G.R. and Scott, D.W. (1992) Variable kernel density estimation. *Annals of Statistics*, **20**, 1236–1265.

Thomas, W. (1991) Influence diagnostics for the cross-validated smoothing parameter in spline smoothing. *Journal of the American Statistical Association*, **86**, 693–698.

Thompson, J.R. and Tapia, R.A. (1990) *Nonparametric Function Estimation, Modeling, and Simulation*, SIAM, Philadelphia.

Thorburn, D. (1986) A Bayesian approach to density estimation. *Biometrika*, **73**, 65–76.

Tibshirani, R. (1988) Estimating optimal transformations for regression via additivity and variance stabilization. *Journal of the American Statistical Association*, **83**, 394–405.

Tibshirani, R. and Hastie, T. (1987) Local likelihood estimation. *Journal of the American Statistical Association*, **82**, 559–568.
Tiede, J.J. and Pagano, M. (1979) The application of robust calibration to radioimmunoassay. *Biometrics*, **35**, 567–574.
Tierney, L. (1990) *LISP–STAT: An Object-Oriented Environment for Statistical Computing and Dynamic Graphics*, John Wiley, New York.
Titterington, D.M. (1980) A comparative study of kernel-based density estimates for categorical data. *Technometrics*, **22**, 259–268.
Titterington, D.M. and Bowman, A.W. (1985) A comparative study of smoothing procedures for ordered categorical data. *Journal of Computational and Graphical Statistics*, **21**, 291–312.
Tran, L.T. (1989) The $L_1$ convergence of kernel density estimates under dependence. *Canadian Journal of Statistics*, **17**, 197–208.
Tran, L.T. (1990) Kernel density estimation under dependence. *Statistics and Probability Letters*, **10**, 193–201.
Trosset, M.W. (1993) Optimal shapes for kernel density estimation. *Communications in Statistics — Theory and Methods*, **22**, 375–391.
Truong, Y.K. (1991) Nonparametric curve estimation with time series errors. *Journal of Statistical Planning and Inference*, **28**, 167–183.
Tseng, C.-T. and Moret, B.M.E. (1990) A method for the choice of smoothing parameter. *Mathematical and Computer Modelling*, **13/9**, 1–16.
Tsybakov, A.B. (1986) Robust reconstruction of functions by the local-approximation method. *Problems of Information Transmission*, **22**, 133–146.
Tukey, J.W. (1977) *Exploratory Data Analysis*, Addison–Wesley, Reading, MA.
Tukey, P.A. and Tukey, J.W. (1981) Data-driven view selection: agglomeration and sharpening. In *Interpreting Multivariate Data*, ed. V. Barnett, John Wiley, Chichester, 215–243.
Turlach, B.A. and Wand, M.P. (1996) Fast computation of auxiliary quantities in local polynomial regression. *Journal of Computational and Graphical Statistics*, **5**, 337–350.
Udina, F. (1994) Notes about KDE objects. Unpublished manuscript.
Ullah, A. and Zinde-Walsh, V. (1992) On the estimation of residual variance in nonparametric regression. *Journal of Nonparametric Statistics*, **1**, 263–265.
Utreras, F. (1981) On computing robust splines and applications. *SIAM Journal on Scientific and Statistical Computing*, **2**, 153–163.
Utreras, F. (1988) Boundary effects on convergence rates for Tikhonov regularization. *Journal of Approximation Theory*, **54**, 235–249.
van der Linde, A. (1994) On cross-validation for smoothing splines in the case of dependent observations. *Australian Journal of Statistics*, **36**, 67–73.
Van Es, A.J. and Hoogstrate, A.J. (1994) Kernel estimators of integrated squared density derivatives in non-smooth cases. In *Asymptotic Statistics, Proceedings of the Fifth Prague Symposium*, eds. P. Mandl and M. Hŭsková, Physica Verlag, Heidelberg, 213–224.
Van Es, A.J. and Hoogstrate, A.J. (1998) How much do plug-in bandwidth selectors adapt to non-smoothness? *Journal of Nonparametric Statistics*, **8**, 185–197.
Van Ryzin, J. (1973) A histogram method of density estimation. *Communications in Statistics*, **2**, 493–506.
Venables, W.N. and Ripley, B.D. (1994) *Modern Applied Statistics with S–Plus*, Springer-Verlag, New York.
Victor, N. (1976) Nonparametric allocation rules (with discussion). In *Decision Making and Medical Care: Can Information Science Help?*, eds. F.T. de Dombal and F. Grémy, North-Holland, Amsterdam, 515–529.
Vieu, P. (1991) Nonparametric regression: optimal local bandwidth choice. *Journal of the Royal Statistical Society, Ser. B*, **53**, 453–464.
Vitale, R.A. (1975) A Bernstein polynomial approach to density estimation. In *Statistical Inference and Related Topics*, Volume 2, ed. M.L. Puri, Academic Press, San Francisco, 87–99.
Wahba, G. (1975a) Interpolating spline methods for density estimation. I. Equispaced knots. *Annals of Statistics*, **3**, 30–48.

Wahba, G. (1975b) Smoothing noisy data with spline functions. *Numerische Mathematik*, **24**, 383–393.

Wahba, G. (1976) Histosplines with knots which are order statistics. *Journal of the Royal Statistical Society, Ser. B*, **38**, 140–151.

Wahba, G. (1978) Improper priors, spline smoothing and the problem of guarding against model errors in regression. *Journal of the Royal Statistical Society, Ser. B*, **40**, 364–372.

Wahba, G. (1983) Bayesian "confidence intervals" for the cross-validated smoothing spline. *Journal of the Royal Statistical Society, Ser. B*, **45**, 133–150.

Wahba, G. (1985) Discussion of "Some aspects of the spline smoothing approach to non-parametric regression curve fitting." *Journal of the Royal Statistical Society, Ser. B*, **47**, 44.

Wahba, G. (1986) Partial and interaction splines for the semiparametric estimation of functions of several variables. In *Computing Science and Statistics: Proceedings of the 18th Symposium on the Interface*, ed. T. Boardman, American Statistical Association, Washington, DC, 75–80.

Wahba, G. (1990) *Spline Models for Observational Data*, SIAM, Philadelphia.

Wahba, G., Wang, Y., Gu, C., Klein, R., and Klein, B. (1995) Smoothing spline ANOVA for exponential families, with application to the Wisconsin epidemiology study of diabetic retinopathy. *Annals of Statistics*, **23**, 1865–1895.

Wahba, G. and Wold, S. (1975) A completely automatic French curve: fitting spline functions by cross-validation. *Communications in Statistics — Theory and Methods*, **4**, 1–17.

Walter, G. and Blum, J.R. (1979) Probability density estimation using delta sequences. *Annals of Statistics*, **7**, 328–340.

Wand, M.P. (1992a) Finite sample performance of density estimators under moving average dependence. *Statistics and Probability Letters*, **13**, 109–115.

Wand, M.P. (1992b) Error analysis for general multivariate kernel estimators. *Journal of Nonparametric Statistics*, **2**, 1–15.

Wand, M.P. (1994) Fast computation of multivariate kernel estimators. *Journal of Computational and Graphical Statistics*, **3**, 433–445.

Wand, M.P. (1995) Personal communication, June 15, 1995.

Wand, M.P. (1997) Data-based choice of histogram bin width. *American Statistician*, **51**, 59–64.

Wand, M.P. and Devroye, L. (1993) How easy is a given density to estimate? *Computational Statistics and Data Analysis*, **16**, 311–323.

Wand, M.P. and Jones, M.C. (1993) Comparison of smoothing parameterizations in bivariate kernel density estimation. *Journal of the American Statistical Association*, **88**, 520–528.

Wand, M.P. and Jones, M.C. (1994) Multivariate plug-in bandwidth selection. *Computational Statistics*, **9**, 97–116.

Wand, M.P. and Jones, M.C. (1995) *Kernel Smoothing*, Chapman and Hall, London.

Wand, M.P., Marron, J.S., and Ruppert, D. (1991) Transformations in density estimation (with discussion). *Journal of the American Statistical Association*, **86**, 343–361.

Wand, M.P. and Schucany, W.R. (1990) Gaussian-based kernels. *Canadian Journal of Statistics*, **18**, 197–204.

Wang, F.T. (1994) Automatic smoothing parameter selection for robust nonparametric regression. *Proceedings of the Statistical Computing Section of the American Statistical Association.*

Wang, F.T. and Scott, D.W. (1994) The $L_1$ method for robust nonparametric regression. *Journal of the American Statistical Association*, **89**, 65–76.

Wang, K. and Gasser, T. (1996) Optimal rate for estimating local bandwidth in kernel estimators of regression functions. *Scandinavian Journal of Statistics*, **23**, 303–312.

Wang, M.-C. and Van Ryzin, J. (1981) A class of smooth estimators for discrete distributions. *Biometrika*, **68**, 301–309.

Wang, Y. and Wahba, G. (1995) Bootstrap confidence intervals for smoothing splines and their comparison to Bayesian "confidence intervals." *Journal of Statistical Computation and Simulation*, **51**, 263–279.

Watson, G.S. (1964) Smooth regression analysis. *Sankhyā, Ser. A*, **26**, 359–372.

Watson, G.S. and Leadbetter, M.R. (1963) On the estimation of the probability density, I. *Annals of Mathematical Statistics*, **34**, 480–491.

Wecker, W. and Ansley, C.F. (1983) The signal extraction approach to nonlinear regression and spline smoothing. *Journal of the American Statistical Association*, **78**, 81–89.

Wegman, E.J. (1969) A note on estimating a unimodal density. *Annals of Mathematical Statistics*, **40**, 1661–1667.

Wegman, E.J. (1970a) Maximum likelihood estimation of a unimodal density function. *Annals of Mathematical Statistics*, **41**, 457–471.

Wegman, E.J. (1970b) Maximum likelihood estimation of a unimodal density function, II. *Annals of Mathematical Statistics*, **41**, 2169–2174.

Wegman, E.J. (1975) Maximum likelihood estimation of a probability density function. *Sankhyā, Ser. A*, **37**, 211–224.

Wei, C.Z. and Chu, C.-K. (1994) A regression point of view toward density estimation. *Journal of Nonparametric Statistics*, **4**, 191–201.

Weisberg, S. (1985) *Applied Linear Regression*, 2nd ed., John Wiley, New York.

Whittaker, E. (1923) On a new method of graduation. *Proceedings of the Edinburgh Mathematical Society*, **41**, 63–75.

Whittle, P. (1958) On the smoothing of probability density functions. *Journal of the Royal Statistical Society, Ser. B*, **20**, 334–343.

Woltring, H.J. (1986) A FORTRAN package for generalized, cross-validatory spline smoothing and differentiation. *Advances in Engineering Software*, **8**, 104–113.

Woodroofe, M. (1970) On choosing a delta-sequence. *Annals of Mathematical Statistics*, **41**, 1665–1671.

World Bank (1994) *World★Data 1994 World Bank Indicators on CD-ROM*, The World Bank, Washington, DC.

Worton, B.J. (1989) Optimal smoothing parameters for multivariate fixed and adaptive kernel methods. *Journal of Statistical Computation and Simulation*, **32**, 45–57.

Yanagimoto, T. and Yanagimoto, M. (1987) The use of marginal likelihood for a diagnostic test for the goodness of fit of the simple linear regression model. *Technometrics*, **29**, 95–101.

Yang, L. (1995) Transformation-density estimation. Unpublished Ph.D. thesis, University of North Carolina, Chapel Hill, NC.

Young, S.G. and Bowman, A.W. (1995) Non-parametric analysis of covariance. *Biometrics*, **51**, 920–931.

Zhang, P. (1991) Variable selection in nonparametric regression with continuous covariates. *Annals of Statistics*, **19**, 1869–1882.

# Author Index

# Subject Index

# Springer Series in Statistics

*(continued from p. ii)*

*Le Cam/Yang:* Asymptotics in Statistics: Some Basic Concepts.
*Longford:* Models for Uncertainty in Educational Testing.
*Manoukian:* Modern Concepts and Theorems of Mathematical Statistics.
*Miller, Jr.:* Simultaneous Statistical Inference, 2nd edition.
*Mosteller/Wallace:* Applied Bayesian and Classical Inference: The Case of *The Federalist Papers.*
*Parzen/Tanabe/Kitagawa:* Selected Papers of Hirotugu Akaike.
*Pollard:* Convergence of Stochastic Processes.
*Pratt/Gibbons:* Concepts of Nonparametric Theory.
*Ramsay/Silverman:* Functional Data Analysis.
*Read/Cressie:* Goodness-of-Fit Statistics for Discrete Multivariate Data.
*Reinsel:* Elements of Multivariate Time Series Analysis, 2nd edition.
*Reiss:* A Course on Point Processes.
*Reiss:* Approximate Distributions of Order Statistics: With Applications to Non-parametric Statistics.
*Rieder:* Robust Asymptotic Statistics.
*Rosenbaum:* Observational Studies.
*Ross:* Nonlinear Estimation.
*Sachs:* Applied Statistics: A Handbook of Techniques, 2nd edition.
*Särndal/Swensson/Wretman:* Model Assisted Survey Sampling.
*Schervish:* Theory of Statistics.
*Seneta:* Non-Negative Matrices and Markov Chains, 2nd edition.
*Shao/Tu:* The Jackknife and Bootstrap.
*Siegmund:* Sequential Analysis: Tests and Confidence Intervals.
*Simonoff:* Smoothing Methods in Statistics.
*Small:* The Statistical Theory of Shape.
*Tanner:* Tools for Statistical Inference: Methods for the Exploration of Posterior Distributions and Likelihood Functions, 3rd edition.
*Tong:* The Multivariate Normal Distribution.
*van der Vaart/Wellner:* Weak Convergence and Empirical Processes: With Applications to Statistics.
*Vapnik:* Estimation of Dependences Based on Empirical Data.
*Weerahandi:* Exact Statistical Methods for Data Analysis.
*West/Harrison:* Bayesian Forecasting and Dynamic Models, 2nd edition.
*Wolter:* Introduction to Variance Estimation.
*Yaglom:* Correlation Theory of Stationary and Related Random Functions I: Basic Results.
*Yaglom:* Correlation Theory of Stationary and Related Random Functions II: Supplementary Notes and References.